Valleytronics in 2D Materials

Other World Scientific Titles by Kuan Eng Johnson Goh

Scanning Probe Microscopy
ISBN: 978-981-4324-76-2

Valleytronics in 2D Materials

Editors

Kuan Eng Johnson Goh
Calvin Pei Yu Wong
Tong Wang

*Agency for Science, Technology and Research (A*STAR), Singapore*

World Scientific

NEW JERSEY • LONDON • SINGAPORE • BEIJING • SHANGHAI • HONG KONG • TAIPEI • CHENNAI • TOKYO

Published by

World Scientific Publishing Co. Pte. Ltd.

5 Toh Tuck Link, Singapore 596224

USA office: 27 Warren Street, Suite 401-402, Hackensack, NJ 07601

UK office: 57 Shelton Street, Covent Garden, London WC2H 9HE

Library of Congress Cataloging-in-Publication Data

Names: Goh, Kuan Eng Johnson, editor. | Wong, Calvin Pei Yu, editor. | Wang, Tong (Tong-Biao), editor.

Title: Valleytronics in 2D materials / editors, Kuan Eng Johnson Goh, Calvin Pei Yu Wong, Tong Wang, A*STAR, Singapore.

Description: Hackensack, NJ : World Scientific, [2023] | Includes bibliographical references and index.

Identifiers: LCCN 2023000263 | ISBN 9789811229091 (hardcover) | ISBN 9789811229107 (ebook for institutions) | ISBN 9789811229114 (ebook for individuals)

Subjects: LCSH: Valleytronics--Materials. | Spintronics--Materials. | Quantum electronics--Materials. | Many-valley semiconductors. | Semiconductor films. | Electron transport. | Crystalline electric field.

Classification: LCC TK7874.887 .V35 2023 | DDC 621.3815/2--dc23/eng/20230411

LC record available at https://lccn.loc.gov/2023000263

British Library Cataloguing-in-Publication Data

A catalogue record for this book is available from the British Library.

For any available supplementary material, please visit
https://www.worldscientific.com/worldscibooks/10.1142/12066#t=suppl

Desk Editor: Shaun Tan Yi Jie

Typeset by Stallion Press
Email: enquiries@stallionpress.com

https://doi.org/10.1142/9789811229107_fmatter

Preface

— Lily of the Valley —

This book started as a labour of love.

It began with the excitement of the editors to share the new common interest and learnings in valleytronics at the end of a fruitful research project at the Institute of Materials Research and Engineering, A*STAR in Singapore. The book proposal was drafted in late 2019 with Wang Tong and Calvin Wong who both kindly agreed to join the effort. Just as the proposal review was completed and approved by the publisher World Scientific in April 2020, Singapore and many parts of the world went into various forms of lockdown due to the COVID-19 pandemic. Like most people, our work was disrupted as we needed to figure out the new norms of operation in the laboratories, offices and every aspect of daily living. Book writing became a luxury as more urgent tasks of just coping with day-to-day living and work became high priority. When writing restarted in mid-2021, the world had just begun a new trajectory to cautiously reopen economies as COVID-19 became endemic in some countries, including Singapore. Although a significant part of the population had been fully vaccinated by then, some variants of COVID-19 took us in turn. The writing effort became sporadic and at some stage, we needed to co-opt additional authors and seek the publisher's understanding for a series of deadline extensions.

Truth be told, what started as a labour of love became a labour of duty.

Thankfully, as the economies reopened, we were fortunate to welcome new colleagues into our midst, and as we started to engage them in our hobby of valleytronics and retold our motivations — of what had been achieved by researchers, and most importantly, of what new opportunities lie ahead — the fire rekindled. I want to specially thank my colleagues Fabio Bussolotti, Ivan Verzhbitskiy, Qian Ling Kee, Sarthak Das, Wei Fu and Zhuojun Jiang for bravely stepping up to help with authoring and vetting the various chapters and bringing to bear each of their specialisations. Understandably, they were all stretched by their individual day jobs in the post-COVID-19 norms of working, having to adapt to and integrate with new colleagues, and the high standards I demanded of each chapter as they laboured over many nights and weekends. We also thank Hiroyo Kawai for helping with some of the band structure illustrations, and Justin Song and Ding Huang for proofreading the theory chapter and providing helpful suggestions for enhancing our presentation. This book would not have been possible without them.

Valleytronics seems a relatively new field, but you might be surprised to find in Chapter 1 that the concept has been pursued for decades. The exciting developments during the last decade bring significant promise for finally having practical valleytronics for new applications.

Our intent is to bring the reader quickly into this field, not just to join the conversation, but hopefully to inspire scientists and engineers to action and bring more hands on deck to speed the development of valleytronics. Chapters 1 and 8 cover the historical overview and future opportunities respectively, and are written with minimal technicalities in the hope of drawing entrepreneurs, managers and policymakers into the fray. Chapters 2 to 7 provide the main technical background from theory to engineering.

— Rising from the Valley —

It is our hope that this labour of both love and duty will inspire future developments in valleytronics, and bring the valley degree of freedom to bear in new practical applications.

Coincidentally, this book is completed as the world emerges from the historical valley of COVID-19 over the last two years. It is time to rise from the valley and take hold of the future.

Kuan Eng Johnson Goh
22 Dec 22

About the Editors

Professor Kuan Eng Johnson Goh is Division Director and Principal Scientist at the Institute of Materials Research and Engineering (IMRE), Agency for Science, Technology and Research (A*STAR), Singapore. He is also Adjunct Professor at the School of Physical and Mathematical Sciences, Nanyang Technological University (NTU), Singapore. He owns 5 patents and has published more than 100 papers including in top journal series under the *Nature*, *Advanced Materials*, *ACS* and the *Physical Review* publishing groups. He has given more than 25 invited talks including keynotes and plenary lectures. He holds a PhD in Physics from University of New South Wales, Australia.

Dr Calvin Pei Yu Wong is Research Scientist and Deputy Head of the Quantum Technologies for Engineering (QTE) department at the Institute of Materials Research and Engineering (IMRE), Agency for Science, Technology and Research (A*STAR), Singapore. His current research involves low temperature (mK) magneto-transport measurements and scanning probe microscopy of 2D valleytronic and quantum materials. He was awarded the NUS Graduate School for Integrative Science and Engineering Scholarship to pursue multidisciplinary research topics for his PhD. He holds a PhD in Integrative Sciences and Engineering from National University of Singapore, Singapore.

Dr Tong Wang is Research Scientist at the Institute for Materials Research and Engineering (IMRE), Agency for Science, Technology and Research (A*STAR), and focuses on strategic research planning, road mapping, and technology analysis. He was awarded the A*STAR National Science Scholarship for his Bachelor and PhD degrees. Having graduated in 2019, he has published 6 first-author papers and contributed to many others. He has also received a number of accolades such as the Springer-Nature Award and the A*STAR Roll of Honour. He holds a PhD in Physics from the University of Oxford, UK.

Contents

https://doi.org/10.1142/9789811229107_0001

Chapter 1

Brief History of Valleytronics

Kuan Eng Johnson Goh

1.1 Introduction

The progress in our understanding of solids has contributed to many of the technological advances in information processing, communication and automation of routines that enabled many of the modern conveniences we now enjoy. They range from the ubiquitous information access on the internet, online economy, social media, and remote access of appliances at the consumer level, to smart working spaces and factories for the industry. These advances owe in no small part to the exquisite control we now have over the "particles" in solids that are fundamental to our technologies for information carriage and storage. The quantum revolution in the 1900s that provided our present insight into the wave-particle nature of matter, albeit uncomfortably at times, allowed us to make the leap from vacuum tube transistors to the modern microelectronics. With that came the coexistence of our current classical and quantum descriptions of, for example, a charge carrier (a type of particle) appropriated to the occasion, whether for education or application. Interestingly, the concept of momentum appears to be the link between the classical and quantum descriptions of particles, and we can perhaps think of them more generally as momentum carriers. If they are tagged on with a charge, they interact with an electric field. If they are tagged on with a spin, they respond to magnetic coercion. We call this the charge or spin degrees of freedom by which we can access and manipulate these momentum carriers for the purpose of information transmission or storage. Despite the conceptual unease, the wave-particle duality remains our current best descriptor of matter.

The band structure, as a characterisation of crystalline solids, is birthed from such a tension, and there are many excellent textbooks [1]–[3] that provide more complete explanations. What is crucial to the discussion in this book, though, are the special insights that the band structure affords us for the purpose of utilising "particles" in solids as information carriers. While the concept of a particle having some position and momentum provides easy classical intuition for how a stream of electrons could behave in a solid, the quantum description of the wave-like nature of electrons in a solid crystal (with atoms in ordered repetition) opens new levels of insight to their behaviour and hence control of such electron streams. The band structure is a visualisation of this advantage in terms of the energy-momentum relationship, $\varepsilon(k)$, as shown in Fig. 1.1, where ε represents energy and k the crystal momentum associated with carrier electrons in the crystal (often labelled in terms of the high symmetry points of the crystal Brillouin zone as in Fig. 1.1(a) and (b)). It reveals many non-classical phenomena possible for the electrons, such as discrete energy levels, tunnelling, quantum interference, correlated transport, and many emergent behaviour remaining to be unearthed.

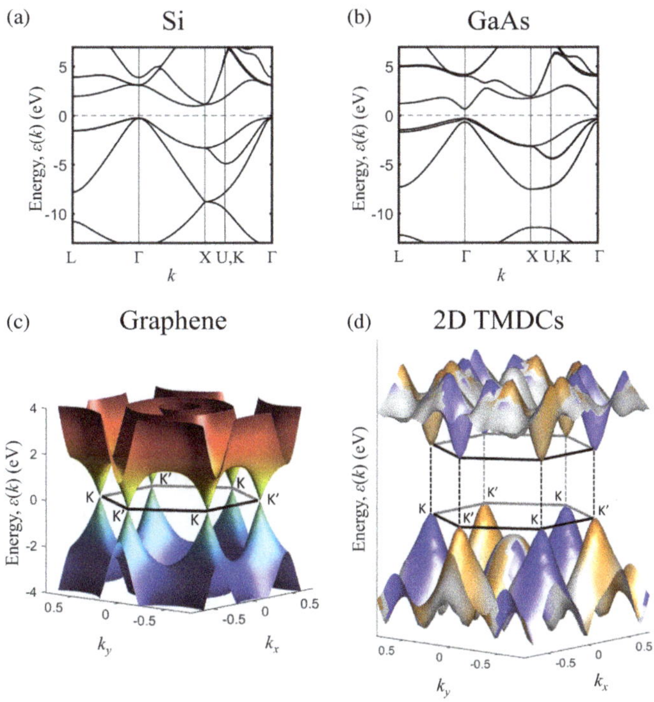

Fig. 1.1. Examples of band structure, where energy is expressed as function of crystal momentum k for semiconductors: (a) Si, (b) GaAs, (c) semi-metal graphene, and (d) 2D semiconductors from the transition metal dichalcogenide family. Band diagrams courtesy of Dr Hiroyo Kawai.

The key features from the band structure that pertains to valleytronics are the energy minima and maxima of $\varepsilon(k)$, commonly referred to as "valleys". Naively, valleys are attractive because near the lowest (highest) points they are simple to describe mathematically using a parabola and hence amenable to analytical solutions. However, it would be remiss to only think about utilising the valleys. Indeed, research in the last decade has drawn our attention strongly towards other exotic features (e.g., Dirac cone-like $\varepsilon(k)$ dispersions with massless fermions [4]–[6], band inversion regions giving rise to topological transport [7], flat-band regions having significant density of states with implication for correlated transport [8]). These exciting developments certainly foreshadow new books that update and consolidate the new understanding. In this book though, we shall focus on describing what we now know about valleys and how we can begin to utilise them like electrons in electronics, hence the portmanteau “valley-tronics”.

It is conceivable that researchers already started thinking about how to utilise $\varepsilon(k)$ valleys for some use early after Bloch’s formulation of the band theory of solids in the late 1920s. The idea of steering a particle with charge via an external field came with the discovery of the electron by J. J. Thompson in 1897. Later, the steering of particles with spin by an external field was shown by O. Stern and W. Gerlach in 1922, even though the concept of spin was still elusive then. However, it was a challenge to access the “valley” in a solid, let alone figure out how to control it. The impact of this challenge is perhaps most keenly felt by the number of devices and applications we now have based on the electron and the magnetic spin, but none based on the valley concept after a century.

Valleys appear naturally in the mathematical conception of the band structure, and because they are a consequence of the periodic nature and constituent atoms that make up the entire crystal, the valley state is expected to be robust and not easily altered. A one-dimensional crystal can have a relatively simple band structure. Increasing the dimension of the crystal and constituent atomic species therein can significantly complicate the band structure. Even if a crystal is identical when viewed along different crystal axes, the different momenta along these axes will conspire to create valleys with multiple degeneracies (i.e., valleys of the same energy). Thus, differentiating these valleys by energetic means would be impossible. Various means of lifting such degeneracies, e.g., by creating anisotropic strain in the crystal, have been attempted. But before we run through the history of such attempts, it is perhaps instructive to first lay down the criteria for a valley state to be considered a practical information carrier.

1.2 Inequivalent Valley States

From the perspective of use, especially for encoding information, most physical systems are expected to have variable states. The simplest implementation of a door, commonly seen as a structural accessory, could encode information such as whether a store was open or closed. Similarly, electrical switches should have at least two or more different states to serve a range of purposes. What is important is that these states are well differentiated or distinct. The binary system of two states being the simplest is surprisingly sufficient to undergird modern information technology and many digitally controlled systems. Such binary systems could be derived from High-Low signals, for example based on two charge/voltage states, or two magnetisation directions for spins. The minimum requirement for information encoding is therefore at least two distinct states. In some literature, the word "contrast" is used, in a similar sense to how the printed words on this page must have a good contrast from the white background to make reading possible. In the same way, for valleys in a band structure to be useful, we must be able to differentiate them. **Measurable valley contrast is a basic criterion** to enable practical access to the contrasting states. The level of contrast is defined in terms of the physical property that is used as a proxy measure of the charge/spin occupancy of one valley *vs* another. Technically, this is referred to in literature as the **degree of valley polarisation**.

It is well known that significant effective mass anisotropy along different crystal axes exists in the valleys of some bulk crystals such as silicon [9], diamond [10] and bismuth [11], and such anisotropy are represented clearly in the cigar-shaped or ellipsoidal Fermi surfaces lying along the high-symmetry axes of these crystals. Such effective mass anisotropy can similarly be found in two-dimensional (2D) transport systems such as the surface state of silicon [9] as well as in AlAs heterostructured quantum wells [11]. Measurable physical transport properties (e.g., relaxation rates, mobilities, etc.) dependent on the effective masses could therefore be prospective proxies for measuring valley contrast. Other researchers also attempted more exotic techniques to create the requisite valley contrast (splitting) by proposing the engineering of atomic steps at the quantum-well interface, such as in Si-SiGe heterostructures [12]. More recently, the 2D material system has re-ignited significant efforts in valleytronics, particularly because valley contrast is already naturally present in some classes of 2D materials and are amenable to direct detection and control by familiar external fields which could be electric, magnetic or optical in nature. We shall trace the history of valleytronics in Section 1.4, but before that, it is instructive to introduce the concept of valley control.

1.3 Valley Control

For a physical system that stores information in contrasting states, practical methods of control are required in order to manipulate the system state as required by the application. Electric fields are used to dynamically turn on or off transistors, and local magnetic fields can switch the polarity of magnetic domains for information storage. In a solid-state device, manipulation of the valley states could involve populating (filling) or depopulating (unfilling) the selected valley subband. The preferential populating of one valley *vs* another valley results in what is known as *valley polarisation*. This control should ideally be implemented by some external field which could be easily varied. If the valleys are distinct, control could also imply a way to channel current only through a specific valley state. **Dynamic valley control is another key criterion** for its utility, and the physics behind such control will be dealt with in Chapter 2.

Attempts to control the valley degree of freedom (also called the valley index or quantum number) were reported for 2D electron-gas systems found in AlAs heterostructure-based devices using a magnetic field [13], as well as employing strain to induce symmetry breaking [14]. In bulk crystal systems, researchers have investigated in 2012 the rotation of an external magnetic field for tuning the degree of valley polarisation in bismuth [11]. Another scheme involving the electric field pulsing to exploit anisotropic effective masses of electrons along orthogonal crystal axes in diamond has also shown the possibility to drive valley repopulation of hot electrons from the light mass valleys to the heavy mass valleys in 2013 [10].

While the above attempts to control the valley degree of freedom clearly show the continued desire to utilise it for applications, they are fraught with practical challenges that hinder rapid advancement.

1.4 Revival of Valleytronics in 2D Crystals

Perhaps the clearest initial efforts to utilise the valley degree of freedom is represented by the attempts in the late 1970s (for example by Ohkawa and Uemura [15] and Sham *et al.* [16]) to explain unusual oscillations observed in the magnetoconductivity of the silicon MOSFET (metal-oxide-semiconductor field-effect transistor). Various theories were forwarded to explain these oscillations in terms of valley splitting and the concomitant formation of minigaps which may be observed as interband absorption using optical conductivity measurements. As required by the criterion for observable or measurable valley contrast in Section 1.2, such distinguishable minigap states may fulfil the first step for potential utilisation of the valleys in the silicon MOSFET. Nevertheless, control over such valley states proved elusive and to date no practical valley devices based on the Si MOSFET abound despite the tremendous advances in that technology for logic applications.

Fortuitously, the present generation has the privilege of witnessing the rise of 2D materials in the early 2000s (heralded by the isolation of graphene in 2004 and entrenched by a Nobel Prize in 2010 to Andre Geim and Kostya Novoselov). A significant body of work on valleytronics in the last two decades has arisen based on 2D materials. We therefore trace the developments of 2D valleytronics here by highlighting key milestones which will be further developed in the respective chapters in the rest of this book.

Typically, there is lag between experimental and theoretical developments. While one expects theoretical predictions to precede experimental verifications, it may sometimes occur that unexpected observations drive theoretical explanations or refinements. In what follows, we shall track the broad historical developments in terms of four categories: (1) theoretical developments; (2) valley contrast detection; (3) valley control demonstration; (4) valley current in devices.

1.4.1 Early Theoretical Developments

The initial theoretical insights for 2D valleytronics are often attributed to a pair of seminal papers published in 2007 by two groups of researchers [17], [18]. In the work by Xiao *et al.* [17], they propose that the breaking of spatial inversion symmetry in the monolayer honeycomb lattice of graphene enables valley-contrasting properties, allowing the direct detection of valley polarisation and observable manifestation of Berry phase-driven topological transport in what is known as the valley Hall effect (more details in Chapter 2). This rides on previous experimental works showing that inversion symmetry in graphene may be broken by interaction with a substrate potential [19] or in a bilayer graphene system subjected to an interlayer voltage [20], [21]. In contrast, the work by Rycerz *et al.* explores the use of a 1D graphene ribbon with zig-zag edges as a valley valve to allow selective spin transfers between two spin reservoirs of 2D graphene whereby the polarity of the spin transfer can be switched by the application of a voltage to the valve via a local electric gate [18]. These initial works motivated many experimental validations in sub-sections below.

Importantly, the insight offered by Xiao *et al.* [17], that the **breaking of inversion symmetry in 2D materials results in Berry curvatures of opposite signs**, led to the fervent investigations of a special class of 2D materials known as the transition metal dichalcogenides (TMDs). While adopting a typical hexagonal lattice like graphene, the two species of atoms within a unit cell deprive TMDs of an inversion centre. Intriguingly, this is the exact condition needed to produce the contrasting Berry curvatures in the K and K' valleys, thus making the valley index a good quantum number. Whereas the valley contrast in graphene is not intrinsic but needs to be interface engineered, TMDs possess this valley contrast intrinsically by virtue of the inherent inversion asymmetry [22], [23]. In addition, the 2H-phase of TMDs such as MoS_2, $MoSe_2$, WS_2 and WSe_2 are direct bandgap semiconductors. This implies that the search for valley physics in such materials could readily start through the use of common photoluminescence experiments with circularly polarised light, which can selectively address the opposite angular momenta in the K and K' valleys.

Indeed, such experiments were reported soon after another theoretical work by Xiao *et al.* in 2012 showed how valley-dependent optical selection rules can be exploited for studying the valley-coupled spins states (or **valley pseudospins**) in TMDs [24].

Experimental efforts quickly accompanied the theoretical treatise by Xiao *et al.*, and the race to demonstrate valley-dependent optical phenomena was marked by no less than the following four papers published in 2012 to 2013. These works are amongst the first to confirm the theoretically predicted valley-dependent optical selection rules based on the valley physics of monolayer TMDs by performing helicity-dependent photoluminescence measurements.

1.4.2 Observations of Valley Selectivity by Helicity-Dependent Optical Excitations

The first three works below employed circularly polarised light excitation using a 633 nm (1.96 eV) HeNe laser to excite monolayer MoS_2. This excitation energy was chosen to closely match the direct bandgap of about 1.9 eV determined for MoS_2 to allow addressing only the topmost valence band state in a specific valley. The experiments were typically performed with the sample at cryogenic temperatures below 100 K.

Cao *et al.* [25] demonstrated that up to 50% valley polarisation was possible in monolayer MoS_2 supported by 90 nm SiO_2 on a Si substrate at a temperature of 83 K, thus indicating a significant degree of selection of one valley excited by the circularly polarised light. Imperfect valley polarisation was ascribed to intervalley relaxation processes, which can be induced by defects and substrate effects.

Zeng *et al.* [26] also showed that for MoS_2 monolayer on Si/SiO_2 substrate under circularly polarised excitation, about 32% valley polarisation was achieved at 10 K using both left- or right-handed circularly polarised light of 633 nm (1.96 eV). For MoS_2 bilayers which are inversion symmetric, they found negligible valley polarisation. These observations were taken as evidence that inversion breaking present in monolayer MoS_2 was responsible for the valley polarisation.

Mak *et al.* [27] impressively alleviated the substrate-related intervalley scattering effects observed by the previous two teams by placing their mono- and bilayer MoS_2 directly on hexagonal boron nitride (hBN) instead of the more disordered SiO_2 substrate. With this improvement, they demonstrated that dynamic control reaching 100% valley polarisation can be achieved in inversion asymmetric monolayer MoS_2 via optical pumping with circularly polarised light. In contrast, they observed that the degree of photoluminescence circular polarisation was significantly suppressed in inversion symmetric bilayer MoS_2, thus providing strong evidence in support of inversion breaking as the requisite condition leading to the valley-selective optical excitation in monolayer MoS_2. They also deduced that valley polarisation persisted for more than 1 ns in their sample at a temperature of 14 K.

Jones *et al.* [28] extended the observation of selective valley optical excitation to another member in the TMD family, namely monolayer WSe_2, using 1.88 eV excitations with the sample at 30 K. Insightfully, their work also shed new light on the phenomenon of valley coherence through an investigation of the linearly polarised luminescence. They found in this study that valley coherence was preserved by valley excitons but not by valley trions (charge-coupled excitons).

Thus far, the first works confirming the optical selectivity of valleys in monolayer TMDs are consistent with theory. However, it is not unusual to find experimental observations that deviate from theoretical expectations in research. With more reports showing similar deviations, researchers would have to reconsider or refine the basic theoretical models to explain the apparent deviations. A balanced survey of the history for valleytronics should therefore present such atypical observations as well. Here, we highlight in particular the anomalous valley polarisation behaviour in bilayer or even multilayer WS_2 that was first reported in 2014 by Zhu *et al.* [29], and further observed by a few other groups including Su *et al.* [30], Nayak *et al.* [31], and also confirmed in the laboratory of this author.

Theory predicts that the non-centrosymmetric TMDs should exhibit strong valley polarisation under circularly polarised excitation which can selectively excite only one of the band edge valleys (i.e., K or K'). Indeed, this was demonstrated experimentally with effectively full valley polarisation for MoS_2 by Mak *et al.* [27] as we saw earlier. One would naturally expect the other members of the TMD family (e.g., $MoSe_2$, WS_2, WSe_2) to behave in the same way. In contrast, bilayers of the TMD family adopt the stable inversion symmetric phase and are expected to have no valley polarisation under circularly polarised illumination. Again, the early works on MoS_2 bilayers appear to concur with this theoretical expectation.

In 2014, Zhu *et al.* [24] surprisingly observed close to 100% valley polarisation in bilayer WS_2 at 10 K when excited close to resonance by photons of energy 2.088 eV, and this strong valley polarisation persisted almost independent of temperature up to 200 K, and thereafter decreases to about 60% at room temperature (300 K). Interestingly, for the inversion asymmetric monolayer WS_2 where one would expect strong valley polarisation, their sample showed less than 50% valley polarisation. Two years later, Nayak *et al.* [31] reported a similar observation showing about 35% and 80% valley polarisations in monolayer and bilayer WS_2, respectively, but this time at room temperature. Su *et al.* [30] followed up in 2017 reporting that multilayer (up to five layers) WS_2 samples showed typically higher degrees of valley polarisation (up to about 80%) compared to a surprisingly low 23% for monolayer WS_2 at 10 K. These unexpected findings throw caution to the rather simple theoretical picture where the degree of valley polarisation could be a straightforward proxy indicator of spin-valley coupling on the basis of breaking inversion symmetry. This author's laboratory repeated the multilayer WS_2 experiments in 2017 and similarly found less than 50% valley polarisation in monolayer WS_2 compared to around 80% or more for 2- to 4-layer WS_2 exfoliated from a bulk crystal. It appears that real 2D TMD systems, e.g., WS_2, could be more complex in character. Several explanations [32] have been proposed by these researchers, including (I) coupling of spin, layer and valley degrees of freedom, (II) layer decoupling with suppressed interlayer hopping events, and (III) suppression phonon-assisted intervalley scattering mechanisms for multilayers. These will be dealt with in more detail in Chapter 3. The actual mechanism leading to the anomalous valley polarisation in WS_2 remains unresolved and will certainly require more work. Some theoretical works [33], [34] have indeed proposed more rigorous ways to model the valley polarisation in 2D TMDs and it remains to be seen if more experiments and theory development could unveil the real reason behind the peculiar observations for 2D WS_2.

1.4.3 Observations of Valley Hall Effect — Spatial Separation of Valley-coupled Spins

For the current transport in a conventional semiconductor, the well-known Hall effect manifests as the migration of positive and negative charges to opposite edges of the transport channel under the application of a magnetic field. There is an analogous **valley Hall effect** (covered in Chapter 2) in valley semiconductors where opposite spins migrate to the opposite edges of the transport channel but this time under the electric field driving the transport current. This effect, which manifests electrically as an anomalous Hall effect (AHE) due to the coupling of the valley degree of freedom to the orbital angular momentum of carriers, was first reported for the TMD monolayer MoS_2 in 2014 by Mak *et al.* [35]. Here they observed photoinduced AHE where a net valley polarisation was induced by elliptically polarised light incident on the monolayer MoS_2. They also showed that AHE was not observed in bilayer MoS_2 with inversion symmetry, confirming the need for inversion breaking to effect the spin-valley coupling.

Although graphene is intrinsically inversion symmetric, this symmetry may be broken, for example, by layer-stacking graphene over hBN with aligned crystallographic axes. Gorbachev *et al.* demonstrated this in 2014 and thereby showed that the valley Hall effect could be induced in graphene by the coupling to hBN [36]. Note that in their experiment, the valley Hall effect is electrically driven, and also electrically detected through the measurement of a non-local voltage in a Hall bar configuration similar to those typically used in spin Hall effect measurements. In high electronic quality graphene, their work suggests that the long-range topological currents originating from the graphene valleys can be controlled by a gate voltage as in typical transistors, and thus could be useful for valley-based information processing.

The above two experiments used electrical methods to detect the valley Hall signature. However, it is also possible to employ optical methods which are sensitive to spin orientations to directly visualise the migration of the opposite spins in the K and K' valleys migrating to opposite edges of the valley Hall device.

In 2016, Lee *et al.* demonstrated this in a bilayer MoS_2 transistor using optical Kerr rotation microscopy which is sensitive to surface spin-polarisation of the material studied [37]. In Kerr rotation measurement, linearly polarised light shone on the material interacts with the net valley (or orbital) magnetisation in the material due to the valley-coupled spins. This results in a small but detectable rotation in the polarisation of the reflected light. This rotation has opposite signs for the K and K' valleys and hence it can be a proxy that allows one to map the spatial accumulation of valley-coupled spins driven by a longitudinal electric field. In addition, the researchers also established that the valley-coupled spin directions are out of plane with respect to the bilayer MoS_2 crystal plane. It is of interest to note that as the bilayer MoS_2 used in their experiment was inversion symmetric, they had to apply an additional vertical electric field to the bilayer to break inversion symmetry in order to enable spin-valley coupling and hence observe the valley Hall effect by Kerr rotation microscopy. The ability to use an electric field to turn on and off the inversion symmetry makes the valley conductivity tunable and hence potentially advantageous for realising valleytronics devices. We will discuss this further in Section 1.4.4.

The valley Hall effect was also observed by spatially resolved polarisation-dependent photoluminescence spectroscopy via the interlayer exciton in bilayer MoS_2/WSe_2 heterostructures in 2020 by Huang *et al.* [38]. Optical observations of the valley Hall effect, while in principle possible, are made challenging by the typically short intralayer exciton lifetimes in monolayer TMDs. By going to a hetero-bilayer system such as MoS_2/WSe_2 with longer interlayer exciton and valley lifetimes, optical detection becomes feasible, and Huang *et al.* showed that this valley Hall signature remained strong up to room temperature [38].

1.4.4 Control of Valley-coupled Spins

Not surprisingly, researchers were already thinking of and exploring valley-related transport phenomena in 2D TMDs even as the proofs of spin and valley coupling from inversion breaking started to emerge in the aforementioned experiments. Theoretical work reported in 2013 by Gong *et al.*, for example, expounded on the possibility to elicit magnetoelectric effects from 2D TMDs as well as the use of valley-controlled spins as qubit building blocks for quantum computing applications [39]. An effort to consolidate the understanding and potential ideas for exploration and applications was epitomised by the review paper of Xu *et al.* in 2014 [22].

Apart from the observations of valley polarisation and coherence, there was also interest to investigate the tunability of valley-related properties by optical means. Using high-intensity circularly polarised light, Sie *et al.* showed in 2015 that it is possible to break time-reversal symmetry and demonstrate valley-selective energy level shifts in practical field used in the laboratory [40]. They showed it was possible to selectively tune the exciton level in each valley by up to 18 meV via the optical Stark effect, and proposed that this may open the way to achieve novel Floquet topological phases in 2D TMDs.

Following the initial demonstrations of the valley Hall effect in monolayer MoS_2 and graphene/hBN heterostructure, a slew of reports showing electrical control of valley transport or current ensued. For bilayer graphene which is intrinsically inversion symmetric, the application of an electric field perpendicular to the bilayer opens a bandgap and breaks inversion symmetry resulting in contrasting Berry curvatures at the K and K' valleys at the corners of the hexagonal Brillouin zone. In 2015, two research teams, Sui *et al.* [41] and Shimazaki *et al.* [42], demonstrated full electrical generation and detection of pure valley currents in the graphene bilayer using Hall-bar devices and 4-terminal measurements which allowed them to electrically generate the valley currents between one terminal pair and electrically detect the valley current via non-local resistance/voltage measurement across the other terminal pair. This was a significant step towards all electrical valleytronic devices.

For TMD materials, this appeared to be the next reasonable goal as well. Hence, Lee *et al.* attempted to apply electrical control to a bilayer MoS_2 transistor in 2016 [37]. Unlike the monolayer MoS_2 they studied earlier to demonstrate the valley Hall effect [35], the bilayer MoS_2 would require an electric field perpendicular to the plane of the bilayer in order to break inversion symmetry and thereby couple spins back to specific valleys. Just as for bilayer graphene devices, a longitudinal in-plane electric field is required to drive the valley current. In this case however, they only used a two-terminal device configuration and hence were not able to electrically detect the valley currents. Instead, they employed optical Kerr rotation measurements to detect opposite spins accumulation at the device edges [37].

The full electrical generation and detection of the valley Hall effect in TMDs had to wait a couple more years. In 2019, Hung *et al.* eventually reported the valley Hall effect in monolayer MoS_2, using the inverse valley Hall effect to electrically detect the presence of the generated valley current [43]. This was a heroic effort nonetheless, because unlike previous scientific studies using exfoliated materials, Hung and co-workers have opted to use materials grown by the more common chemical vapour deposition (CVD) method which has better prospects for process scalability. Not only was their MoS_2 CVD grown, their capping dielectric (hBN) layer was also CVD grown. We note here that these CVD processes are still non-standard and have to be self-developed in research laboratories as of this writing. Impressively they showed that the valley currents generated in their devices persisted over a distance of 4 microns at room temperature, thus demonstrating the viability to operate a valley device without the overheads of cryogenic cooling.

Soon after, Li *et al.* reported in 2020 the implementation of a full valley current-based monolayer MoS_2 transistor that works at room temperature and improved the valley current drift distance to 16 microns [44]. In contrast to conventional charge current-driven transistors, they demonstrated that the transistor effect was not driven by a charge current, but by a valley current instead. They also showed how this valley Hall output could be cascaded to drive the next valley transistor using purely valley-based input or output signals. This opens ways for future development of full valley-based logic circuits.

Finally, we mention briefly here the work reported in 2018 by Cha *et al.* [45] showing how valley-locked spin photocarriers in WSe_2 can be exploited to effect spin diffusion into a laterally connected graphene layer over submicron length scale. While they observed a sizable non-local spin polarisation of about 50% in their integrated devices, their observed valley spin injection efficiency is $<1\%$. This nevertheless demonstrates the unique opportunity to engineer devices requiring novel integration of optical, spin and valley degrees of freedom for applications. This work will be discussed in more detail in Chapter 7.

1.4.5 Magnetic Control of Valley Pseudospins

It is well known that the application of a magnetic field *B* lifts the energy degeneracy in opposite spin states by the Zeeman effect resulting an energy splitting of $-\mu B$ where μ is the associated magnetic moment. In addition, the spin orbiting in the plane of a monolayer TMD also possesses an additional magnetic dipole moment due to the orbit. As the spin-orbit interaction in TMDs are typically strong, the total magnetic moment (comprising the spin and orbital contributions) normally describes the energy spectrum splitting due to an external *B* field. Applying a perpendicular magnetic field *B* lifts the degeneracy between the opposite magnetic moments coupled to the K and K' valleys resulting in an associated energy splitting or difference between the valley states in the TMD. This is known as the valley Zeeman effect.

Srivastava *et al.* [46] and Aivazian *et al.* [47] reported the valley Zeeman effect in 2015 by studying the optical excitations of monolayer WSe_2 in the presence of a magnetic field. In their circular polarisation-resolved photoluminescence studies, they found that the spectral peaks from the K and K' valleys are separated by ~0.1 to 0.25 meV T^{-1} under out-of-plane magnetic fields of several teslas. In that same year, MacNeill *et al.* used the same technique to measure the valley Zeeman splitting in monolayer $MoSe_2$, reporting a linear splitting of –0.22 meV T^{-1} [48].

The measurement of valley Zeeman splitting marks a first step towards using magnetic control to tune valley-associated states. In

2016, Wang *et al.* went another step further to demonstrate that the exciton valley coherence in monolayer WSe_2 could be controlled by an external magnetic field [49]. Using a photoluminescence setup, they first prepared a coherent superposition of valley states, and then demonstrated the possibility to rotate the coherent superposition of valley states by as much as 30° using perpendicular fields up to 9 T. For applications requiring quantum state manipulation (e.g., qubits), this represents the essential capability to rotate the exciton spin around the equatorial plane of the Bloch sphere, albeit using a rather large external magnetic field.

1.5 Summary

In this chapter, we have given a brief historical treatment of the developments attempting to harness the valley degree of freedom for devices toward information processing. In particular, we trace the fervent developments in 2D materials which appear to be naturally imbued with the desirable properties for valley accessibility and control, and thus potentially attractive for technology engineering. This short whirlwind tour does not do justice to the multitude of works that represent decades of attempts to make a practical valley device. Our main goal is to provide the reader with a general appreciation of the difficulties that delayed the access and control of the valley degree of freedom for a few decades and highlight the key milestones that represent the recent uptrend in the area of valleytronics. Fig. 1.2 provides a corresponding pictorial historical timeline. For the motivated reader, the key developments will be expanded in the subsequent chapters, and we provide our perspective for future developments in the concluding Chapter 8. In the next chapter, we start with an introduction to the valley physics that undergird valleytronics.

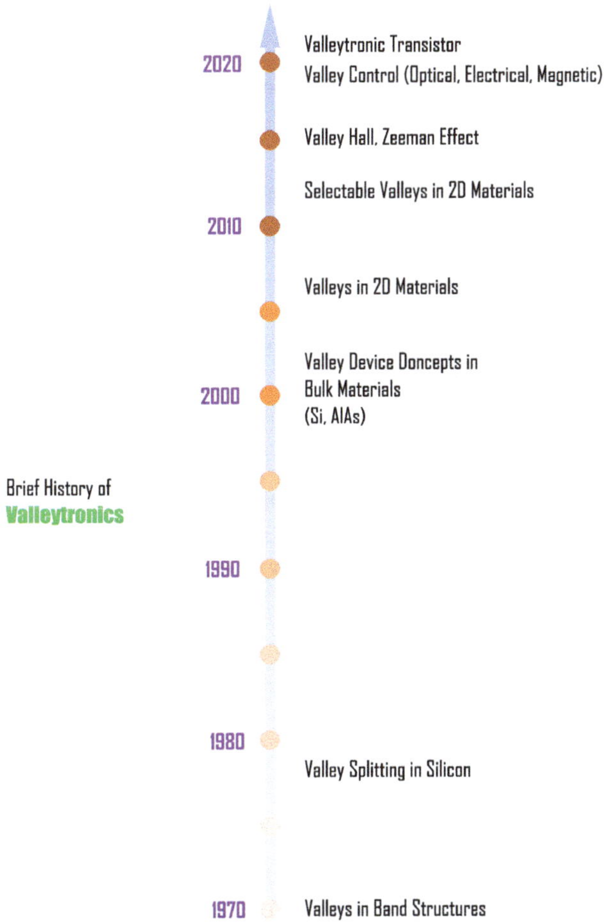

Fig. 1.2. A brief graphical history of valleytronics.

References

[1] J. M. Ziman, *Principles of the Theory of Solids*. Cambridge University Press, 1972, doi: 10.1017/CBO9781139644075.

[2] N. W. Ashcroft and N. D. Mermin, *Solid State Physics*. Saunders College Publishing, 1976.

[3] C. Kittel, *Introduction to Solid State Physics*, 8th ed. John Wiley & Sons, 2004.

[4] K. Novoselov, "Graphene: Materials in the Flatland," *Noble Lecture*. https://www.nobelprize.org/prizes/physics/2010/novoselov/lecture/ (accessed Sep. 27, 2022).

[5] A. K. Geim and K. S. Novoselov, "The rise of graphene," *Nat. Mater.*, vol. 6, no. 3, pp. 183–191, 2007, doi: 10.1038/nmat1849.

[6] A. H. Castro Neto, F. Guinea, N. M. R. Peres, K. S. Novoselov, and A. K. Geim, "The electronic properties of graphene," *Rev. Mod. Phys.*, vol. 81, no. 1, pp. 109–162, 2009, doi: 10.1103/RevModPhys.81.109.

[7] X.-L. Li and S.-C. Zhang, "Topological insulators and superconductors," *Rev. Mod. Phys.*, vol. 83, no. 4, p. 1057, 2011, doi: 10.1103/RevModPhys.83.1057.

[8] E. Y. Andrei *et al.*, "The marvels of moiré materials," *Nat. Rev. Mater.*, vol. 6, no. 3, pp. 201–206, 2021, doi: 10.1038/s41578-021-00284-1.

[9] K. Takashina, Y. Ono, A. Fujiwara, Y. Takahashi, and Y. Hirayama, "Valley polarization in Si(100) at zero magnetic field," *Phys. Rev. Lett.*, vol. 96, no. 23, p. 236801, 2006, doi: 10.1103/PhysRevLett.96.236801.

[10] J. Isberg, M. Gabrysch, J. Hammersberg, S. Majdi, K. K. Kovi, and D. J. Twitchen, "Generation, transport and detection of valley-polarized electrons in diamond," *Nat. Mater.*, vol. 12, no. 8, pp. 760–764, 2013, doi: 10.1038/nmat3694.

[11] Z. Zhu, A. Collaudin, B. Fauqué, W. Kang, and K. Behnia, "Field-induced polarization of Dirac valleys in bismuth," *Nat. Phys.*, vol. 8, no. 1, pp. 89–94, 2012, doi: 10.1038/nphys2111.

[12] S. Goswami *et al.*, "Controllable valley splitting in silicon quantum devices," *Nat. Phys.*, vol. 3, no. 1, pp. 41–45, 2007, doi: 10.1038/nphys475.

[13] Y. P. Shkolnikov, E. P. De Poortere, E. Tutuc, and M. Shayegan, "Valley splitting of AlAs two-dimensional electrons in a perpendicular magnetic field," *Phys. Rev. Lett.*, vol. 89, no. 22, p. 226805, 2002, doi: 10.1103/PhysRevLett.89.226805.

[14] O. Gunawan, Y. P. Shkolnikov, K. Vakili, T. Gokmen, E. P. De Poortere, and M. Shayegan, "Valley susceptibility of an interacting two-dimensional electron system," *Phys. Rev. Lett.*, vol. 97, no. 18, pp. 1–4, 2006, doi: 10.1103/PhysRevLett.97.186404.

[15] F. J. Ohkawa and Y. Uemura, "Theory of valley splitting in an N-channel (100) inversion layer of Si I. Formulation by extended zone effective mass theory," *J. Phys. Soc. Japan*, vol. 43, pp. 907–916, 1977, doi: 10.1143/JPSJ.43.907.

[16] L. J. Sham, S. J. Allen, A. Kamgar, and D. C. Tsui, "Valley-valley splitting in inversion layers on a high-index surface of silicon," *Phys. Rev. Lett.*, vol. 40, no. 7, pp. 472–475, 1978, doi: 10.1103/PhysRevLett.40.472.

[17] D. Xiao, W. Yao, and Q. Niu, "Valley-contrasting physics in graphene: Magnetic moment and topological transport," *Phys. Rev. Lett.*, vol. 99, no. 23, p. 236809, 2007, doi: 10.1103/PhysRevLett.99.236809.

[18] A. Rycerz, J. Tworzydło, and C. W. J. Beenakker, "Valley filter and valley valve in graphene," *Nat. Phys.*, vol. 3, no. 3, pp. 172–175, 2007, doi: 10.1038/nphys547.

[19] S. Gwo and C. K. Shih, "Site-selective imaging in scanning tunneling microscopy of graphite: The nature of site asymmetry," *Phys. Rev. B*, vol. 47, no. 19, pp. 13059–13062, 1993, doi: 10.1103/PhysRevB.47.13059.

[20] T. Ohta, A. Bostwick, T. Seyller, K. Horn, and E. Rotenberg, "Controlling the electronic structure of bilayer graphene," *Science*, vol. 313, no. 5789, pp. 951–954, 2006, doi: 10.1126/science.1130681.

[21] E. McCann and V. I. Fal'ko, "Landau-level degeneracy and quantum hall effect in a graphite bilayer," *Phys. Rev. Lett.*, vol. 96, no. 8, pp. 1–4, 2006, doi: 10.1103/PhysRevLett.96.086805.

[22] X. Xu, W. Yao, D. Xiao, and T. F. Heinz, "Spin and pseudospins in layered transition metal dichalcogenides," *Nat. Phys.*, vol. 10, no. 5, pp. 343–350, 2014, doi: 10.1038/nphys2942.

[23] J. R. Schaibley *et al.*, "Valleytronics in 2D materials," *Nat. Rev. Mater.*, vol. 1, no. 11, pp. 1–15, 2016, doi: 10.1038/natrevmats.2016.55.

[24] D. Xiao, G. Bin Liu, W. Feng, X. Xu, and W. Yao, "Coupled spin and valley physics in monolayers of MoS_2 and other group-VI dichalcogenides," *Phys. Rev. Lett.*, vol. 108, no. 19, p. 196802, 2012, doi: 10.1103/PhysRevLett.108.196802.

[25] T. Cao *et al.*, "Valley-selective circular dichroism of monolayer molybdenum disulphide," *Nat. Commun.*, vol. 3, no. May, p. 887, 2012, doi: 10.1038/ncomms1882.

[26] H. Zeng, J. Dai, W. Yao, D. Xiao, and X. Cui, "Valley polarization in MoS_2 by optical pumping," *Nat. Nanotechnol.*, vol. 7, no. 8, pp. 490–493, 2012, doi: 10.1038/nnano.2012.95.
[27] K. F. Mak, K. He, J. Shan, and T. F. Heinz, "Control of valley polarization in monolayer MoS_2 by optical helicity," *Nat. Nanotechnol.*, vol. 7, no. 8, pp. 494–498, 2012, doi: 10.1038/nnano.2012.96.
[28] A. M. Jones *et al.*, "Optical generation of excitonic valley coherence in monolayer WSe_2," *Nat. Nanotechnol.*, vol. 8, no. 9, pp. 634–638, 2013, doi: 10.1038/nnano.2013.151.
[29] B. Zhu, H. Zeng, J. Dai, Z. Gong, and X. Cui, "Anomalously robust valley polarization and valley coherence in bilayer WS_2," *Proc. Natl. Acad. Sci.*, vol. 111, no. 32, pp. 11606–11611, 2014, doi: 10.1073/pnas.1406960111.
[30] H. Su, C. Wei, A. Deng, D. Deng, C. Yang, and J. F. Dai, "Anomalous enhancement of valley polarization in multilayer WS_2 at room temperature," *Nanoscale*, vol. 9, no. 16, pp. 5148–5154, 2017, doi: 10.1039/c7nr00554g.
[31] P. K. Nayak, F.-C. Lin, C.-H. Yeh, J.-S. Huang, and P.-W. Chiu, "Robust room temperature valley polarization in monolayer and bilayer WS_2," *Nanoscale*, vol. 8, no. 11, pp. 6035–6042, 2016, doi: 10.1039/C5NR08395H.
[32] F. Bussolotti *et al.*, "Roadmap on finding chiral valleys: screening 2D materials for valleytronics," *Nano Futur.*, vol. 2, no. 3, p. 032001, 2018, doi: 10.1088/2399-1984/aac9d7.
[33] X. Zhang, Q. Liu, J. W. Luo, A. J. Freeman, and A. Zunger, "Hidden spin polarization in inversion-symmetric bulk crystals," *Nat. Phys.*, vol. 10, no. 5, pp. 387–393, 2014, doi: 10.1038/nphys2933.
[34] Q. Liu, X. Zhang, and A. Zunger, "Intrinsic circular polarization in centrosymmetric stacks of transition-metal dichalcogenide compounds," *Phys. Rev. Lett.*, vol. 114, no. 8, p. 087402, 2015, doi: 10.1103/PhysRevLett.114.087402.
[35] K. F. Mak, K. L. McGill, J. Park, and P. L. McEuen, "The valley Hall effect in MoS_2 transistors," *Science*, vol. 344, no. 6191, pp. 1489–92, 2014, doi: 10.1126/science.1250140.

[36] R. V. Gorbachev *et al.*, "Detecting topological currents in graphene superlattices," *Science*, vol. 346, no. 6208, pp. 448–451, 2014, doi: 10.1126/science.1254966.
[37] J. Lee, K. F. Mak, and J. Shan, "Electrical control of the valley Hall effect in bilayer MoS_2 transistors," *Nat. Nanotechnol.*, vol. 11, no. 5, pp. 421–425, 2016, doi: 10.1038/nnano.2015.337.
[38] Z. Huang *et al.*, "Robust room temperature valley Hall effect of interlayer excitons," *Nano Lett.*, vol. 20, no. 2, pp. 1345–1351, 2020, doi: 10.1021/acs.nanolett.9b04836.
[39] Z. Gong *et al.*, "Magnetoelectric effects and valley-controlled spin quantum gates in transition metal dichalcogenide bilayers," *Nat. Commun.*, vol. 4, no. May, pp. 1–6, 2013, doi: 10.1038/ncomms3053.
[40] E. J. Sie, J. W. McLver, Y. H. Lee, L. Fu, J. Kong, and N. Gedik, "Valley-selective optical Stark effect in monolayer WS_2," *Nat. Mater.*, vol. 14, no. 3, pp. 290–294, 2015, doi: 10.1038/nmat4156.
[41] M. Sui *et al.*, "Gate-tunable topological valley transport in bilayer graphene," *Nat. Phys.*, vol. 11, no. 12, pp. 1027–1031, 2015, doi: 10.1038/nphys3485.
[42] Y. Shimazaki, M. Yamamoto, I. V. Borzenets, K. Watanabe, T. Taniguchi, and S. Tarucha, "Generation and detection of pure valley current by electrically induced Berry curvature in bilayer graphene," *Nat. Phys.*, vol. 11, no. 12, pp. 1032–1036, 2015, doi: 10.1038/nphys3551.
[43] T. Y. T. Hung, K. Y. Camsari, S. Zhang, P. Upadhyaya, and Z. Chen, "Direct observation of valley-coupled topological current in MoS_2," *Sci. Adv.*, vol. 5, no. 4, p. eaau6478, 2019, doi: 10.1126/sciadv.aau6478.
[44] L. Li *et al.*, "Room-temperature valleytronic transistor," *Nat. Nanotechnol.*, vol. 15, pp. 743–749, 2020, doi: 10.1038/s41565-020-0727-0.
[45] S. Cha *et al.*, "Generation, transport and detection of valley-locked spin photocurrent in WSe_2-graphene-Bi_2Se_3 heterostructures," *Nat. Nanotechnol.*, vol. 13, no. 10, pp. 910–914, 2018, doi: 10.1038/s41565-018-0195-y.
[46] A. Srivastava, M. Sidler, A. V. Allain, D. S. Lembke, A. Kis, and A. Imamoɋlu, "Valley Zeeman effect in elementary optical excitations

of monolayer WSe_2," *Nat. Phys.*, vol. 11, no. 2, pp. 141–147, 2015, doi: 10.1038/nphys3203.
[47] G. Aivazian *et al.*, "Magnetic control of valley pseudospin in monolayer WSe_2," *Nat. Phys.*, vol. 11, no. 2, pp. 148–152, 2015, doi: 10.1038/nphys3201.
[48] D. MacNeill *et al.*, "Breaking of valley degeneracy by magnetic field in monolayer $MoSe_2$," *Phys. Rev. Lett.*, vol. 114, p. 037401, 2015, doi: 10.1103/PhysRevLett.114.037401.
[49] G. Wang *et al.*, "Control of exciton valley coherence in transition metal dichalcogenide monolayers," *Phys. Rev. Lett.*, vol. 117, no. 18, p. 187401, 2016, doi: 10.1103/PhysRevLett.117.187401.

https://doi.org/10.1142/9789811229107_0002

Chapter 2

Valley Physics in 2D Materials

Ivan Verzhbitskiy and Kuan Eng Johnson Goh

2.1 2-Dimensional Materials

2-Dimensional (2D) materials are a modern 21st-century discovery led by the isolation of graphene from graphite and the demonstration of its unique electronic properties [1]. Layered materials like graphite show unique crystalline order: the layers of covalently bonded atoms are stacked on top of each other, held only by a weak van der Waals force. Hence, these are often called van der Waals materials. Single and few-layer materials can be isolated from the stack through mechanical exfoliation, solution processing, or grown as single crystals or polycrystalline films. In the extreme case of a single isolated layer, they exhibit quite different physical properties compared to their bulk counterparts, and some are extraordinary and unexpected. Graphene, for example, has many unique properties not commonly observed in the bulk (i.e., the stack). These properties have earned graphene many superlatives, including stronger than steel, lightweight, flexible, transparent electrical and thermal conductor, and the two scientists (Andre Geim and Konstantin Novoselov) credited with its exploration won the 2010 Nobel Prize in Physics [2], [3].

Electrically, graphene is considered a semiconductor but with zero bandgap and hence is also often referred to as a semi-metal. However, there are many other 2D monolayers that do have finite bandgaps ranging from elemental 2D layers such as silicene, germanene, phosphorene, and boronene, to the multi-elemental materials like the family of transition metal dichalcogenides (TMDs) with the general chemical formula of MX_2 (where M represents a transition metal atom and X stands for a chalcogen atom). It was quickly realised that the bandgap of the van der Waals materials can be tuned with the thickness of the stack. Moreover, certain materials, like MoS_2, revealed the transition from indirect to direct band transport as the stack is reduced to a single layer. All this inspired the breadth of possibilities for electronic device engineering with TMDs [4].

Crucially, the insight that monolayers can be a good valleytronics candidate opened many routes to control the valley degree of freedom for information processing based on valley-coupled spins [5], [6]. Since the discovery of graphene, over a hundred 2D materials have been isolated with thousands waiting to be discovered [7]. Unique properties of single-layer materials have boosted research in many areas of fundamental physics and applied science. For example, monolayers of WTe_2 or $MnBi_2Te_4$ have been praised for their impact on the physics of topological matter [8], [9], while 2D magnetism in layers of CrI_3 and $CrGeTe_3$ has sparked continued research in spintronics [10], [11].

Apart from the exciting physical properties of single layers, 2D van der Waals materials brought to light a plethora of novel phenomena. Among these are the evolution of these properties with strain [12] or the number of layers, interlayer dynamics in homo- and hetero-layered vertical stacks [13], and superconductivity and flat-band physics in stacks of misaligned or twisted layers [14]. These works have led to the genesis of even newer fields known as 'straintronics', 'layertronics', and 'twistronics'.

In this book, our main focus, however, is on 'valleytronics' and in the remaining sections of this chapter, our main objective is to establish the physics associated with the valleys in 2D semiconductors that is amenable to practical access and control by external fields so that meaningful devices could be realised. In Section 2.2, we shall motivate the idea of spin-valley coupling from the perspective of symmetries in space and time. We first discuss the 'simpler' system based on monolayer graphene with only a single type of atom in its crystal lattice. We extend this to the case of TMDs with two atomic species in each unit cell that tessellates the 2D crystal. In Section 2.3, we cover the concept of the Berry phase (and the associated Berry curvature) and how it leads to the valley Hall effect. The section further details the notion of the magnetic orbital moment and its impact on the optical selection rules in 2D TMDs. Next, we dissect the valley Zeeman effect which parallels the traditional Zeeman effect that breaks Kramer's degeneracy associated with the time-reversed spin-up and spin-down states. Section 2.4 delves into the valley dynamics in bilayer systems. Finally, we summarise with a reiteration of the defining physics for valleytronics in 2D systems with broken inversion symmetry.

2.2 Electronic and Optical Properties of 2-Dimensional Materials

In solid-state materials, atoms, ions, or molecules arrange themselves in a highly ordered, periodic structure known as a crystal lattice.

The periodicity of a lattice defines many properties of the material, from the crystal shape and cleavage to the electronic and optical response. Thinning down crystals into single layers may lead to a change of symmetry and corresponding changes in electronic and optical behaviours. In the following, we will discuss the lattice structures and symmetries of 2D materials and how they affect the crystal properties.

2.2.1 Graphene Lattice and Valleys

2.2.1.1 *Graphene Lattice*

The crystal lattice of graphene is formed by closely packed carbon atoms arranged in an infinite hexagonal (often called "honeycomb") structure. Owing to a particular bonding, the graphene lattice is ultimately 2D flat with no buckling, and thus offers a material of a single-atom thickness.

The graphene's unit cell, a fundamental unit of a lattice, has two carbon atoms, usually denoted as **A** and **B**, and is rotationally symmetric with a 120° angle around any point in the lattice, Fig. 2.1(a). In the momentum space, the fundamental unit of the reciprocal lattice is the Brillouin zone [15]. Graphene's Brillouin zone is also hexagonal with several high-symmetry points: Γ (the centre), M (the middle of the hexagon's side), and two consecutive, inequivalent corner points K and K′, Fig. 2.1(b). The other corner points can be expressed through one of these two by adding the reciprocal lattice vector.

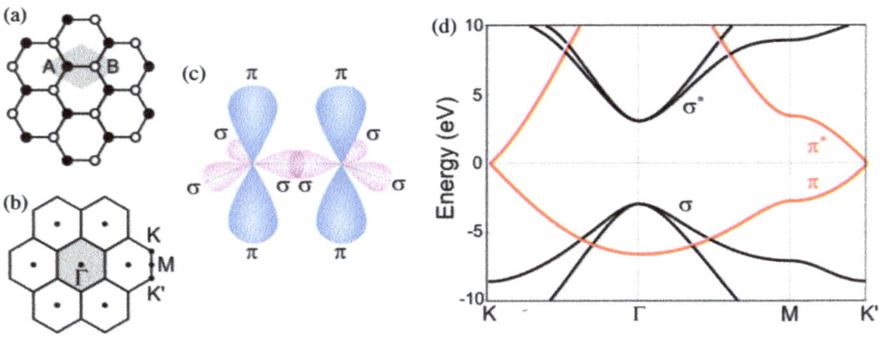

Fig. 2.1. (a) Crystal lattice (real space) of graphene; (b) Brillouin zone (reciprocal space) of graphene with high-symmetry points Γ, K and K′; (c) carbon orbitals form in-plane σ and out-of-plane π bonds; (d) 2D electronic structure of graphene calculated by tight-binding approximation along K-Γ-K′ direction.

2.2.1.2 *Electronic Structure of Graphene*

Carbon atoms in graphene are bonded through sp^2 orbital hybridisation. Hybridised orbitals (s, p_x, and p_y) form σ bonds with the three nearest neighbours, orienting themselves in the same plane with a 120° angle between each orbital, Fig. 2.1(c). The only unhybridised, pure orbitals (p_z) form a π bond which is oriented perpendicular to the plane. Two graphene sublattices form two bands, π and π*. From a simple tight-binding model that assumes overlapping of the wavefunctions of the neighbouring atoms, these two bands fully describe the low-energy electronic excitations in graphene [16] as well as graphite [17]. These orbitals produce valence and conduction bands, Fig. 2.1(d), which intersect at the corners of the Brillouin zone (K, K′ points). Calculation of the energy spectrum near these corner points yields a linear dispersion [16]:

$$\varepsilon_{\mathbf{K},\mathbf{K}'}(\mathbf{k}) = \pm\hbar\nu_{\mathrm{F}}\left|\mathbf{k}\right| \tag{2.1}$$

where $\hbar = h/2\pi$ and h is Planck constant, $\mathbf{k} = (k_x, k_y)$ is the electron wavevector and v_F is the Fermi velocity. The latter is the group velocity of the electrons at the Fermi energy, a chemical potential at $T = 0$. The negative (positive) energy band in Eq. (2.1), i.e., π (π^*), is fully occupied (empty), representing the valence (conduction) band of graphene. In pristine graphene, the Fermi energy level is set at the zero reference, illustrating the charge neutrality condition, where the Fermi surface (a surface separating filled and empty states) is 0D, i.e., represented by the K and K′ points only. This makes graphene a zero-gap material, often called "semi-metallic".

2.2.1.3 *Massless Particles of Graphene*

The linear energy-momentum relation of graphene, Eq. (2.1), near the K and K′ points resemble the relativistic energy-momentum relation of massless particles, e.g., phonons. Indeed, the energy, ε, of a relativistic free particle of mass m_0 in motion is $\varepsilon^2 = (m_0c^2)^2 + (cp)^2$, where c is the speed of light and $p = |\mathbf{p}|$ is the particle's momentum. In the special case of massless particles, $m_0 = 0$, this relation reduces to

$$\varepsilon = c|\mathbf{p}| \tag{2.2}$$

Expressing the momentum via wavevector, $p = \hbar k$, we can see that Eq. (2.2) mimics Eq. (2.1) with c replaced by v_F. This analogy concludes that electrons and holes in graphene at low energies move like massless particles! The electronic group velocities near the K and K′ points are as high as 8.5×10^5 m·s^{-1}, representing the effective "speed of light" of the massless fermions [18].

2.2.1.4 *Graphene Pseudospin and Chirality*

A more rigorous description of particles' behaviour involves solving the Schrödinger equation, a ubiquitous tool of quantum mechanics [19] that has the following form:

$$H\Psi = \varepsilon\Psi \tag{2.3}$$

where Ψ is the wavefunction describing the positions and spins of the particles (electrons, nuclei), while H is the Hamiltonian operator. The wavefunction, $\Psi(x,t)$, is a quantum mechanical concept that describes the quantum state (e.g., particle's position at a given time) of a system. The generally accepted wavefunction interpretation relates the integral of squared amplitude, $\int_a^b |\Psi(x,t)|^2 \, dx$, to the probability of finding the system at the region (between locations a and b) at time t [18]. This interpretation, however, makes us think that the phase of the wavefunction does not matter. On the contrary, the phase is responsible for a series of very important physical phenomena, like the Berry phase and Berry curvature that we will discuss in detail in Section 2.3.

The Hamiltonian operator corresponds to the total energy of the system and describes all forces that act on the system as well as all the interactions between the particles. In the remaining part of this chapter, we will often use the Hamiltonian representation as a very intuitive and descriptive approach.

The two inequivalent corner points of the Brillouin zone, K and K′, are described by individual *effective* Hamiltonians, H_K and $H_{K'}$. For simplicity, we will treat the H_K while the $H_{K'}$ can be obtained by the transposed operator: $H_{K'} = -H_K^t$. In the low-energy approximation, the effective Hamiltonian of the K point reads as:

$$H_K = \hbar \nu_F \begin{pmatrix} 0 & k_x - ik_y \\ k_x + ik_y & 0 \end{pmatrix} \tag{2.4}$$

Replacing the wavevector notation with the momentum, $p = \hbar k$, Eq. (2.4) can be rewritten as follows:

$$H_K = \hbar \nu_F \left(p_x \sigma_x + p_y \sigma_y \right) \tag{2.5}$$

where σ_x and σ_y are the Pauli matrices [19] which are generally defined as

$$\sigma_x = \begin{pmatrix} 0 & 1 \\ 1 & 0 \end{pmatrix}, \quad \sigma_y = \begin{pmatrix} 0 & -i \\ i & 0 \end{pmatrix}, \quad \sigma_z = \begin{pmatrix} 1 & 0 \\ 0 & -1 \end{pmatrix} \tag{2.6}$$

Using the notation $\hat{\sigma} = \left(\sigma_x, \sigma_y, \sigma_z \right)$, Eq. (2.5) can be written in a short form:

$$H_K = \nu_F \hat{\sigma} \cdot \mathbf{p} \tag{2.7}$$

Remarkably, the effective Hamiltonian in Eq. (2.7) is a 2D equivalent of the Dirac Hamiltonian for massless particles [20]. The Dirac Hamiltonian follows from the Dirac equation that was developed [21] to treat the relativistic electrons as waves and was one of the first theories that understood the need for the spin degree of freedom. Henceforth, the electrons and holes in graphene are often called massless Dirac fermions, while the linear energy-momentum dispersion at low energies is referred to as Dirac "cones" at Dirac points (K and K′) [22]. Fig. 2.2(a) shows a 3D view of the electronic band structure of graphene with prominent "conical" dispersion at Dirac points. From a more general standpoint, the extrema of the

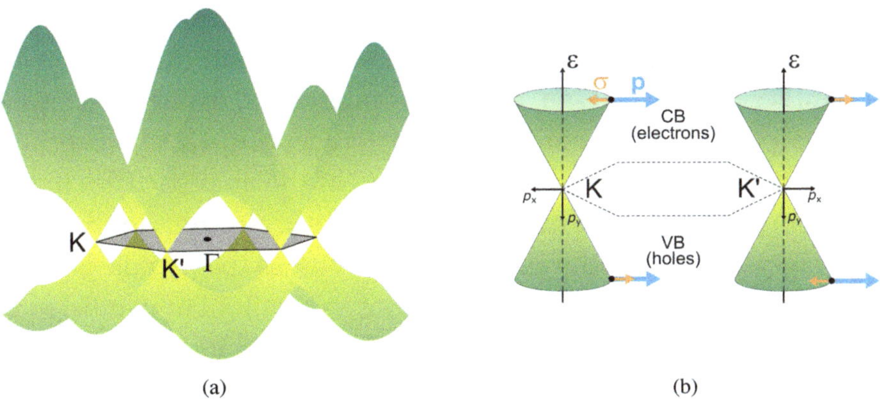

Fig. 2.2. (a) 3D electronic band structure of graphene (at low energy) calculated by the tight-binding approach; (b) zoom-in structure of the two inequivalent Dirac cones at the K and K′ points of the Brillouin zone (outlined by a dashed line). The directions of sublattice pseudospin, $\boldsymbol{\sigma}$, and momentum, $\boldsymbol{p}$, are shown to highlight the parallel and antiparallel relative orientations at conduction (CB) and valence (VB) bands.

electronic band structure are often called *valleys*. In graphene, conduction band minima and valence band maxima thus form valleys at Dirac points, K and K′.

Unlike in the Dirac representation, however, in graphene the Pauli matrices in Eq. (2.6) do not represent the "spin up" or "spin down" character of quasiparticles. Instead, the spin character is viewed through the sublattice parameter, e.g., "sublattice A" or "sublattice B", Fig. 2.1(a). Like the electron spin, this parameter takes one of the two values and is thus often referred to as a sublattice *pseudospin*.[a]

---

[a]Generally, a pseudospin of a system represents a two-level physical quantity acting like an effective spin-1/2. Originally, the term was introduced by Heisenberg to describe an atomic nucleus consisting of protons and neutrons, which he modelled as two states of the same particle [84]. In the case of monolayer TMD, we invoke valley pseudospin, while in bilayer materials, the layer index is a good quantum number, keying the layer pseudospin [85].

Interestingly, the projection of the pseudospin onto the momentum is a well-defined quantum mechanical quantity that can have a positive or negative sign, and this is interpreted as the pseudospin can be parallel or antiparallel to the momentum [23]. This brings in the notion of *chirality*, or handedness, of the quasiparticles in the momentum space. For example, the chirality of electrons in the K valley is opposite to the chirality of K′ electrons, as the pseudospin and momentum are antiparallel for electrons in K valley but parallel in K′ valleys, Fig. 2.2(b). The chirality introduces several intriguing properties of graphene. For example, backscattering, i.e., a change of a sign of the momentum of electrons in graphene, will require flipping the pseudospin. However, the direction of the pseudospin is locked to the direction of the momentum, which makes backscattering impossible for pristine, unperturbed graphene at low energies [24].

2.2.1.5 *Spin-orbit Interaction*

Another important relativistic effect that electrons experience in the presence of a local electric field is the spin-orbit coupling. Accelerated by the potential V, electrons move with velocity $\mathbf{v}$ under the electric field $\boldsymbol{E}$ obeying $-e\boldsymbol{E} = -\nabla V$. In the rest frame, however, electrons experience an effective (self-inflicted) magnetic field $B = -\mathbf{v} \times \boldsymbol{E}/c$. This effective magnetic field acts on the electron's spin through the Zeeman effect. The Hamiltonian of the spin-orbit interaction, H_{SOC}, then follows:

$$H_{\text{SOC}} \propto \boldsymbol{s} \cdot \boldsymbol{p} \times \nabla V \tag{2.8}$$

The notation $\boldsymbol{s}$ for spin is used to distinguish it from the graphene's pseudospin, σ. In an atom, V is the central potential while spin-orbit interaction is $\propto \boldsymbol{L} \cdot \boldsymbol{S}$, where $\boldsymbol{L}$ and $\boldsymbol{S}$ are the orbital angular momentum and spin, respectively. Clearly, as atomic potential increases with atomic number, spin-orbit interaction becomes stronger for heavier atoms. In a crystal, V stems from the periodic potential of the lattice as well as all the electric fields that act on the material. Due to the low mass of carbon atoms, however, the spin-orbit coupling in graphene is relatively weak, inducing a small spin splitting of ~40 μeV only [25]. In contrast, strong spin-orbit interactions in monolayer TMDs induce spin splittings of several hundreds of meV [26] (see Section 2.2.2.3).

2.2.2 Transition Metal Dichalcogenide Lattice

2.2.2.1 *Polymorphs and Polytypes*

Akin to graphene, TMDs are layered materials, where layers are bound by weak van der Waals forces. Monolayer TMD is a three-atom-thick material: the layer of transition metal atoms is sandwiched between two layers of chalcogen atoms. Depending on the metal coordination geometry within the layer, TMDs form different polymorphs, whereas the stacking order defines the polytype. For example, TMDs from group VIB transition metals, such as MoS_2 or WSe_2, are stable in trigonal prismatic coordination with a 2H (hexagonal, 2-layer unit cell) or 3R (rhombohedral, 3-layer unit cell) crystal structure, Fig. 2.3 [27], [28]. Naturally, TMD monolayers of 2H and 3R polytypes are identical. Often, these monolayers are identified as 1H-TMD, as the unit cell is limited by the monolayer thickness. Notably, TMDs formed by metals of other groups often occur in perfect octahedral coordination, 1T (trigonal, 1-layer unit cell), e.g., TiS_2 or VSe_2, or various distorted octahedral arrangements, e.g., 1T' in WTe_2, as shown in Fig. 2.3.

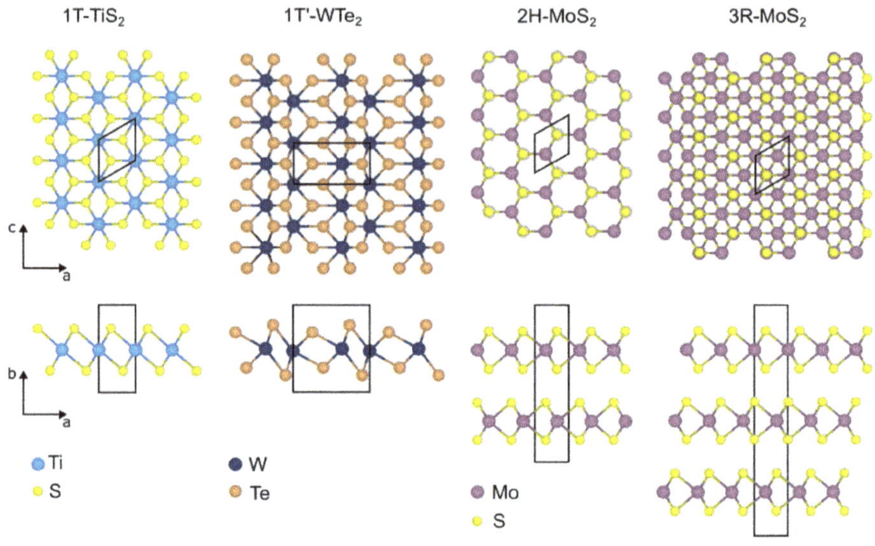

Fig. 2.3. Common crystal structures of transition metal dichalcogenides. The unit cell is outlined in black.

While TMD is a large family of layered materials, accounting for more than forty different compounds [29], including metals, semiconductors, superconductors [30], and quantum spin-Hall insulators [31], [32], we will focus our attention on MX_2 (M = Mo, W; X = S, Se), i.e., MoS_2, $MoSe_2$, WS_2, and WSe_2. Owing to their unique properties discussed below, these materials are an ideal platform for addressing and manipulating the valley degree of freedom, paving the way for new research in *valleytronics*.

2.2.2.2 *The Electronic Band Structure of 2H-MX_2*

The electronic structure of MX_2 can be inferred using a simple filling rule of the *d*-orbitals that are split according to the crystal field theory.

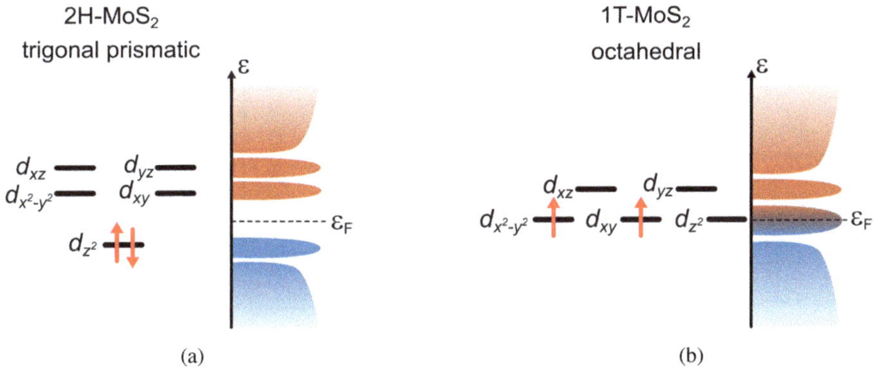

Fig. 2.4. Filling of the *d*-orbitals of a typical d^2 TMD (left panel) alongside the sketched band structure for (a) trigonal prismatic (2H-MoS_2), and (b) octahedral (1T-MoS_2) atomic arrangement. Adapted with permission from Ref. [86].

For example, MoS_2 is in a naturally occurring trigonal prismatic configuration (2H or 3R) with a completely filled d_{z^2} and fully empty d_{xy} and $d_{x^2-y^2}$ orbitals, Fig. 2.4(a). This makes 2H-MoS_2 exhibit properties of an insulator, as the Fermi energy, ε_F, is located inside the gap. In practice, 2H-MoS_2 often demonstrates semiconducting properties as the Fermi level can be tuned into the onset of the conduction or valence bands. In contrast, the octahedral 1T-MoS_2, Fig. 2.4(b), shows metallic behaviour with partially filled energy-degenerate bands (d_{xy}, d_{xz}, and d_{yz}). Such a simple band-filling method, however, can only provide a qualitative picture. A more rigorous electronic band structure can be obtained from density functional theory (DFT) calculations.

Figure 2.5 shows a DFT-calculated dispersion of the electronic bands of a monolayer, Fig. 2.5(b), and bulk 2H-MoS_2, Fig. 2.5(c). The band structure is represented by several valleys associated with high-symmetry points of the Brillouin zone, e.g., the centre (Γ) and corners (K and K′), as well as midpoints like Λ and M, Fig. 2.5(a). When monolayer

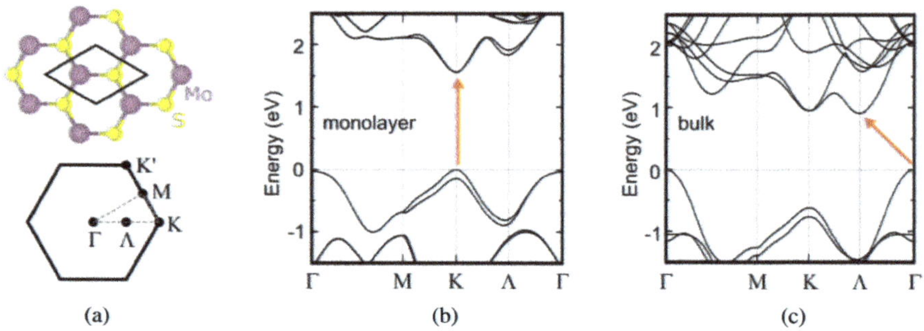

Fig. 2.5. (a) Crystal lattice (top) and the Brillouin zone of monolayer MoS_2 (bottom), highlighting high-symmetry points; Band structure of (b) bulk and (c) monolayer MoS_2. The arrow depicts the lowest energy transition; (b) and (c) adapted with permission from Ref. [87].

and bulk band dispersions are compared side by side, the most salient feature is the evolution of the lowest conduction and highest valence bands that lead to the transition from an indirect gap in bulk to a direct gap in the monolayer. Indeed, the valence band maximum (VBM) and conduction band minimum (CBM) of bulk MoS_2 are located at Γ and Λ points, respectively. In monolayer, however, both VBM and CBM are at the K point. This remarkable transition is reproduced in similar MX_2 materials, such as WS_2 and $MoSe_2$ [33].

The direct bandgap is a characteristic feature of the monolayer MX_2 that vanishes abruptly with increasing thickness. Indeed, the indirect bandgap nature is already restored in bilayers, which rules out the quantum confinement effect as the sole reason behind the indirect-to-direct bandgap transition. Detailed evolution of the MX_2 electronic band dispersion [34], [35] reveals that valleys at Λ and Γ points are strongly affected by the number of layers. The electronic bands at these points consist of an admixture of the metal *d*-orbitals

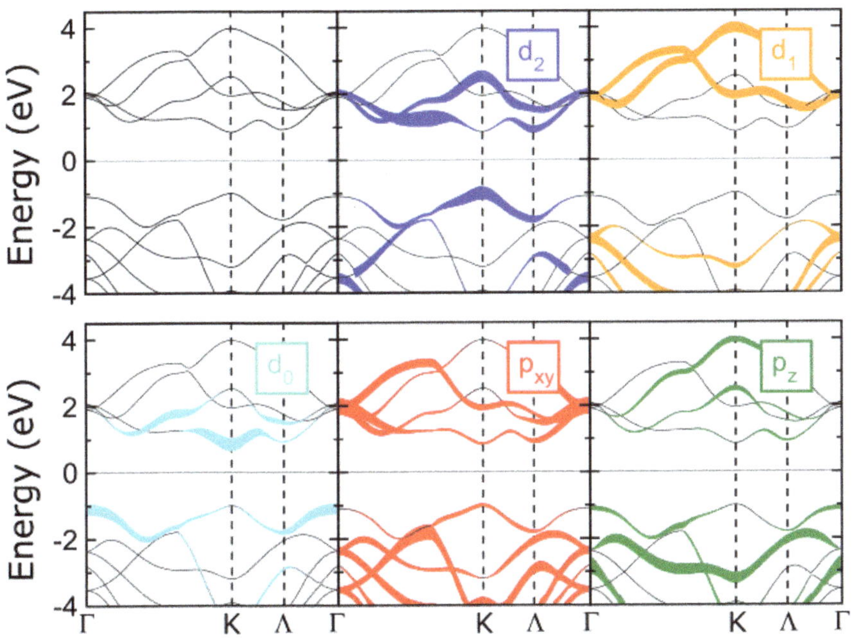

Fig. 2.6. Band structure of monolayer MoS_2 highlighting the contributions from different Mo *d*-orbitals ($\mathbf{d}_2 = d_{x^2-y^2}, d_{xy}$, $\mathbf{d}_1 = d_{xz}, d_{yz}$, $\mathbf{d}_0 = d_{z^2}$) and *p*-orbitals of S ($p_{xy} = p_x, p_y$; $p_z = p_z$). The coloured belt represents the strength of contributions. Reproduced with permission from Ref. [34].

and chalcogen *p*-orbitals, thus spreading the wavefunctions across the whole three-atom-thick X-M-X slab. In contrast, wavefunctions associated with K points stem predominantly from the metal *d*-orbitals, Fig. 2.6, tightly confining the states to the middle of the layer. This makes the indirect gap between Λ and Γ points change faster with the thickness than the states at K, triggering the indirect-to-direct transition. Notably, a similar evolution has been reported for other 2D semiconductors like 2H-$MoTe_2$, where a direct bandgap is established in a few-layer thickness already [36].

2.2.2.3 *Spin-orbit Interaction and Spin-valley Locking*

Owing to the strong spin-orbit coupling of MX_2, the band dispersion exhibits strong spin-orbit splitting predominantly in the valence band, where it can reach several hundreds of meV at the K point (Δ_v in Fig. 2.7) in monolayers. The spin splitting of the conduction band at the K point is notably smaller (Δ_c in Fig. 2.7), varying in different materials from a few meV to several tens of meV [37]. This clearly distinguishes monolayer MX_2 from graphene, which exhibits an intrinsic spin-orbit gap of about four orders of magnitude smaller [25]. Besides the large masses of atoms like Mo or W, strong spin-orbit coupling in TMDs benefits from the broken inversion symmetry in a monolayer limit [26].

Similar to graphene, K and K′ points in monolayer MX_2 are connected by time-reversal symmetry, which produces the opposite spin splitting in these valleys, i.e., degenerate in energy, but non-degenerate in spin. For example, the spin direction of the top valence band in K is the opposite to the one at K′, Fig. 2.7. This inherent

Fig. 2.7. Schematic of the spin-orbit splitting of the conduction and valence bands in time-reversal-symmetric K and K′ valleys of (a) Mo-based and (b) W-based TMDs. Arrows ↑ and ↓ denote spin-up and spin-down bands, respectively.

coupling of spin and momentum allows us to access the valley degree of freedom of charge carriers, which is sometimes dubbed as *spin-valley locking* (see Section 2.3.4). Interestingly, as the spin-orbit interaction scales with the atomic number (Section 2.2.1), Mo-based and W-based TMDs experience a different strength of the spin-orbit coupling. This leads to a peculiar order swap of the spin-split conduction bands at the K point, Fig. 2.7. Remarkably, top valence and bottom conduction bands in MoX_2 exhibit the same spin character, while in WX_2 the order is reversed, and the nearest conduction and valence bands are of the opposite spin.

2.2.3 Optical Properties of Transition Metal Dichalcogenide Monolayers

As discussed previously, layered MX_2 (M = Mo, W; X = S, Se) materials are semiconductors with electronic properties that strongly depend on thickness. When thinned down to a monolayer limit, MX_2 exhibits a direct bandgap with the energy matching the visible spectrum. Interestingly, the dielectric screening in monolayer materials is substantially weaker than that in bulk slabs. This leads to the strong geometric confinement of the charge carriers within the layer and, consequently, strong Coulomb interactions. Such conditions favour the formation of excitons, i.e., strongly bound electron-hole pairs, when the material is illuminated with light of energy equal to or higher than the bandgap. The binding energy of these excitons falls in the range of ~ 0.5 eV, which is one to two orders of magnitude higher than in conventional materials such as GaAs [38], and thus exceeds thermal energy ($k_BT \approx 26$ meV at 300 K), stabilising excitons at room temperature and beyond.

The natural 2D confinement in monolayer MX_2 governs some of the most intriguing properties of excitons in condensed matter physics. For example, since both electrons and holes are tightly confined within the layer, excitons exhibit vanishingly small out-of-plane polarisability, highlighting their robustness in the presence of strong out-of-plane electric fields [39]. Coulomb interactions in monolayer MX_2 promote a formation of a plethora of different many-body quasiparticles. Besides neutral excitons, hydrogen atom-like electron-hole pairs, monolayer semiconductors host charged excitons (or trions), comprised of an exciton bound to a free charge charrier (electron or hole); biexcitons — four-particle quasiparticles (two holes and two electrons); charged biexcitons, formed by a trion and a neutral exciton, and many more [40]. It is not a surprise that excitonic species dominate the optical response of MX_2 monolayers.

The strong layer dependence of the electronic band structure MX_2 is firmly imprinted in the optical properties of MX_2. For example, excitonic photoluminescence from monolayer MoS_2 is more than four orders of magnitude higher compared with the bulk [41], [42]. The bright photoluminescence is attributed to the direct bandgap, Fig. 2.8(a), and features two peaks, A at ~1.9 eV and B at ~2.03 eV, Fig. 2.8(b), with energy difference matching the spin-orbit splitting of the bands at the K point. In multilayers and bulk MX_2, the photoluminescence quantum yield is hampered by the radiative recombination of indirect excitons, i.e., the phonon-assisted second-order process, which is significantly slower than non-radiative decay processes.

Concurrently, the optical absorption spectrum of MX_2 monolayers is dominated by excitonic resonances. Fig. 2.8(c) shows the

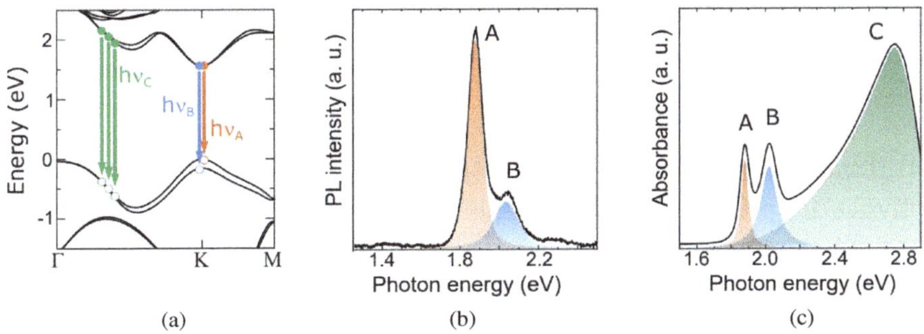

Fig. 2.8. Optical response of the monolayer MoS_2. (a) Schematic of the electronic band structure of monolayer MoS_2; optical transitions responsible for peaks in (b) and (c) are highlighted. (b) Photoluminescence (PL) excited at 2.33 eV and (c) optical absorbance of monolayer MoS_2 at room temperature. Labels A and B correspond to the direct optical transitions at the K (and K′) point highlighted in (a). The C peak originates from the band-nesting effect [43] near the Γ point.

room-temperature absorption spectrum of monolayer MoS_2. Alongside the A and B peaks, the third prominent feature at ~2.8 eV (C peak) dominates the spectrum. The C peak has been attributed to the band-nesting effect near the Γ point, which leads to the divergence in the joint density of states in 2D materials [43]. Despite being ultimately thin, monolayer MX_2 absorbs typically more than 15% of incident photons at the direct excitonic transition energy and even higher at the C peak resonance, highlighting the strength of light-matter interactions at the 2D limit. For the given thickness, monolayer MX_2 exhibits sunlight absorption an order of magnitude higher than Si or GaAs [44]. In contrast, monolayer graphene absorbs only 2.3% of incident light across the visible spectral range [45]. Interestingly, excitonic peaks obscure the onset of the free-carrier absorption in monolayer MX_2, which is known to dominate the room-temperature absorption in conventional semiconductor quantum wells.

2.3 Valley Physics in Broken Inversion Symmetry Lattices

In the previous sections of this chapter, we discussed the pseudospin physics in pristine graphene and the pseudospin-momentum locking effect observed at the time-reversal symmetric K and K′ points. Unlike pristine graphene, however, monolayer TMD exhibits inherently broken inversion symmetry. The following subsections introduce the consequences of broken inversion symmetry in 2D lattices, such as the finite Berry curvature and orbital magnetic moment, and examine the valley Hall and valley Zeeman effects.

2.3.1 Berry Phase and Berry Curvature

Formally, the Berry phase is a phase angle that can take values from 0 to 2π and represents the evolution of the (complex) vector as it is carried along a certain path. Vector here may represent a ground state of a molecule or the electrons in a solid. For example, consider a triatomic molecule with continuous "looped" deformation that shortens one of the bonds, shown as a double bond in Fig. 2.9(a). This deformation travels from bond to bond, shortening and lengthening the distance between two atoms as shown in snapshots 1 to 4. Importantly, the molecule is fixed and only the pattern of deformation rotates around its axis. The Berry phase then represents the phase evolution of the ground state of that molecule.

Often, the Berry phase is called the geometric phase in the sense that its value depends only on the geometry of the path and the state that is being evaluated. The concept of the geometric phase has been developed through the early work of Pancharatnam [46] and

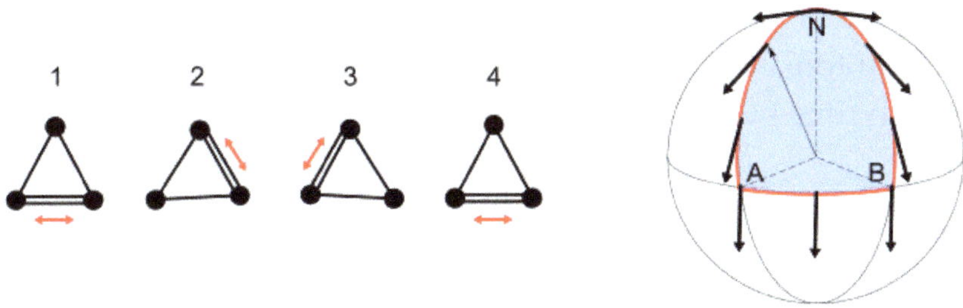

(a) (b)

Fig. 2.9. (a) Interatomic bond distortion sequence of the triatomic molecule, from step 1 to step 4. The first and the last steps are identical. (b) A thought experiment of transporting a javelin following the N-A-B-N path (see the text).

generalised by Sir Michael Berry in a seminal paper [47]. Intuitively, the geometric phase can be understood through the example of parallel transport along a curved surface. Imagine holding a javelin and taking a walk from the north pole along the path shown in Fig. 2.9(b) (red solid line). While following the path N-A-B-N, keep constant the rotation angle of the javelin around local vertical. Surprisingly, at the end of the journey, the javelin will accumulate an angle difference. The javelin never rotated, yet it turned! The accumulated angle is the geometric phase analogous to the Berry phase.

Note that both examples illustrate a closed-loop path where the initial and final states of the system are the same. This is an important condition that is imposed on the behaviour of the Berry phase. Throughout this chapter, we will discuss only closed-loop path phenomena and the Berry phase associated with it. A more general treatment can be found elsewhere [48].

Understanding the role of the Berry phase of charged quasiparticles in crystalline solids ultimately appeals to the quantum mechanical representation. Here, we are interested in tracing the temporal evolution of the system that experiences a looped (periodic) perturbation, i.e., the Hamiltonian of the initial state, $H(t = 0)$, returns to its value after a certain time $t = T$:

$$H(0) = H(T) \tag{2.9}$$

This is an analogy to a closed-loop path condition discussed earlier. Importantly, the temporal evolution of the Hamiltonian must be slow, often called '*adiabatic*'. As the system is exposed to a perturbation, the Hamiltonian of the system changes accordingly. The adiabatic condition then requires that the system in the initial state $|n\rangle$ remains in the same (instantaneous) state throughout the process [19].

Generally, the temporal evolution of the wavefunction can be solved for a slow periodic perturbation in an arbitrary parameter space [47]. In favour of further discussion, we will treat the Hamiltonian as it evolves in momentum space,[b] i.e., $H = H(\mathbf{k}(t))$. The question is then what happens to the wavefunction when the Hamiltonian returns to its initial value, $H(\mathbf{k}(0)) = H(\mathbf{k}(T))$?[c] To answer this question, we must appeal to the time-dependent Schrödinger equation:

$$i\hbar\frac{\partial}{\partial t}|\Psi(t)\rangle = H(k(t))|\Psi(t)\rangle \tag{2.10}$$

The conditions discussed earlier demand that the form of the wavefunction for an eigenstate $|n\rangle$ to be as follows:

$$|\Psi(t)\rangle = e^{i\phi(t)}|n,t\rangle \tag{2.11}$$

---

[b] Remember that the crystal momentum, p, relates to the wavevector, k, as $p = \hbar k$. Hereafter the momentum- and k-space notions are used interchangeably.

[c] In crystalline solids, this requirement is fulfilled for the momentum space since the reciprocal lattice is periodic in nature. Thus, the energy band in the Brillouin zone can be treated as a loop in momentum space.

The only unknown parameter here is the phase of the wavefunction, $\phi(t)$, which then takes the following form:

$$\phi(t) = -\frac{1}{\hbar}\int_0^t \varepsilon_n(\boldsymbol{k}(t'))dt' + \gamma_n \tag{2.12}$$

Here, the solution contains two terms: the conventional dynamical phase and the additional term γ_n. The dynamical phase retains its form ($\sim\varepsilon t/\hbar$) from the solution of the time-independent Schrödinger equation. The second term is the Berry phase that takes the following form as the system completes the cycle at $t = T$:

$$\gamma_n(T) = i\int_0^T \left\langle n,t \middle| \frac{\partial}{\partial t} \middle| n,t \right\rangle dt \tag{2.13}$$

As our Hamiltonian depends on time through the momentum, $\hbar\boldsymbol{k}(t)$, we can rewrite Eq. (2.13) with the exchange of variables:

$$\gamma_n = i\oint_C \left\langle n,\boldsymbol{k} \middle| \frac{\partial}{\partial \boldsymbol{k}} \middle| n,\boldsymbol{k} \right\rangle d\boldsymbol{k} \tag{2.14}$$

where C is the closed-loop path in the momentum space. Thus, the Berry phase is the closed-path integral in the momentum space and is independent of how the system travels along that path (hence the name 'geometric'), assuming that it travels slowly enough for the adiabatic approximation to hold. Even though we have invoked the time-dependent Schrödinger equation, the resulting Berry phase is independent of how $\boldsymbol{k}(t)$ varies in time! Reflecting on the complete form of the phase, Eq. (2.12), one may conclude that the dynamical phase answers the question "how long was the journey?", while the Berry phase indicates "where has it been?".

The closed-path condition of the Berry phase draws a vague analogy with the magnetic field, $\boldsymbol{B}$, acting on an electron and being associated with the closed-loop trajectories as the Lorentz force, $\boldsymbol{F}_B$, is a centrifugal force that pushes the electrons into circular motion, $\boldsymbol{F}_B = e\boldsymbol{\upsilon} \times \boldsymbol{B}$ (where $\boldsymbol{\upsilon}$ is the electrons' velocity). This intuitive thinking relates the phenomenon of the Berry phase with the magnetic field. Indeed, one can show that Eq. (2.14) can be rewritten as the flux of a vector field through a surface area, S, that the path C encircles:

$$\gamma_n = i\iint_C dS\boldsymbol{\Omega}_n \tag{2.15}$$

This vector field is called the *Berry curvature*, $\boldsymbol{\Omega}_n$. The derivation of the Berry curvature then requires solving a complex differential equation. Often, however, it is substituted by the summation over the eigenstates that yields numerically identical results:

$$\boldsymbol{\Omega}_{n,z} = i\sum_{n\neq n'} \frac{\langle n|\partial_x H|n'\rangle\langle n'|\partial_y H|n\rangle - \langle n|\partial_y H|n'\rangle\langle n'|\partial_x H|n\rangle}{\left(\varepsilon_n - \varepsilon_{n'}\right)^2} \tag{2.16}$$

where $\partial_a H = \partial H/\partial k_a$ and ε_n is the energy of the n-th state.

The Berry curvature modulates the motion of charged particles in the same way as the magnetic field, with the only difference being that it is defined over the momentum space! This can be traced through the semiclassical equations of motion:

$$\hbar\dot{k} = -e(E + \dot{r} \times B) \tag{2.17}$$

$$\hbar\dot{r} = \frac{\partial \varepsilon_n}{\partial k} - \hbar\dot{k} \times \boldsymbol{\Omega}_n \tag{2.18}$$

where $\boldsymbol{r}$ and $\hbar k$ are the position and the momentum of the charged particle, and the dot represents the first time derivate [49]. The first equation, Eq. (2.17), encapsulates the combined effect of the Coulomb and Lorentz forces acting on the particle of a charge e (i.e., electron). The second equation, Eq. (2.18), describes the velocity of the particle, where the Berry curvature enters in the second term, which is called *anomalous velocity*. The similarity between the anomalous velocity term in Eq. (2.18) and Lorentz force in Eq. (2.17) underlines the analogy between the magnetic field and the Berry curvature. In the absence of a magnetic field, the Lorentz force is zero and thus the Eq. (2.18) can be simplified as:

$$\hbar\dot{r} = \frac{\partial \varepsilon_n}{\partial k} + eE \times \boldsymbol{\Omega}_n \tag{2.19}$$

Interestingly, the second term in Eq. (2.19), the anomalous velocity, is always transverse to the electric field, giving rise to the transverse current in the absence of a magnetic field, i.e., the anomalous Hall effect that will be discussed later.

2.3.1.1 *Implications of Symmetry on the Berry Curvature*

Historically, the motion of charged particles has been successfully described by the conventional form of Eq. (2.18), consisting only of the first term, the band dispersion contribution (i.e., the group velocity). Thus, it is imperative to understand the conditions under which the Berry curvature cannot be neglected. These conditions can be drawn from a symmetry analysis of Eq. (2.18).

If the host system of charged particles is symmetric under the time-reversal ($\mathcal{T}$) and spatial inversion ($\mathcal{P}$) operations, Eq. (2.18) should be invariant under these transformations as well. Under $\mathcal{T}$ symmetry, $\dot{r}$ and $\boldsymbol{k}$ change sign while $\boldsymbol{E}$ is fixed, implying that the Berry curvature must be an odd function of the particle's momentum, i.e.,

$$\boldsymbol{\Omega}_n(-k) = -\,\boldsymbol{\Omega}_n(k) \tag{2.20}$$

Intuitively, this condition is supported by a path-dependent origin of the Berry curvature. Under $\mathcal{P}$ symmetry, however, $\dot{r}$, $\boldsymbol{k}$ and $\boldsymbol{E}$ change sign, requiring the Berry curvature to be an even function of the momentum:

$$\boldsymbol{\Omega}_n(-k) = \boldsymbol{\Omega}_n(k) \tag{2.21}$$

Consequently, if the system possesses both time-reversal and spatial inversion symmetries, i.e., Eq. (2.20) and Eq. (2.21) hold simultaneously, then the Berry curvature can only be zero. Therefore, the Berry curvature is finite in systems where either $\mathcal{T}$ or $\mathcal{P}$ symmetry is broken.

2.3.2 Valley Hall Effect

The anomalous velocity that enters as a second term in the semiclassical Eq. (2.18) immediately hints at the possibility of the electric-field-transverse currents generated by the Berry curvature. Indeed, anomalous Hall conductivity is a direct consequence of the finite Berry curvature:

$$\sigma_{\mathrm{AH}} = \frac{e^2}{\hbar}\int \frac{dk}{(2\pi)^2} f(k)\boldsymbol{\Omega}_n(k) \tag{2.22}$$

where $f(\boldsymbol{k})$ is the Fermi-Dirac distribution function [50].

For such currents to exist, however, one needs to identify a system that lacks either time-reversal ($\mathcal{T}$) or inversion ($\mathcal{P}$) symmetry. Broken $\mathcal{T}$ symmetry condition, for example, is fulfilled by magnetisation in ferromagnetic materials. Here, the Berry curvature stems from the spin-orbit interactions and gives rise to the intrinsic anomalous Hall effect [50]. In non-magnetic systems, the Berry curvature may arise from broken $\mathcal{P}$ symmetry, leading to the valley Hall effect (VHE), one of the key concepts of valleytronics.

In the following, we will consider an example of a monolayer graphene with broken $\mathcal{P}$ symmetry. Experimentally, this can be achieved through interlayer interactions with the substrate [51], [52]. Here, the two sublattices of graphene, A and B in Fig. 2.1(a), are exposed to different electrostatic potentials, thus breaking the sublattice symmetry. At the same time, this opens a bandgap at the K and K′ points, Fig. 2.10(a). The two valleys can be distinguished through the valley index, $\tau = \pm 1$, which changes the sign when valleys are switched. We can write the VHE acting in presence of the in-plane electric field, $\boldsymbol{E}_{\parallel}$, as follows:

$$\boldsymbol{j}_{\text{VH}} = \sigma_{\text{VH}} \times \boldsymbol{E}_{\parallel} \tag{2.23}$$

where σ_{VH} is the valley Hall conductivity pseudovector. Thus, the valley-contrasting current is the average of the product of the valley index and velocity operator, $\boldsymbol{v}, \boldsymbol{j}_{\text{VH}} = \langle \tau \boldsymbol{v} \rangle$ With broken $\mathcal{P}$ symmetry,

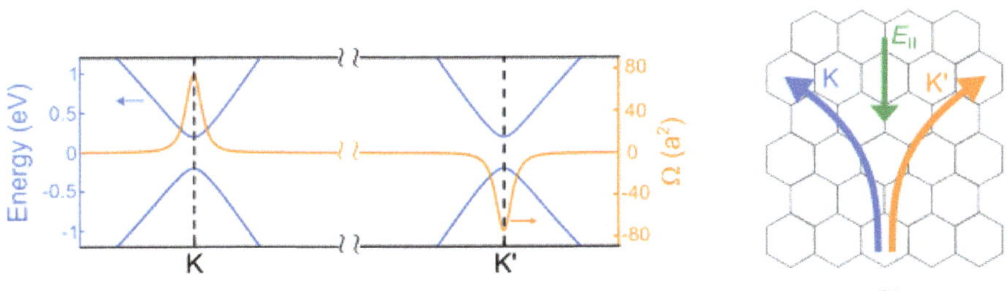

Fig. 2.10. (a) Berry curvature, **Ω**, (orange) of the conduction band in the $\mathcal{P}$-broken graphene superimposed with the electronic band structure. (b) Schematics of valley Hall effect, where in-plane electric field generates a transverse valley-contrasting current that leads to valley polarization on the sample edges. Adapted with permission from Refs. [50], [56].

valley current is invariant while the electric field changes its sign. As the $\mathcal{T}$ symmetry is preserved, the VHE generates a non-zero current in the $\mathcal{P}$-broken monolayer graphene.

In analogy to the pristine monolayer graphene, the $\mathcal{P}$-broken graphene can be modelled with the tight-binding approximation. We need to modify the massless Dirac Hamiltonian, Eq. (2.5), by introducing the site of potential difference between two sublattices, Δ, and the nearest-neighbour hopping energy, $t_{\parallel}$. Then, the Hamiltonian of the systems at low doping (near the Dirac points K and K′) reads as:

$$H = at_{\parallel}\left(\tau k'_x \sigma_x + k'_y \sigma_y\right) + \frac{\Delta}{2}\sigma_z \tag{2.24}$$

where a is the lattice constant, $\boldsymbol{k}'$ is measured from the Dirac point, and $\sigma_i = (\sigma_x, \sigma_y, \sigma_z)$ is the Pauli matrix element for lattice pseudospin, Eq. (2.6). The Berry curvature of the conduction band, $\Omega_c(k')$, is given as follows[d]:

$$\Omega_c(\tau, k') = \tau \frac{2a^2 t_{\parallel}^2 \Delta}{\left(\Delta^2 + 4a^2 t_{\parallel}^2 k'^2\right)^{3/2}} \tag{2.25}$$

Interestingly, the Berry curvature at the K and K′ valleys is identical in magnitude but opposite in sign as shown in Fig. 2.10(a), fulfilling the time-reversal symmetry condition. If we reduce the energy gap Δ to zero, gradually recovering the $\mathcal{P}$ symmetry, the Berry curvature

[d]The expression for the Berry curvature of the valence band, $\Omega_v(k')$, is straightforward due to the particle-hole symmetry of graphene: $\Omega_c(k') = -\Omega_v(k')$.

vanishes except for the Dirac points where it diverges. At the same time, the Berry phase at the Dirac points reaches the exact value of $\pm\pi$ [50].

In the absence of intervalley scattering, charged particles from different valleys accumulate at opposite edges of the sample, manifesting the VHE, Fig. 2.10(b). The valley transport is a general phenomenon and can be observed in many systems with broken $\mathcal{P}$ symmetry. For example, Eqs. (2.24) and (2.25), derived for gapped graphene, can be applied to monolayers of TMDs where the $\mathcal{P}$ symmetry is broken intrinsically [53]. However, strong spin-orbit coupling in the valence band of TMDs requires an additional term in the Hamiltonian:

$$H = at_{\parallel}\left(\tau k'_x \sigma_x + k'_y \sigma_y\right) + \frac{\Delta}{2}\sigma_z - \tau\lambda\frac{\sigma_z - 1}{2}s_z \tag{2.26}$$

where 2λ is the spin splitting at the top of the valence band, s_z is the Pauli matrix for spin, while σ_i is the Pauli matrix element defined on the basis of two *d*-orbitals of metal atoms with magnetic quantum numbers $m = 0$ and $m = 2\tau$. Berry curvature in the same valley depends on the spin via the spin-dependent bandgap, $\Delta' = \Delta - \tau s_z \lambda$. Following Eq. (2.22), one can calculate the valley Hall conductivity (in units of $e/\hbar$):

$$\sigma^n_{\mathrm{VH}} = 2\int \frac{dk}{(2\pi)^2}\left(f_{n,\uparrow}(k)\mathbf{\Omega}_{n,\uparrow}(k) + f_{n,\downarrow}(k)\mathbf{\Omega}_{n,\downarrow}(k)\right) \tag{2.27}$$

where arrows $\uparrow$ and $\downarrow$ denote spin-up and spin-down, respectively, and integration is performed around the K (or K′) point. For moderate hole doping, where the Fermi energy, ε_F, is located between spin-split bands, $\varepsilon_F << \Delta - \lambda$, the valley Hall conductivity becomes:

$$\sigma^h_{VH} = \frac{1}{\pi}\frac{\varepsilon_F}{\Delta - \lambda} \tag{2.28}$$

In the case of monolayer MoS_2, the spin splitting in the conduction band is considerably small (scales as k^2), and we can assume that both sub-bands are spin degenerate at the band bottom:

$$\sigma^e_{VH} = \frac{1}{\pi}\frac{\Delta}{\Delta^2 - \lambda^2}\varepsilon_F \tag{2.29}$$

Evidently, no VHE is observed in pristine bilayer TMDs samples, where $\mathcal{P}$ symmetry is restored, and no finite Berry curvature exists. Indeed, a 2H-phase bilayer TMD consists of two identical monolayers stacked with a relative rotation of 180°, and thus possesses the $\mathcal{P}$ symmetry in the stack. However, the $\mathcal{P}$ symmetry of bilayer TMD can be broken by an external out-of-plane (vertical) electric field, $E_\perp$.

Just like the interactions with the substrate create a potential difference between two sublattices in monolayer graphene, the vertical electric field generates the potential difference, $U(E_\perp)$, between two layers in a bilayer. A continuously tuned electric field gradually breaks the $\mathcal{P}$ symmetry. As the charge carriers from K and K′ valleys now possess finite but opposite-in-sign Berry curvature, VHE manifests itself through the valley Hall conductivity, σ_{VH}, which now depends on the interlayer potential difference:

$$\sigma_{VH} \propto n\left(U/\sqrt{(2t_\perp)^2 + U^2}\right) \tag{2.30}$$

where n and $t_\perp$ are the total carrier density and interlayer hopping constant of the bilayer, respectively [54]. It is worth noting that neither valley polarisation nor VHE shows any electric-field dependence in a monolayer TMD with inherently broken $\mathcal{P}$ symmetry.

2.3.3 Orbital Magnetic Moment

Alongside the Berry curvature, valley states can be distinguished by the *orbital magnetic moment*. This physical quantity is an intrinsic property of electrons in crystalline materials and can be understood if we treat the electron's dynamics in crystals through the motion of a *wave packet*. The wave packet representation stems from the uncertainty principle ($\Delta r \Delta k > 1$) and can be viewed as a superposition of the electron's band states, localised in a certain area of the real space, Fig. 2.11(a) [55]. Narrowly localised wave packet in the real space possesses a centre of mass, r_c, and wavevector, k_c, in the momentum space.

The motion of a wave packet is described through the motion of its centre of mass, r_c, and due to finite spread in the real space, a wave packet may exhibit a self-rotation around the centre of mass, Fig. 2.11(b). This self-rotation results in an orbital magnetic moment of the wave packet, in addition to the spin moment. The orbital magnetic moment, $\mathfrak{m}$, shows exactly the same symmetry transformations as the Berry curvature. Namely, it exists only in $\mathcal{T}$-broken or $\mathcal{P}$-broken symmetry systems [56].

Taking the previous example of $\mathcal{P}$-broken graphene, Eq. (2.24), $\mathfrak{m}$ in K and K′ valleys can be calculated as follows:

$$\mathfrak{m}(\tau,k') = \tau \frac{2a^2 t_{\parallel}^2 \Delta}{\Delta^2 + 4a^2 t_{\parallel}^2 k'^2} \frac{e}{2\hbar} \tag{2.31}$$

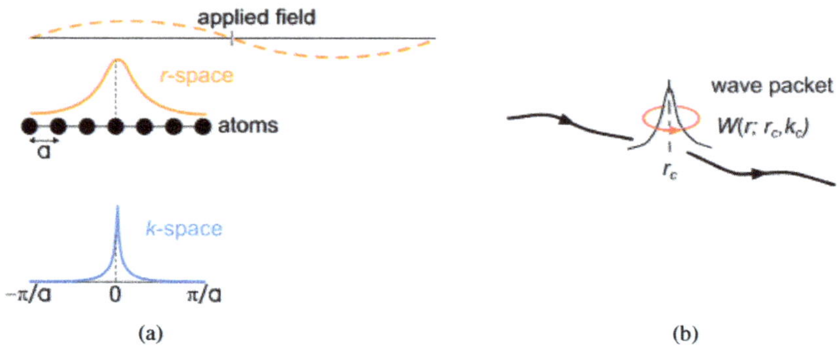

Fig. 2.11. The wave packet construction and dynamics. (a) A wave packet consists of a set of wavevectors in the *k*-space much smaller than the size of the Brillouin zone, making the real space (*r*-space) representation much larger than the lattice constant, ***a***. When the size of the perturbation, e.g., a wavelength of the applied field, is much larger than the size of the wavepacket, it is possible to model the system semiclassically, i.e., treat the position and the wavevector of an electron simultaneously [55]; (b) the motion of the wave packet generally is represented by the motion of its centre of mass, r_c, and self-rotation around this centre. (b) is reproduced with permission from Ref. [88].

Similar to the Berry curvature, $\mathfrak{m}$ values around K and K′ points are equal in magnitude but opposite in sign, as required by $\mathcal{T}$ symmetry. Interestingly, exactly at the band bottom, the magnetic orbital moment is reduced to:

$$\mathfrak{m}(\tau) = \tau\mu_B^*, \quad \mu_B^* = \frac{e\hbar}{2m^*}, \quad m^* = \frac{2\Delta\hbar^2}{3a^2t_{\parallel}^2} \tag{2.32}$$

where m^* is the effective mass at the band bottom. Clearly, the orbital magnetic moment is analogous to the Bohr magneton of the electron's spin. In real systems, however, it may exceed the value of the spin magneton by ~30 times [56]. As the orbital magnetic moment is derived directly from the self-rotation of the wave packet, it is an intrinsic property of the band structure and can be used to distinguish between valleys for use in valleytronics.

2.3.4 Valley-selective Optical Rules

From the previous sections, we have identified that the Berry curvature and the orbital magnetic moment around K and K′ points are finite and opposite in sign, taking that the system lacks $\mathcal{P}$ symmetry. In the following, we will derive optical selection rules for monolayer TMDs, MX_2, with inherently broken $\mathcal{P}$ and sizable optical gap (see Section 2.2.3).

The finite valley-contrasting orbital magnetic moment at K and K′ points hints that the valleys can be optically coupled to the circularly polarised excitations. To derive this coupling, let's first consider interband transitions with the matrix element of the canonical momentum operator of the following form:

$$P_\alpha \equiv m_0 \left\langle u_c(\boldsymbol{k}) \left| \frac{1}{\hbar} \frac{\partial H}{\partial k_\alpha} \right| u_v(\boldsymbol{k}) \right\rangle \tag{2.33}$$

where m_0 is the free electron mass and $u(\boldsymbol{k})$ is the periodic part of the electron's wavefunction (Bloch function) in the crystal. The coupling strength with the circularly polarised light, $\sigma^\pm$, is then given by:

$$P^\pm(\boldsymbol{k}) \equiv P_x(\boldsymbol{k}) \pm iP_y(\boldsymbol{k}) \tag{2.34}$$

For transitions near the K and K′ points, one can find [53]:

$$\left|P^{\pm}(\boldsymbol{k})\right|^2 = \frac{m_0^2 a^2 t_{\parallel}^2}{\hbar^2}\left(1 \pm \tau \frac{\Delta'}{\sqrt{\Delta'^2 + 4a^2 t_{\parallel}^2 k^2}}\right)^2 \tag{2.35}$$

where Δ' is the bandgap modified by the spin-orbit coupling, $\Delta' = \Delta - \tau s_z \lambda$, as was introduced earlier. Near the K point, $\Delta' >> a t_{\parallel} k$ and the coupling strength can be further reduced:

$$\left|P^{\pm}(\boldsymbol{k})\right| = \frac{m_0 a t_{\parallel}}{\hbar}\left|1 \pm \tau\right| \tag{2.36}$$

Hence, one can see that the interband transitions are exclusively coupled to the valleys, defined by valley index τ. Due to the strong spin-orbit splitting in the valence band of MX_2, the polarisation of light and valley-locked spins are strongly coupled. Simply put, the right-handed (σ^+) or left-handed (σ^-) circularly polarised excitation near the excitonic resonance induces the optical transitions only in the K or K′ valley, respectively. Since the spin-valley locking effect occurs in monolayer MX_2, a valley polarisation results in a concurrent spin polarisation. These selection rules provide a direct and easy way to read and write the valley polarisation by addressing the individual valleys.

We can identify the degree of circular polarisation as follows:

$$\eta(\boldsymbol{k}) \equiv \frac{\left|P^{+}(\boldsymbol{k})\right|^2 - \left|P^{-}(\boldsymbol{k})\right|^2}{\left|P^{+}(\boldsymbol{k})\right|^2 + \left|P^{-}(\boldsymbol{k})\right|^2} \tag{2.37}$$

This quantity can be directly expressed through the orbital magnetic moment, or the Berry curvature as follows:

$$\eta(\boldsymbol{k}) = \frac{\mathbf{m}(\tau, k')}{\mu_{\mathrm{B}}^{*}} = \frac{\Omega(k')}{\mu_{\mathrm{B}}^{*}} \frac{e}{2\hbar} \sqrt{\Delta'^2 + 4a^2 t_{\parallel}^2 k^2} \tag{2.38}$$

At the energy minima of K or K′ points, the depolarisation ratio reduces to $\eta(\boldsymbol{k}) = \tau$.

Experimentally, this effect can be demonstrated through polarisation-dependent photoluminescence measurements. To this end, excitation must be conditioned to either σ^+ or σ^- circularly polarised light, with detection locked to circularly polarised emission, i.e., $I(\sigma^+)$ and $I(\sigma^-)$, Fig. 2.12(a, b). Then, valley polarisation, η, reads as:

$$\eta = \frac{I(\sigma^+) - \mathrm{I}(\sigma^-)}{I(\sigma^+) + \mathrm{I}(\sigma^-)} \tag{2.39}$$

Remarkably, valley polarisation in monolayer MX_2 can reach 50–100% highlighting the robustness of the optical addressing of the valley degree of freedom [57].

Consequently, linearly polarised excitation can be used to generate and probe valley coherence. Indeed, linear polarisation is a coherent superposition of σ^+ and σ^- and thus excites both K and K′ valleys *simultaneously*, imprinting the optical coherence to valley quantum coherence. Hence, linearly polarised photoluminescence, parallel to the linear polarisation of the excitation, detects the valley coherence state, Fig. 2.12(c). Following the Bloch vector representation that is often used to depict an evolution of a quantum two-level system [58], the K and K′ states correspond to the north, Fig. 2.12(g), and south poles, Fig. 2.12(h), of the unit sphere, respectively. The linearly polarised light creates a superposition of the two states, pinning the Bloch vector to the sphere's equator, Fig. 2.12(i). Linear polarisation of the photoluminescence from monolayer WSe_2 confirms that the

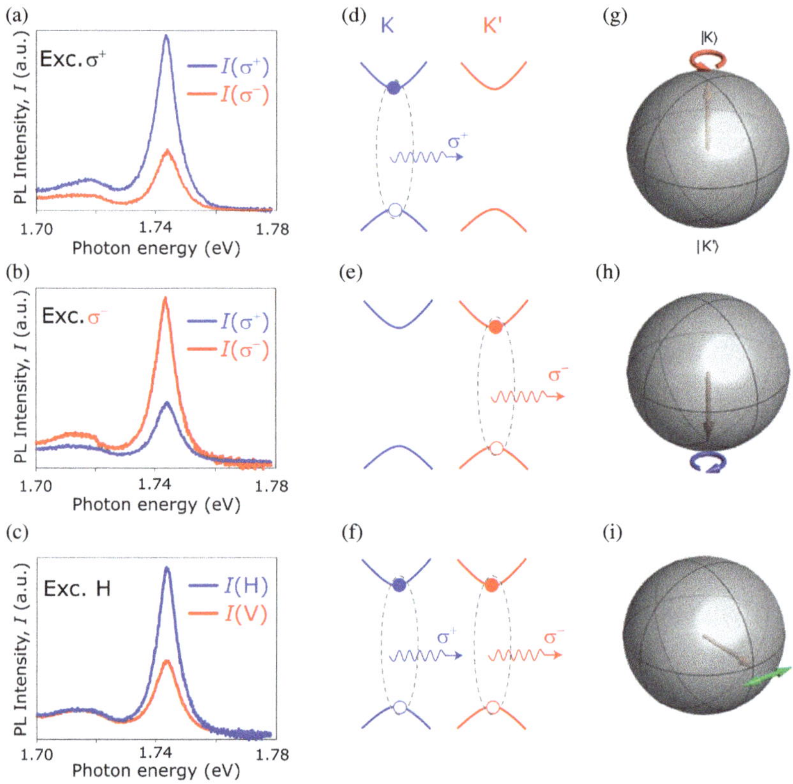

Fig. 2.12. Valley polarisation and valley coherence. (a–c) Polarisation-resolved photoluminescence (PL) of neutral exciton from monolayer WSe_2, excited with circularly polarised (a, b) and linearly polarised light (c). (d–f) Schematic energy diagrams for optical transitions at K and K′ valleys that correspond to the PL depicted in (a–c). (g–i) Bloch-sphere representations for the processes in (a–c). Reproduced with permission from Ref. [85].

valley polarisation coherence is preserved for neutral excitons but is destroyed in trions due to inherent quasiparticle asymmetry and excitonic exchange interactions [59].

While the degree of valley polarisation may reach unity, the lifetime of valley polarisation is surprisingly short, lasting only on the timescale of a few picoseconds [60]. Although the valley degree of freedom is protected by the large separation in the momentum space, other mechanisms have been suggested to gradually reduce the valley polarisation and lifetime [40]. For example, Coulomb exchange interaction assisted by disorder, biexcitons, and lattice phonons effectively couples K and K′ valleys, promoting the intervalley scattering that gives rise to the valley polarisation decay on a picosecond timescale [61], [62]. Similarly, valley polarisation can be destroyed by the intervalley transfer of an excitonic population from A to B excitonic states with the same spin in different valleys [63]. This mechanism implies an effective coupling between two valleys akin to Dexter coupling [64] in spatially separated systems. Consequently, it leads to zero-valley-polarisation for an A exciton if the B exciton is resonantly excited and vice versa [63]. These depolarisation mechanisms limit the application of optically addressed valley degree of freedom in monolayer MX_2 and stimulate the development of new devices and concepts that can improve the degree of valley polarisation and extend the valley lifetime.

2.3.5 Valley Zeeman Effect

While the Berry curvature responds to the electric field via the VHE, the orbital magnetic moment manifests itself through the magnetic field effects, such as the *valley Zeeman effect.*

In a 2D system with $\mathcal{P}$-broken symmetry, the orbital motion is restricted in-plane, thus the orbital magnetic moment points out-of-plane and couples to the out-of-plane magnetic field, $\boldsymbol{B}_z$. As the orbital magnetic moments in K and K′ valleys are opposite in sign, $\boldsymbol{B}_z$ causes the opposite energy shifts in the order of $-\boldsymbol{m} \cdot \boldsymbol{B}_z$. This lifts the valley energy degeneracy in monolayer TMDs. The valley Zeeman effect is the direct analogue of the real spin Zeeman effect, i.e., the spin-dependent energy shift of the electronic states in external magnetic field.

Given the optical selection rules discussed earlier, monolayer TMD is the ideal platform for experimental investigation of the valley Zeeman effect. With circularly polarised excitation, excitonic photoluminescence reflects the carrier population of the individual valleys, Eq. (2.35). Assuming the right- and left-circularly polarised emission correspond to the K and K′ valleys respectively, both emission peaks overlap at 0T magnetic field, thus showing a pristine, energy-degenerate state, in full accordance with time-reversal symmetry of the valleys, Fig. 2.13(a). When the strong out-of-plane magnetic field is present, the peaks split in energy, Fig. 2.13(b). The splitting is linear with the magnetic field strength and is in the order of $\Delta_{VZ} \approx 0.1 - 0.2$ meV/T, Fig. 2.13(c) [40].

As the magnetic field couples to the orbital magnetic moment as well as the spin magnetic moment, one may expect to observe

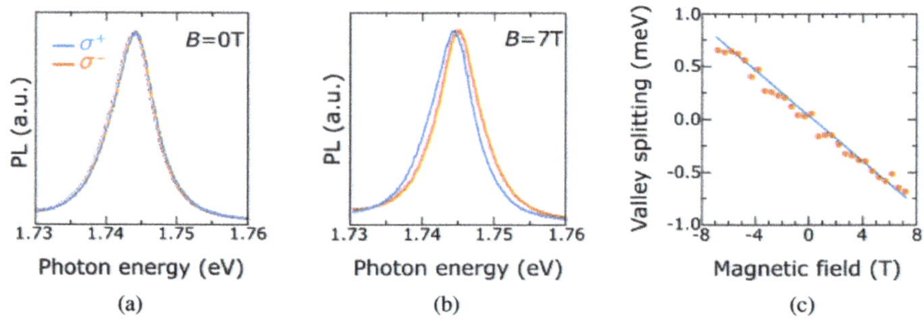

Fig. 2.13. Valley Zeeman effect. Polarisation-resolved photoluminescence (PL) excited with circularly polarized light (σ^+) at the magnetic field (a) $B = 0$ and (b) $B = 7$ T. The PL intensity is normalized for clarity. (c) The linear shift of the energy splitting, Δ_{VZ}, as a function of the external magnetic field. Reproduced with permission from Ref. [68].

contributions from both effects. However, photoluminescence probes the difference between the conduction band and valence band Zeeman shifts, and thus the spin contribution cancels out due to spin conservation for bright excitons. Therefore, the energy shift reflects the orbital magnetic moment contribution. Following the analysis of the orbital character of the band structure of monolayer MoS_2 (see Section 2.2.2), the conduction band at K and K′ valleys is formed by hybridised d_{z^2} orbitals that carry zero angular momentum, $l_z = 0$, Fig. 2.6(d_0). However, the valence band at K and K′ comprises hybridisation of $d_{x^2-y^2}$ and d_{xy} with angular momentum $l_z = 2$, Fig. 2.6(d_2). This difference in the orbital magnetic moment between conduction and valence bands is responsible for the exciton energy shift in the valley Zeeman effect. Indeed, the shift, Δ_{VZ}, is calculated as the difference between energy shifts of conduction (ε_c) and valence (ε_v) bands, $\Delta_{VZ} = 2(\varepsilon_c - \varepsilon_v)$, thus resulting precisely in $\Delta_{VZ} = 4\mu_B B_Z$, where μ_B is the Bohr magneton. This can be rewritten as $\Delta_{VZ} = g\mu_B B_Z$, where $g = 4$ is the g-factor of bright neutral excitons in monolayer TMDs [65]. Remarkably, this linear Zeeman shift extends up to the magnetic fields as high as B_Z~65 T [66], [67], highlighting the robust nature of the valley Zeeman effect.

Interestingly, several works report a deviation from the $g = 4$ scenario [68], [69], suggesting additional contributions to the excitonic Zeeman shift. One of the contributions may originate from the unequal effective masses of electrons and holes in the valley, $m_e \neq m_h$, which results in a finite *valley* magnetic moment, as a counterpart to the *orbital* magnetic moment discussed earlier. While $4\mu_B B_Z$ is the result of a simple two-band approximation, more general models of Zeeman shift should include electron-hole mass asymmetry [70], [71] as well as other higher-order effects [72], [73].

External magnetic field can induce a finite energy gap between K and K′ valleys through the valley Zeeman effect. However, it requires a large field strength (~ few tesla) which comes with a substantial energy cost, defeating the purpose of *valleytronics* being an energy-efficient computing principle. The energy degeneracy of valleys can be broken via proximity effects in heterostructures of monolayer semiconductors with magnetic insulators. Through the interaction with the ferromagnetic substrate, also known as exchange coupling, 2D materials experience an additional, effective magnetic field. Theoretical estimations predict that such fields can be as high as a few thousand teslas [74], quickly surpassing the strengths of state-of-the-art laboratory magnets. Early experiments with magnetic insulating substrates like yttrium iron garnet or EuS demonstrated proximity-induced anomalous Hall effect in graphene [75] and effective magnetic fields reaching ~14 T [76]. One of the limiting factors of the exchange coupling is the poor interface quality in heterostructures of 3D substrates and 2D van der Waals materials [77]. Recently re-discovered van der Waals magnets like CrI_3 and $Cr_2Ge_2Te_6$ offer ultimately smooth surfaces and ideal coupling for 2D-on-2D van der Waals heterostructures. A monolayer WSe_2 placed on top of bulk CrI_3 has demonstrated the zero-external-field valley Zeeman effect in an effective magnetic field reaching ~13 T [78]. Furthermore, an ideal van der Waals coupling allows studying of newly discovered ultra-thin magnets through the optical response of TMD materials and a better understanding of the properties of all-2D magnetic heterostructures [79]. As a new material platform, van der Waals magnets aim to expand the toolkit of *valleytronics* in pursuit of energy-efficient computing.

2.4 Valleytronics in 2D Heterostructures

One of the most important advantages of van der Waals materials is their ability to form vertical heterostructures with an ideal, atomically sharp interface. This opens a new direction for valleytronic device implementation. While in pristine 2H-bilayers of TMDs the inversion symmetry is restored and no valley-contrasting effects exist, a bilayer comprising of two different monolayer TMDs will form a vertical heterostructure where excitons can be formed across the layers, providing intriguing prospects for *valleytronics*.

When two different TMD monolayers form a vertical stack, their Fermi energies re-adjust and their relative band alignment forms one of three possible combinations: type I (straddling gap), type II (staggering gap) and type III (broken gap), Fig. 2.14(a). Early DFT calculations [80] have shown that several combinations of TMD monolayers form staggering Type II band alignment, Fig. 2.14(b), the material system that can host interlayer excitons. The latter is the quasiparticle that comprises an electron and a hole localised in different layers.

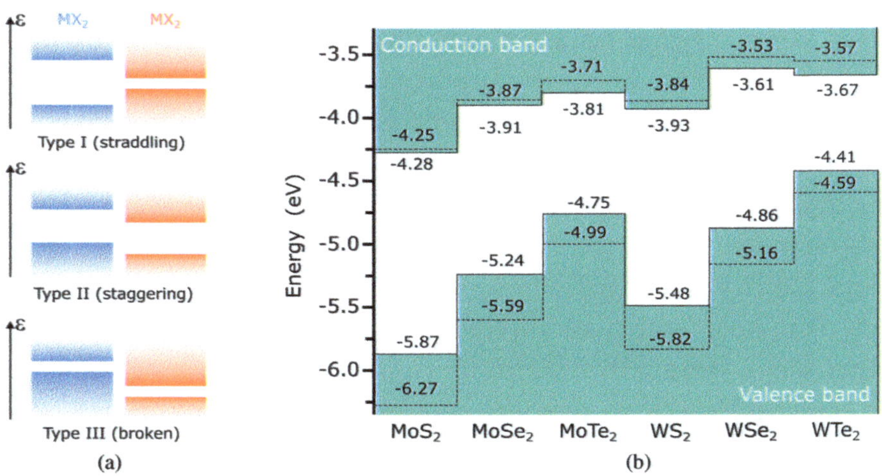

Fig. 2.14. Van der Waals heterostructures. (a) Band offset for straddling, staggering or broken type heterostructures. (b) DFT-calculated band alignment for MX_2 monolayers. Solid and dashed lines are obtained by different DFT models, PBE and HSE06. (b) is reproduced with permission from Ref. [80].

Upon light excitation with a large enough energy, the electrons from the valence bands of both monolayers get promoted to the corresponding conduction bands, thus forming spatially direct, intralayer excitons. Due to the ideal interface between two layers, however, quick interlayer charge transfer (in the order of a few hundred femtoseconds) leaves the carriers only in the lowest conduction and highest valence bands, which are localised in different layers. These electrons and holes form *spatially indirect*, interlayer excitons that cannot easily recombine. The lifetime of interlayer excitons may reach several nanoseconds, while spatially direct intralayer excitons usually recombine within a few picoseconds.

In interlayer excitons, electrons and holes from different layers still occupy K and K′ valleys, thus allowing us to evaluate the ability to address and control the valley degree of freedom of hetero-bilayers. The optical selection rules for such heterostructures strongly depend on the relative twist between the constituent layers [81], [82]. For nearly perfect alignment, however, the interlayer excitons are nearly direct in momentum, making their polarisation selection rules follow the case of monolayer TMDs, discussed earlier (see Section 2.2.3). The polarisation of the photoluminescence of the heterostructure is thus expected to be strongly co-polarised with the handedness of the laser excitation. Remarkably, as the holes and electrons are spatially separated by the van der Waals gap, electron-hole exchange interactions are suppressed, and valley lifetime is gradually improved from picosecond time scale for monolayer TMDs to tens of nanoseconds in hetero-bilayers. The experiments on WSe_2-$MoSe_2$ heterostructure [83] demonstrated that the degree of polarisation and the valley lifetime can be controlled electrically by changing the carrier densities *in situ*.

2.5 Summary

Throughout this chapter, we have discussed the basic theoretical aspects of *valleytronics*. After a brief introduction in Section 2.1, we explored the fundamental electronic and optical properties of graphene and TMDs through Section 2.2. We started from the symmetry aspects of the crystalline lattice of graphene, where we introduced electronic structure and the valleys, discussed massless charge transport, identified the important characteristics of graphene electrons, like pseudospin and chirality, and touched upon the spin-orbit interaction in pristine graphene. Next, we introduced the 2D monolayers of TMDs, their crystal lattice, electronic band structure, and the well-pronounced spin-orbit coupling. Armed with this knowledge, we extended the discussion of the optical properties of TMDs and emphasised the importance of the excitonic resonances and the impact of the reduced dimensionality on the many-body interactions in these materials.

Section 2.3 thoroughly covered the physics of materials with broken inversion symmetry. We presented the foundational concept of Berry phase and Berry curvature for charged particles in a crystalline solid and its intuitive analogy to the magnetic field. This was followed by a discussion of crystalline symmetry and how it affects the Berry curvature. Following the previous discussion, the fundamental physics of the VHE was derived. The experimental investigation of VHE will be thoroughly discussed in Chapter 5 of this book. We then introduced the notion of the magnetic orbital moment and its intuitive understanding from the wave packet dynamics. This opened the important consequence of the coupling of the valleys to circularly polarised light and its implications on the optical valley selectivity which was discussed in detail. The experimental implications of these phenomena will be comprehensively covered in Chapters 3 and 5 of this book. Following the discussion of the magnetic orbital moment, we examined the valley Zeeman effect and how it can be used to lift the energy degeneracy of the valleys.

In the final Section 2.4, we briefly investigated the physics of valleys in bilayer heterostructures of TMDs and how interlayer physics may contribute to the field of *valleytronics*.

References

[1] K. S. Novoselov *et al.*, "Two-dimensional atomic crystals," *Proc. Natl. Acad. Sci.*, vol. 102, no. 30, pp. 10451–10453, 2005, doi: 10.1073/pnas.0502848102.

[2] K. S. Novoselov, "Graphene: Materials in the Flatland," *Noble Lecture*. https://www.nobelprize.org/prizes/physics/2010/novoselov/lecture/ (accessed Sep. 27, 2022).

[3] A. Geim, "Random Walk to Graphene," *Noble Lecture*. https://www.nobelprize.org/prizes/physics/2010/geim/lecture/ (accessed Sep. 27, 2022).

[4] M. Chhowalla, D. Jena, and H. Zhang, "Two-dimensional semiconductors for transistors," *Nat. Rev. Mater.*, vol. 1, no. 11, pp. 1–15, 2016, doi: 10.1038/natrevmats.2016.52.

[5] J. R. Schaibley *et al.*, "Valleytronics in 2D materials," *Nat. Rev. Mater.*, vol. 1, no. 11, pp. 1–15, 2016, doi: 10.1038/natrevmats.2016.55.

[6] S. A. Vitale *et al.*, "Valleytronics: opportunities, challenges, and paths forward," *Small*, vol. 14, no. 38, p. 1801483, 2018, doi: 10.1002/smll.201801483.

[7] N. Mounet *et al.*, "Two-dimensional materials from high-throughput computational exfoliation of experimentally known compounds," *Nat. Nanotechnol.*, vol. 13, no. 3, pp. 246–252, 2018, doi: 10.1038/s41565-017-0035-5.

[8] Q. L. He, T. L. Hughes, N. P. Armitage, Y. Tokura, and K. L. Wang, "Topological spintronics and magnetoelectronics," *Nat. Mater.*, vol. 21, no. 1, pp. 15–23, 2022. doi: 10.1038/s41563-021-01138-5.
[9] J. Wang and S. C. Zhang, "Topological states of condensed matter," *Nat. Mater.*, vol. 16, no. 11, pp. 1062–1067, 2017, doi: 10.1038/NMAT5012.
[10] M. Gibertini, M. Koperski, A. F. Morpurgo, and K. S. Novoselov, "Magnetic 2D materials and heterostructures," *Nat. Nanotechnol.*, vol. 14, no. 5, pp. 408–419, 2019, doi: 10.1038/s41565-019-0438-6.
[11] K. S. Burch, D. Mandrus, and J. G. Park, "Magnetism in two-dimensional van der Waals materials," *Nature*, vol. 563, no. 7729, pp. 47–52, 2018, doi: 10.1038/s41586-018-0631-z.
[12] S. Deng, A. V. Sumant, and V. Berry, "Strain engineering in two-dimensional nanomaterials beyond graphene," *Nano Today*, vol. 22, pp. 14–35, 2018, doi: 10.1016/j.nantod.2018.07.001.
[13] J. C. W. Song and N. M. Gabor, "Electron quantum metamaterials in van der Waals heterostructures," *Nat. Nanotechnol.*, vol. 13, no. 11, pp. 986–993, 2018, doi: 10.1038/s41565-018-0294-9.
[14] K. F. Mak and J. Shan, "Semiconductor moiré materials," *Nat. Nanotechnol.*, vol. 17, no. 7, pp. 686–695, 2022, doi: 10.1038/s41565-022-01165-6.
[15] C. Kittel, *Introduction to Solid State Physics*, 8th ed. John Wiley & Sons, 2004.
[16] P. R. Wallace, "The band theory of graphite," *Phys. Rev.*, vol. 71, no. 9, pp. 622–634, 1947, doi: 10.1103/PhysRev.71.622.
[17] J. C. Charlier, J. P. Michenaud, X. Gonze, and J. P. Vigneron, "Tight-binding model for the electronic properties of simple hexagonal graphite," *Phys. Rev. B*, vol. 44, no. 24, pp. 13237–13249, 1991, doi: 10.1103/PhysRevB.44.13237.
[18] K. S. Novoselov *et al.*, "Two-dimensional gas of massless Dirac fermions in graphene," *Nature*, vol. 438, no. 7065, pp. 197–200, 2005, doi: 10.1038/nature04233.
[19] D. J. Griffiths and D. F. Schroeter, *Introduction to Quantum Mechanics*, 3rd ed. Cambridge University Press, 2018. doi: 10.1017/9781316995433.
[20] J. J. Sakurai and J. Napolitano, *Modern Quantum Mechanics*, 3rd ed. Cambridge University Press, 2020. doi: 10.1017/9781108587280.

[21] P. A. M. Dirac, *The Principles of Quantum Mechanics*, 4th ed. Clarendon Press, 1981. [Online]. Available: https://global.oup.com/academic/product/the-principles-of-quantum-mechanics-9780198520115.

[22] A. K. Geim and K. S. Novoselov, "The rise of graphene," *Nat. Mater.*, vol. 6, no. 3, pp. 183–191, 2007, doi: 10.1038/nmat1849.

[23] L. E. F. Foa Torres, S. Roche, and J.-C. Charlier, *Introduction to Graphene-Based Nanomaterials*. Cambridge University Press, 2020. doi: 10.1017/9781108664462.

[24] T. Ando, T. Nakanishi, and R. Saito, "Berry's phase and absence of back scattering in carbon nanotubes," *J. Phys. Soc. Japan*, vol. 67, no. 8, pp. 2857–2862, 1998, doi: 10.1143/JPSJ.67.2857.

[25] J. Sichau *et al.*, "Resonance microwave measurements of an intrinsic spin-orbit coupling gap in graphene: a possible indication of a topological state," *Phys. Rev. Lett.*, vol. 122, no. 4, p. 046403, 2019, doi: 10.1103/PhysRevLett.122.046403.

[26] Z. Y. Zhu, Y. C. Cheng, and U. Schwingenschlögl, "Giant spin-orbit-induced spin splitting in two-dimensional transition-metal dichalcogenide semiconductors," *Phys. Rev. B*, vol. 84, no. 15, p. 153402, 2011, doi: 10.1103/PhysRevB.84.153402.

[27] J. A. Wilson and A. D. Yoffe, "The transition metal dichalcogenides discussion and interpretation of the observed optical, electrical and structural properties," *Adv. Phys.*, vol. 18, no. 73, pp. 193–335, 1969, doi: 10.1080/00018736900101307.

[28] R. A. Bromley, R. B. Murray, and A. D. Yoffe, "The band structures of some transition metal dichalcogenides. III. Group VIA: trigonal prism materials," *J. Phys. C*, vol. 5, no. 7, pp. 759–778, 1972, doi: 10.1088/0022-3719/5/7/007.

[29] S. Z. Butler *et al.*, "Progress, challenges, and opportunities in two-dimensional materials beyond graphene," *ACS Nano*, vol. 7, no. 4, pp. 2898–2926, 2013, doi: 10.1021/nn400280c.

[30] M. M. Ugeda *et al.*, "Characterization of collective ground states in single-layer NbSe2," *Nat. Phys.*, vol. 12, no. 1, pp. 92–97, 2016, doi: 10.1038/nphys3527.

[31] S. Tang *et al.*, "Quantum spin Hall state in monolayer 1T'-WTe 2," *Nat. Phys.*, vol. 13, no. 7, pp. 683–687, 2017, doi: 10.1038/nphys4174.

[32] S. Wu *et al.*, "Observation of the quantum spin Hall effect up to 100 kelvin in a monolayer crystal," *Science*, vol. 359, no. 6371, pp. 76–79, 2018, doi: 10.1126/science.aan6003.

[33] W. Zhao, R. M. Ribeiro, and G. Eda, "Electronic structure and optical signatures of semiconducting transition metal dichalcogenide nanosheets," *Acc. Chem. Res.*, vol. 48, no. 1, pp. 91–99, 2015, doi: 10.1021/ar500303m.

[34] E. Cappelluti, R. Roldán, J. A. Silva-Guillén, P. Ordejón, and F. Guinea, "Tight-binding model and direct-gap/indirect-gap transition in single-layer and multilayer MoS2," *Phys. Rev. B*, vol. 88, no. 7, p. 075409, 2013, doi: 10.1103/PhysRevB.88.075409.

[35] C. H. Chang, X. Fan, S. H. Lin, and J. L. Kuo, "Orbital analysis of electronic structure and phonon dispersion in MoS2, MoSe2, WS2, and WSe2 monolayers under strain," *Phys. Rev. B*, vol. 88, no. 19, p. 195420, 2013, doi: 10.1103/PhysRevB.88.195420.

[36] I. G. Lezama *et al.*, "Indirect-to-direct band gap crossover in few-layer MoTe2," *Nano Lett.*, vol. 15, no. 4, pp. 2336–2342, 2015, doi: 10.1021/nl5045007.

[37] A. Kormányos *et al.*, "$k \cdot p$ theory for two-dimensional transition metal dichalcogenide semiconductors," *2D Mater.*, vol. 2, no. 2, p. 022001, 2015, doi: 10.1088/2053-1583/2/2/022001.

[38] M. M. Ugeda *et al.*, "Giant bandgap renormalization and excitonic effects in a monolayer transition metal dichalcogenide semiconductor," *Nat. Mater.*, vol. 13, no. 12, pp. 1091–1095, 2014, doi: 10.1038/nmat4061.

[39] I. Verzhbitskiy, D. Vella, K. Watanabe, T. Taniguchi, and G. Eda, "Suppressed out-of-plane polarizability of free excitons in monolayer WSe2.," *ACS Nano*, vol. 13, no. 3, pp. 3218–3224, 2019, doi: 10.1021/acsnano.8b08905.

[40] G. Wang *et al.*, "Colloquium: Excitons in atomically thin transition metal dichalcogenides," *Rev. Mod. Phys.*, vol. 90, no. 2, p. 021001, 2018, doi: 10.1103/RevModPhys.90.021001.

[41] K. F. Mak, C. Lee, J. Hone, J. Shan, and T. F. Heinz, "Atomically thin MoS_2: A new direct-gap semiconductor," *Phys. Rev. Lett.*, vol. 105, no. 13, pp. 2–5, 2010, doi: 10.1103/PhysRevLett.105.136805.

[42] A. Splendiani *et al.*, "Emerging photoluminescence in monolayer MoS2," *Nano Lett.*, vol. 10, no. 4, pp. 1271–1275, 2010, doi: 10.1021/nl903868w.
[43] A. Carvalho, R. M. Ribeiro, and A. H. Castro Neto, "Band nesting and the optical response of two-dimensional semiconducting transition metal dichalcogenides," *Phys. Rev. B*, vol. 88, no. 11, p. 115205, 2013, doi: 10.1103/PhysRevB.88.115205.
[44] M. Bernardi, M. Palummo, and J. C. Grossman, "Extraordinary sunlight absorption and one nanometer thick photovoltaics using two-dimensional monolayer materials," *Nano Lett.*, vol. 13, no. 8, pp. 3664–3670, 2013, doi: 10.1021/nl401544y.
[45] R. R. Nair *et al.*, "Fine structure constant defines visual transparency of graphene," *Science*, vol. 320, no. 5881, p. 1308, 2008, doi: 10.1126/science.1156965.
[46] S. Pancharatnam, "Generalized theory of interference, and its applications — Part I. Coherent pencils," *Proc. Indian Acad. Sci. A*, vol. 44, no. 5, pp. 247–262, 1956, doi: 10.1007/BF03046050.
[47] M. V. Berry, "Quantal phase factors accompanying adiabatic changes," *Proc. R. Soc. Lond. Aes*, vol. 392, no. 1802, pp. 45–57, 1984, doi: 10.1098/rspa.1984.0023.
[48] D. Vanderbilt, *Berry Phases in Electronic Structure Theory: Electric Polarization, Orbital Magnetization and Topological Insulators*. Cambridge University Press, 2018. doi: 10.1017/9781316662205.
[49] G. Sundaram and Q. Niu, "Wave-packet dynamics in slowly perturbed crystals: Gradient corrections and Berry-phase effects," *Phys. Rev. B*, vol. 59, no. 23, pp. 14915–14925, 1999, doi: 10.1103/PhysRevB.59.14915.
[50] D. Xiao, M. C. Chang, and Q. Niu, "Berry phase effects on electronic properties," *Rev. Mod. Phys.*, vol. 82, no. 3, pp. 1959–2007, 2010, doi: 10.1103/RevModPhys.82.1959.
[51] S. Y. Zhou *et al.*, "Substrate-induced bandgap opening in epitaxial graphene," *Nat. Mater.*, vol. 6, no. 10, pp. 770–775, 2007, doi: 10.1038/nmat2003.
[52] R. V. Gorbachev *et al.*, "Detecting topological currents in graphene superlattices," *Science*, vol. 346, no. 6208, pp. 448–451, 2014, doi: 10.1126/science.1254966.

[53] D. Xiao, G.-B. Liu, W. Feng, X. Xu, and W. Yao, "Coupled spin and valley physics in monolayers of MoS2 and other group-VI dichalcogenides," *Phys. Rev. Lett.*, vol. 108, no. 19, p. 196802, 2012, doi: 10.1103/PhysRevLett.108.196802.

[54] J. Lee, K. F. Mak, and J. Shan, "Electrical control of the valley Hall effect in bilayer MoS_2 transistors," *Nat. Nanotechnol.*, vol. 11, no. 5, pp. 421–425, 2016, doi: 10.1038/nnano.2015.337.

[55] N. W. Ashcroft and N. D. Mermin, *Solid State Physics*. Saunders College Publishing, 1976. [Online]. Available: https://www.bibsonomy.org/bibtex/27759092df316587a126f200107b24a85/brouder.

[56] D. Xiao, W. Yao, and Q. Niu, "Valley-contrasting physics in graphene: Magnetic moment and topological transport," *Phys. Rev. Lett.*, vol. 99, no. 23, p. 236809, 2007, doi: 10.1103/PhysRevLett.99.236809.

[57] K. F. Mak, K. He, J. Shan, and T. F. Heinz, "Control of valley polarization in monolayer MoS2 by optical helicity," *Nat. Nanotechnol.*, vol. 7, no. 8, pp. 494–498, 2012, doi: 10.1038/nnano.2012.96.

[58] F. T. Arecchi, E. Courtens, R. Gilmore, and H. Thomas, "Atomic coherent states in quantum optics," *Phys. Rev. A*, vol. 6, no. 6, pp. 2211–2237, 1972, doi: 10.1103/PhysRevA.6.2211.

[59] A. M. Jones *et al.*, "Optical generation of excitonic valley coherence in monolayer WSe_2," *Nat. Nanotechnol.*, vol. 8, no. 9, pp. 634–638, 2013, doi: 10.1038/nnano.2013.151.

[60] Y. Ye *et al.*, "Electrical generation and control of the valley carriers in a monolayer transition metal dichalcogenide," *Nat. Nanotechnol.*, vol. 11, no. 7, pp. 598–602, 2016, doi: 10.1038/nnano.2016.49.

[61] M. M. Glazov, T. Amand, X. Marie, D. Lagarde, L. Bouet, and B. Urbaszek, "Exciton fine structure and spin decoherence in monolayers of transition metal dichalcogenides," *Phys. Rev. B*, vol. 89, no. 20, p. 201302, 2014, doi: 10.1103/PhysRevB.89.201302.

[62] D. Y. Qiu, T. Cao, and S. G. Louie, "Nonanalyticity, valley quantum phases, and lightlike exciton dispersion in monolayer transition metal dichalcogenides: theory and first-principles calculations," *Phys. Rev. Lett.*, vol. 115, no. 17, p. 176801, 2015, doi: 10.1103/PhysRevLett.115.176801.

[63] G. Berghäuser *et al.*, "Inverted valley polarization in optically excited transition metal dichalcogenides," *Nat. Commun.*, vol. 9, no. 1, p. 971, 2018, doi: 10.1038/s41467-018-03354-1.

[64] D. L. Dexter, "A theory of sensitized luminescence in solids," *J. Chem. Phys.*, vol. 21, no. 5, pp. 836–850, 1953, doi: 10.1063/1.1699044.

[65] D. MacNeill *et al.*, "Breaking of valley degeneracy by magnetic field in monolayer $MoSe_2$," *Phys. Rev. Lett.*, vol. 114, no. 3, p. 037401, 2015, doi: 10.1103/PhysRevLett.114.037401.

[66] A. A. Mitioglu *et al.*, "Optical investigation of monolayer and bulk tungsten diselenide (WSe_2) in high magnetic fields," *Nano Lett.*, vol. 15, no. 7, pp. 4387–4392, 2015, doi: 10.1021/acs.nanolett.5b00626.

[67] A. V. Stier, K. M. McCreary, B. T. Jonker, J. Kono, and S. A. Crooker, "Exciton diamagnetic shifts and valley Zeeman effects in monolayer WS2 and MoS2 to 65 Tesla," *Nat. Commun.*, vol. 7, no. 1, p. 10643, 2016, doi: 10.1038/ncomms10643.

[68] G. Aivazian *et al.*, "Magnetic control of valley pseudospin in monolayer WSe_2," *Nat. Phys.*, vol. 11, no. 2, pp. 148–152, 2015, doi: 10.1038/nphys3201.

[69] A. Srivastava, M. Sidler, A. V. Allain, D. S. Lembke, A. Kis, and A. Imamoglu, "Valley Zeeman effect in elementary optical excitations of monolayer WSe2," *Nat. Phys.*, vol. 11, no. 2, pp. 141–147, 2015, doi: 10.1038/nphys3203.

[70] H. Rostami, A. G. Moghaddam, and R. Asgari, "Effective lattice Hamiltonian for monolayer MoS2: Tailoring electronic structure with perpendicular electric and magnetic fields," *Phys. Rev. B*, vol. 88, no. 8, p. 085440, 2013, doi: 10.1103/PhysRevB.88.085440.

[71] G.-B. Liu, W.-Y. Shan, Y. Yao, W. Yao, and D. Xiao, "Three-band tight-binding model for monolayers of group-VIB transition metal dichalcogenides," *Phys. Rev. B*, vol. 88, no. 8, p. 085433, 2013, doi: 10.1103/PhysRevB.88.085433.

[72] T. Deilmann, P. Krüger, and M. Rohlfing, "Ab initio studies of exciton g factors: monolayer transition metal dichalcogenides in magnetic fields," *Phys. Rev. Lett.*, vol. 124, no. 22, p. 226402, 2020, doi: 10.1103/PhysRevLett.124.226402.

[73] T. Woźniak, P. E. Faria Junior, G. Seifert, A. Chaves, and J. Kunstmann, "Exciton g factors of van der Waals heterostructures from first-principles calculations," *Phys. Rev. B*, vol. 101, no. 23, p. 235408, 2020, doi: 10.1103/PhysRevB.101.235408.
[74] J. Qi, X. Li, Q. Niu, and J. Feng, "Giant and tunable valley degeneracy splitting in MoTe2," *Phys. Rev. B*, vol. 92, no. 12, p. 121403, 2015, doi: 10.1103/PhysRevB.92.121403.
[75] Z. Wang, C. Tang, R. Sachs, Y. Barlas, and J. Shi, "Proximity-induced ferromagnetism in graphene revealed by the anomalous Hall effect," *Phys. Rev. Lett.*, vol. 114, no. 1, p. 016603, 2015, doi: 10.1103/PhysRevLett.114.016603.
[76] P. Wei *et al.*, "Strong interfacial exchange field in the graphene/EuS heterostructure," *Nat. Mater.*, vol. 15, no. 7, pp. 711–716, 2016, doi: 10.1038/nmat4603.
[77] C. Zhao *et al.*, "Enhanced valley splitting in monolayer WSe2 due to magnetic exchange field," *Nat. Nanotechnol.*, vol. 12, no. 8, pp. 757–762, 2017, doi: 10.1038/nnano.2017.68.
[78] D. Zhong *et al.*, "Van der Waals engineering of ferromagnetic semiconductor heterostructures for spin and valleytronics," *Sci. Adv.*, vol. 3, no. 5, p. e1603113, 2017, doi: 10.1126/sciadv.1603113.
[79] L. Ciorciaro, M. Kroner, K. Watanabe, T. Taniguchi, and A. Imamoglu, "Observation of magnetic proximity effect using resonant optical spectroscopy of an electrically tunable MoSe2/CrBr3 heterostructure," *Phys. Rev. Lett.*, vol. 124, no. 19, p. 197401, 2020, doi: 10.1103/PhysRevLett.124.197401.
[80] J. Kang, S. Tongay, J. Zhou, J. Li, and J. Wu, "Band offsets and heterostructures of two-dimensional semiconductors," *Appl. Phys. Lett.*, vol. 102, no. 1, p. 012111, 2013, doi: 10.1063/1.4774090.
[81] J. Kang, J. Li, S.-S. Li, J.-B. Xia, and L.-W. Wang, "Electronic structural Moiré pattern effects on MoS2/MoSe2 2D heterostructures," *Nano Lett.*, vol. 13, no. 11, pp. 5485–5490, 2013, doi: 10.1021/nl4030648.
[82] H. Yu, G.-B. Liu, J. Tang, X. Xu, and W. Yao, "Moiré excitons: From programmable quantum emitter arrays to spin-orbit–coupled artificial lattices," *Sci. Adv.*, vol. 3, no. 11, p. e1701696, 2017, doi: 10.1126/sciadv.1701696.

[83] P. Rivera *et al.*, "Valley-polarized exciton dynamics in a 2D semiconductor heterostructure," *Science*, vol. 351, no. 6274, pp. 688–691, 2016, doi: 10.1126/science.aac7820.
[84] W. Heisenberg, "Über den Bau der Atomkerne. I," *Zeitschrift für Physik*, vol. 77, no. 1–2, pp. 1–11, 1932, doi: 10.1007/BF01342433.
[85] X. Xu, W. Yao, D. Xiao, and T. F. Heinz, "Spin and pseudospins in layered transition metal dichalcogenides," *Nat. Phys.*, vol. 10, no. 5, pp. 343–350, 2014, doi: 10.1038/nphys2942.
[86] M. Chhowalla, H. S. Shin, G. Eda, L. J. Li, K. P. Loh, and H. Zhang, "The chemistry of two-dimensional layered transition metal dichalcogenide nanosheets," *Nat. Chem.*, vol. 5, no. 4, pp. 263–275, 2013, doi: 10.1038/nchem.1589.
[87] T. Brumme, M. Calandra, and F. Mauri, "First-principles theory of field-effect doping in transition-metal dichalcogenides: Structural properties, electronic structure, Hall coefficient, and electrical conductivity," *Phys. Rev. B*, vol. 91, no. 15, p. 155436, 2015, doi: 10.1103/PhysRevB.91.155436.
[88] D. Xiao, Y. Yao, Z. Fang, and Q. Niu, "Berry-phase effect in anomalous thermoelectric transport," *Phys. Rev. Lett.*, vol. 97, no. 2, p. 026603, 2006, doi: 10.1103/PhysRevLett.97.026603.

https://doi.org/10.1142/9789811229107_0003

Chapter 3

Valley Addressability and Controllability

Fabio Bussolotti, Sarthak Das, Wang Tong, and Kuan Eng Johnson Goh

3.1 Introduction

In Chapter 2, we have discussed the theoretical background needed to understand the "valley" properties of 2D materials. The unique valley properties in layered semiconducting materials inspire us to investigate the underlying physics of valley-coupled spins in these material systems. Furthermore, the time-reversal symmetry and the intrinsic broken inversion symmetry in semiconducting 2D transition metal dichalcogenides (TMDs) make it feasible to explore the valley-selective light-matter interaction in those materials. In this chapter, we will first focus on *valley addressability* where we will discuss three key methods to access valley-selective information from 2D materials with reference to relevant examples:

3.2 Spin- and angle-resolved photoemission spectroscopy,
3.3 Circular dichroic micro-photoluminescence, and
3.4 Kerr rotation spectroscopy.

Thereafter, we will discuss *valley controllability* and cover the following concept demonstrations that have been reported in the literature:

3.6 Electrical control of valley carriers,
3.7 Magnetic control via the valley Zeeman effect,
3.8 Optical field control via AC Stark effect, and
3.9 Temperature-induced valley control.

All the above control options could be implemented individually or in combination, depending on the device action needed to the realise practical valleytronic devices. The addressability techniques in Sections 3.2 to 3.4 are often used in experiments to confirm that the valley control has been successful.

3.2 Spin- and Angle-resolved Photoemission Spectroscopy of TMD

One of the methods that can directly probe the spin character of the charges carriers in valleys of TMD materials is spin- and angle-resolved photoemission spectroscopy. This technique involves shining light on a material and then inferring the band structure information from the photoelectrons ejected from the material — the well-known photoelectric effect. We shall first present the more commonly used angle-resolved photoemission spectroscopy for measuring crystal band structures before extending to the spin-resolved counterpart.

3.2.1 Angle-resolved Photoemission Spectroscopy (ARPES)

Angle-resolved Photoemission Spectroscopy (ARPES) is an advanced experimental technique relying on the photoelectric effect to trigger the emission of electrons from a solid into vacuum [1]. Fig. 3.1 shows a schematic of a typical ARPES experimental setup. When monochromatised radiation produced by a gas discharge lamp, synchrotron beamline and/or laser source is incident on a sample shown in Fig. 3.1(a), electrons are emitted *into the vacuum*, provided that the photon energy ($h\nu$) is sufficiently large to overcome the energy barrier determined by the sample work function (WF). The as-emitted electrons are then collected by an electron analyser in Fig. 3.1(a).

By measuring and analysing their kinetic energy (ε_{Kin}) and emission direction, detailed information on the electron energies and momentum inside a single crystalline solid (i.e., electronic band structure) can be obtained [1]–[3]. Under the final state free-particle assumption [1], the photoelectron momentum $\boldsymbol{P} = \hbar\boldsymbol{K}$ (where $\boldsymbol{K}$ is the corresponding wavevector) can be entirely related to its kinetic energy and emission direction: its modulus is given by $P = \sqrt{2m\varepsilon_{Kin}}$ while its components parallel ($\boldsymbol{P}_{\parallel} = \hbar\boldsymbol{K}_{\parallel}$) and perpendicular ($\boldsymbol{P}_{\perp} = \hbar\boldsymbol{K}_{\perp}$) to the sample surface are determined from the measured emission angles θ and φ as shown in Fig. 3.1(a). Assuming total energy and momentum conservation during the photoemission process, the electron binding energy ε_B referenced to the Fermi level (ε_F) is given by

$$\varepsilon_{Kin} = h\nu - WF - \varepsilon_B \tag{3.1}$$

and the crystal momentum $\hbar\boldsymbol{k}$ can be related to the kinetic energy and momentum of photoelectrons as follows

$$\hbar\boldsymbol{K}_{\parallel} = \hbar\boldsymbol{k}_{\parallel} = \sqrt{2m\varepsilon_{Kin}}\sin\theta \tag{3.2}$$

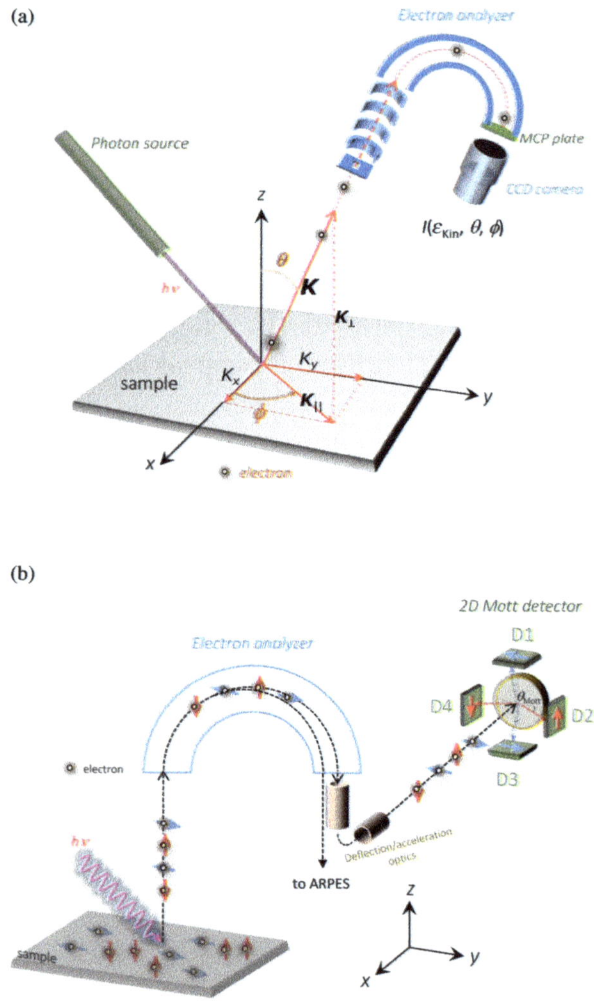

Fig. 3.1. (a) Schematic of a typical setup of ARPES measurements. (b) Schematic of a typical setup of SARPES measurements, equipped with a 2D Mott detector, i.e., only the spin component along two reference axes can be measured (P_x and P_z). Part of the electrons detected by ARPES are deflected and accelerated (25–100 keV) towards a target (Au, Th) and photoemission intensity measured by separated detectors (polarimeters D1–D4) located at the scattering angle (θ_{Mott}).

where m is the electron mass, and $\hbar \boldsymbol{k}_{\|} = \hbar \boldsymbol{K}_{\|}$ is the crystal momentum parallel to the surface which is preserved across the solid/vacuum interface ($\boldsymbol{K}_{\|} = \boldsymbol{k}_{\|}$), $\boldsymbol{k}_{\|}$ being the parallel electron momentum inside the solid.

Expressing $\boldsymbol{k}_{\|} = (k_x, k_y)$ and $\boldsymbol{K}_{\|} = (K_x, K_y)$ where k_x, k_y, K_x and K_y are the respective components in the xy plane of the sample surface in Fig. 3.1(a), one obtains:

$$\hbar K_x = \hbar k_x = \sqrt{2m\varepsilon_{Kin}} \sin\theta \cos\varphi \tag{3.3}$$

$$\hbar K_y = \hbar k_y = \sqrt{2m\varepsilon_{Kin}} \sin\theta \sin\varphi \tag{3.4}$$

Note that the component of the electron wavevector $\boldsymbol{k}_{\perp}$ is not preserved upon crossing of the sample surface ($\boldsymbol{K}_{\perp} \neq \boldsymbol{k}_{\perp}$) due to the lack of translational symmetry along the surface normal. As a result, the *in-plane* band dispersion $\varepsilon_B(\boldsymbol{k}_{\|})$ along the high-symmetry direction of the Surface Brillouin Zone (SBZ) of a 2D layered material can be completely mapped by measuring the kinetic energy position ε_{Kin} of the photoemission peaks at various emission angles θ and φ (see Eqs. (3.2)–(3.4)). Moreover, due to the limited inelastic mean free path (< 2 nm) of photoelectrons at the kinetic energy range (< 100 eV) acquired in typical ARPES measurement [4], the ARPES signal is dominated by the contribution of the outer layers of the crystalline sample. A natural consequence of such a high surface sensitivity is that ARPES acquisition needs to be performed in ultrahigh vacuum (UHV) environment (pressure $\approx 10^{-10}$~10^{-11} mbar) on a clean surface, which is obtained by materials grown in UHV, or bulk crystals cleaved where possible. Otherwise, some UHV surface cleaning (e.g., outgas annealing, or controlled ion sputtering) might be needed to reduce the adsorption of surface contaminants.

3.2.2 Spin-resolved ARPES

Spin-resolved ARPES (SARPES) is an extension of the above technique that allows a complete mapping of the direction and magnitude of the spin-polarisation in the ensemble of emitted photoelectrons. Aside from *final state* photoemission effects [5], i.e., electron-electron correlation, electron-phonon scattering, etc., SARPES also reflects the spin polarisation in the electronic bands of the investigated materials.

A standard SARPES experimental setup involves multiple pairs of Mott detector (polarimeter), which utilises the spin-orbit interaction in the process of electron scattering (Mott scattering) by heavy elements targets (Au, Th, etc.) [5]. The signal detection process is inefficient when compared to the regular ARPES since the intensity of the signal detected is typically a few orders of magnitude lower [5]. As a result, SARPES requires synchrotron-based photoemission facilities, where high photon flux can be achieved relative to a standard lab source. Recent developments of multi-channel spin detectors using different and more efficient detection schemes promise a better understanding of spin textures from materials as well as the spin-ARPES process [6], [7].

The scattering geometry defines the spin-sensitive axes of the polarimeter, onto which the spin polarisation vector $\boldsymbol{P}$ is projected. For a particular spin quantisation axis i, two spin-dependent scattered photoemission intensities $I_i^{\uparrow}$ ("up") and $I_i^{\downarrow}$ ("down") are measured. The spin polarisation component P_i along this axis is then calculated from

$$P_i = \frac{1}{S}\frac{\left(I_i^{\uparrow} - I_i^{\downarrow}\right)}{\left(I_i^{\uparrow} + I_i^{\downarrow}\right)} \tag{3.5}$$

where S is the Sherman function indicating the sensitivity of the detector [5].

A schematic of a typical SARPES experimental setup is described in Fig. 3.1(b). For clarity, a 2D Mott detection system is illustrated, where only two spin polarisation components (P_y, P_z) can be directly measured. After energy/angular filtering by the hemispherical analyser, some fraction of photoelectrons is deflected and accelerated along normal incidence direction towards a heavy elements target (Au, Th). Here the electrons can be "left"/ "right" backscattered at a given angle (θ_{Mott}) with respect to the target surface plane depending on their original spin direction. The scattering mechanism is based on the spin-orbit interaction between the incoming electron and atoms of the target (Mott scattering). The scattering plane is defined by the direction of incidence of accelerated electrons and by their spin direction, i.e., xz plane for spin polarisation along z axis (P_z) and xy for electron spin-polarised along the y axis (P_y). After scattering, the electrons will be detected by a series of single-channel detectors (D1–D4) positioned along the scattering direction determined by θ_{Mott}, and typically presented in the form of a photoemission energy distribution curve. The intensity and energy position of the photoemission peaks $I^{\uparrow}$ and $I^{\downarrow}$ is determined by the degree of spin polarisation and the binding energy of the initial electronic states.

3.2.3 Band Structure and Spin-Polarised States in 2H TMDs (MX_2)

Electronic band structures. ARPES investigations of TMDs were originally focused on the electronic band structure of small-sized, high-quality TMD flakes obtained by direct exfoliation from single crystal bulk and then transferred on a conductive substrate required for photoemission measurements [8]–[12]. For exfoliated TMDs samples, typical measurements constraints are (i) limited lateral size (~μm) of the flakes, (ii) low density of flakes on the supporting substrate and (iii) difficulties in accurate control of the number of layers, ranging from monolayer to multilayer thickness [8]–[12]. Band structure investigations on these samples were therefore mainly conducted by micro-ARPES techniques at designated synchrotron radiation facilities [8]–[14] where the reduced photon beam size (~μm) combined with microscopy techniques allow an accurate selection of the probed sample area. The micro-ARPES technique was successfully employed in mapping the electronic band structure of TMDs exfoliated layers of various thickness. Varying the number of layers changes the band structure in a few ways, namely, (i) the change in the number of valence band branches at the Γ point (following the number of layers) and (ii) the concomitant change of the valence band maximum position in the reciprocal space from K point to Γ point when the layer thickness increases from monolayer to bilayer as shown in Fig. 3.2(a). Finally, the same experimental approach was applied to the study of suspended TMDs flakes [9] or at the interface with the high dielectric materials [12] to clarify the impact of substrate-induced lattice strain [9] or electronic screening on the valence band structure of TMDs [12].

Fig. 3.2(a) illustrates the experimental results from Ref. [8], in which the layer-by-layer band structure evolution of MoS_2 flakes is directly mapped by using micro-ARPES technique on precisely selected areas of the exfoliated flakes. As evidenced in Fig. 3.2(a), a major drawback of the micro-ARPES technique is its intrinsically limited energy resolution (> 0.2 eV) [8]–[12]. This hinders measurements that reveal important details of the electronic band structures of the TMD layers,

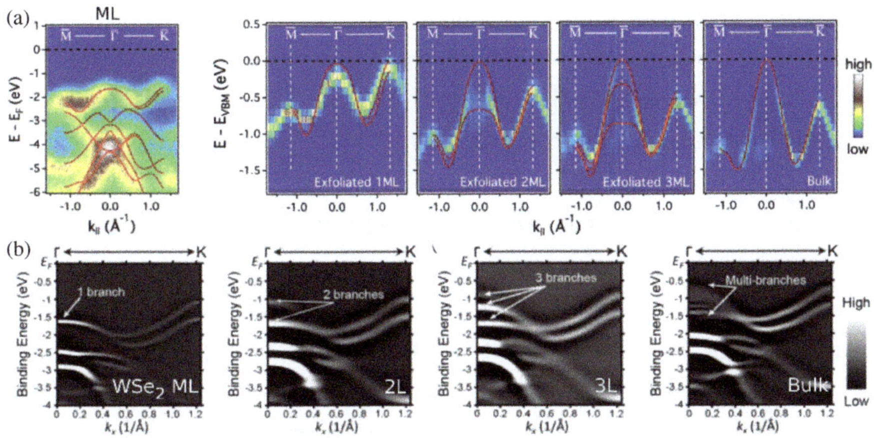

Fig. 3.2. (a) Micro-ARPES spectra from exfoliated MoS_2 monolayer (ML) (left panel). Experimental band dispersion along the MΓK high-symmetry direction of exfoliated monolayer, bilayer, trilayer and multilayer (bulk-like) of MoS_2 (right panels), as obtained by second derivative filter of ARPES data. Adapted with permission from Ref. [8]. (b) Experimental band dispersion along the ΓK high-symmetry direction of monolayer, bilayer, trilayer and multilayer (bulk-like) WSe_2 on bilayer graphene (second derivative filtering). Adapted with permission from Ref. [20].

i.e., number of valence band branches at the Γ point and spin-orbit energy splitting at the K point, which is relevant for valleytronics applications.

Collecting more photoelectrons over larger area samples could increase the ARPES energy resolution. Towards this end, several large-scale growth methods of TMDs were alternatively developed based on both physical [15]–[17] and chemical deposition [14], [18]–[20] processes. For large-scale growth of TMDs, single crystal substrates are generally used as templates, and a conductive substrate is required to avoid charging effects during the photoemission measurements. Single crystal substrates may favour the formation of larger TMD domains (of up to ~100 μm of lateral size) via epitaxy, thus allowing a detailed characterisation by more conventional ARPES systems with large beam-spot sizes [14]–[20], where higher energy/momentum resolution limit can be generally achieved.

The electronic band structures of TMD mono- and multilayers deposited large scale on epitaxial graphene were mapped in great detail by high energy-resolved ARPES measurements [15]–[20]. The peculiar change of the valence band structure with varying layer thickness, i.e., direct-to-indirect band transition and multiple splitting of the valence bands at the Γ point as layer number is increased are highlighted in Fig. 3.2(b). Studies have also found that the post-growth deposition of alkaline metals (Na, K) resulted in *n*-type doping of the TMD layers, and the excess charges populated the lower conduction band states as observed by ARPES [17], [19], [20]. ARPES studies conducted on *n*-doped TMD mono- and multilayers demonstrated the thickness-dependent direct-to-indirect bandgap transition confirmed by theoretical simulations [17], [19], [20]. With regard to the pristine TMD case, reduction of the electronic bandgap (~0.2–0.3 eV) and valence band distortion (i.e., different binding energy shift observed at various high-symmetry points of the SBZ) were reported in ARPES studies of doped MoS_2 [17], [19] and WSe_2 films [20]. They were mainly attributed to the substrate- and/or doping-enhanced electronic screening as mediated by the difference in the wave function localisation at the high-symmetry point, i.e., out-of-plane orbital character of the valence band (conduction band) at the Γ(K) point *vs* in-plane orbital character of the valence band at the K point [16], [17]. In addition, electronic wavefunctions with out-of-plane character of the layer are expected to strongly interact with the supporting substrate (i.e., hybridisation and/or electronic screening effect), thus resulting in large energy broadening and valence band distortion around the Γ point of MoS_2 and WS_2 monolayers reported for metallic and semi-metallic substrates [16], [17].

More recently, ARPES investigation on single layers of MoS_2 and WS_2 on HOPG substrate [21], [22] suggests that the charge and locked spin dynamics at the K valley valence band states are both essentially "protected", even at cryogenic temperatures, from local change in the interfacial potential landscape originating from tiny lateral variation or layer/substrate distance and/or substrate impurities as commonly introduced in any real interface fabrication process. This demonstration of "immunity" of the valley-locked spins to interfacial landscape may have crucial implications for hole transport in single-layer devices seeking to exploit valley pseudospins and the control thereof in TMD-based heterointerfaces.

Spin polarisation. With respect to ARPES investigation, comparatively fewer experimental attempts were made to determine the spin polarisation of electronic bands of TMD layered materials, mostly due to the lower detection efficiency of SARPES [23]–[27].

SARPES measurements on monolayer films of $MoSe_2$ and WSe_2 grown on bilayer graphene substrate were reported [23], [24] showing out-of-plane spin polarisation of the K valley states with respect to the sample surface. The amount of spin polarisation (P_z = 0.1~0.2) of valley states was found to be much smaller than theoretical expectation, i.e., P_z = 1, (corresponding to full out-of-plane spin polarisation). The authors of the report attributed this to (i) low experimental resolution and (ii) overlap of signals coming from the twin domains (relative rotation of 60°) in the epitaxially grown graphene when probed by the relative beam spot of ~100 μm size.

Interestingly, a theoretical study [28] later proposed that the spin texture of the TMD could be probed by SARPES even in the inversion-symmetric bulk crystals ("hidden" spin polarisation), due to the localisation of two spin-degenerated valence band maxima on various layers of the unit cell and the finite penetration depth of the photoemission process probing preferentially the uppermost layer [4]. Experimentally, this effect was observed for WSe_2, which has the strongest spin-orbit coupling among the TMDs [25], by reducing the photoelectron mean free path via tuning of the incident photon energy (Fig. 3.3). The existence of spin-polarised electronic states was

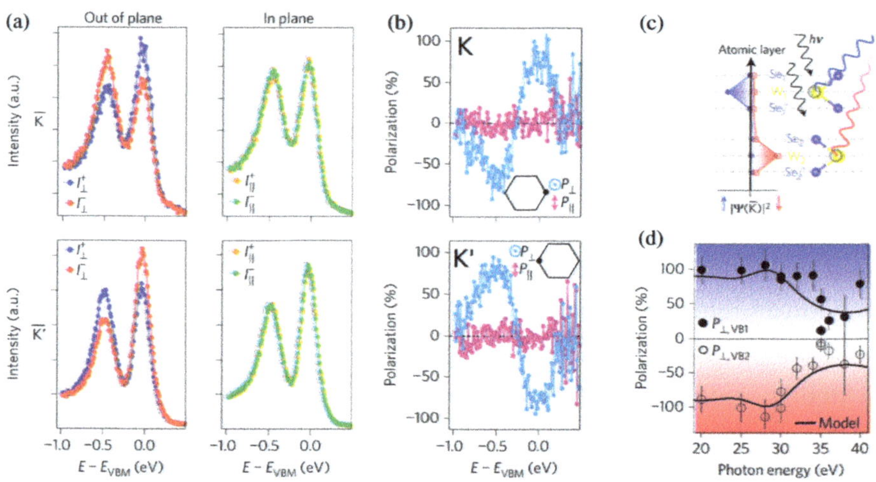

Fig. 3.3. (a) Energy distribution curves (in arbitrary units (a.u.)) from spin-resolved ARPES measurements (**hυ** = 25 eV, T = 300 K) at the K point (upper panels) and K′ point (lower panels) of WSe_2 bulk single crystal measured by the out-of-plane ($I_\perp$) and in-plane ($I_\parallel$) detectors in the Mott scattering chamber. (b) Extracted out-of-plane and in-plane spin polarisation at K and K′. (c) Schematic of the layer spin polarisation measurements by the photoemission technique. (d) Measured spin polarisation at the K point exhibiting a strong photon energy dependence. By increasing photon energy, photoelectrons are emitted at higher kinetic energy. Due to the increase of the electron mean free path, ARPES signals from both 1st and 2nd layers with opposite spin polarisation can be detected, thus leading to an overall reduction of the measured spin polarisation. Adapted with permission from Ref. [26].

also demonstrated in surface-sensitive SARPES study on high-quality 2H-MoS_2 single crystal [26]. More recently, Razzoli *et al.* [27] also reported on a large out-of-plane spin polarisation in the bulk dichalcogenide MoS_2, with its sign depending on the handedness (left/right) of the incident circularly polarised light. Theoretical calculations showed that (i) the above findings are related to initial state effect of photoemission influenced by dipole selection rules, and (ii) the valley- and spin-polarised states of the outermost and innermost layer can be independently addressed by tuning circular polarisation of the incident light. In this context, this method represents an alternative approach to the investigation of the layer-locked spin-polarised states, other than taking only advantage of the high ARPES surface sensitivity to selectively probe the outer layer.

In summary, SARPES can be an effective tool for directly probing the valley-coupled spins in TMD band structures, especially when one can elicit a strong SARPES signal from the TMD under study, either through a large single crystal or using a high intensity source, e.g., at a synchrotron.

3.3 Circular Dichroic Photoluminescence

3.3.1 Micro-photoluminescence

Broadly, photoluminescence (PL) is the process where photons incident on a material excite the material and in turn trigger the emission of new photons from that material. Loosely speaking, light comes in and PL goes out. Since PL is a method that examines the process of light-matter interaction, it carries a signature that is the result of the said interaction, which is why we sometimes call it a "final state" characterisation method. To trigger PL, the excitation source usually needs to be coherent, e.g., lasers, and the wavelength and power of excitation need to be specifically chosen depending on the material's optical properties. Only when the right excitation conditions are met can we promote electrons from the valence band to the conduction band and form electron-hole pairs. These electron-hole pairs are also bound by Coulomb interactions with specific binding energies related to the material and the mesoscopic environment of the point of excitation. As these electron-hole pairs recombine, they will release photons of a very specific energy, or a range of energies, which carry information of the binding energies, bandgap, perturbation from the local environment, and so on.

By collecting these emitted photons and analysing their properties, we can then obtain valuable information about the material, some of which we may not be able to access with other means. By slightly altering the excitation conditions, e.g., temperature, excitation wavelength, or laser power, the optical properties of PL will also change. The amount and trend of change can also provide us with important information of the material, such as the degree of electron-phonon coupling. As the PL process makes use of the quantum nature of light, a proper understanding of the PL data would require the quantum description of light.

The area of PL analysis on the material is determined by the spot-size of the excitation laser beam. Often, using appropriate lensing, the laser spot-size may be reduced to the order of only a few microns, thus allowing samples of only several microns in dimension to be studied. The term μPL is used when areas of material excitation and PL collection are of the order of microns in dimension.

In practice, PL requires tailoring the wavelengths of the excitation and collection instruments to trigger the desired excitation and effectively detect the PL signal. These are typically achieved through highly stable optical benchtop setups that incorporate high-precision optical components that are carefully aligned [29]. A general μPL setup is shown in Fig. 3.4, which illustrates the basic principle of how a PL experiment is conducted.

To excite the samples, a suitable laser source is usually controlled to generate the desired power at a wavelength that is comparable to the material's emission energy of interest (generally the energy of laser excitation is above the bandgap of the material ranging from μW to mW of power). The laser can be a continuous-wave one, where the power is continuous at any time instance, or a pulsed one, where

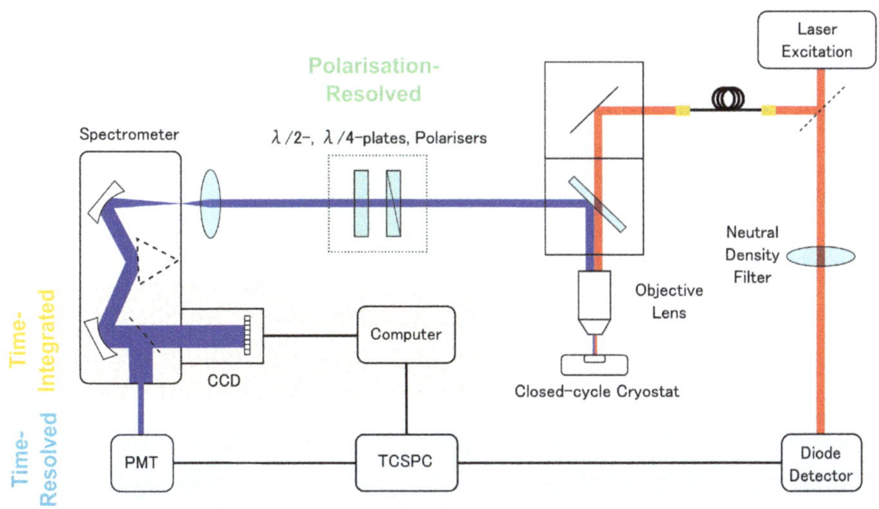

Fig. 3.4. A schematic of PL setup that can perform time-integrated, time-resolved, and polarisation-resolved μPL. Here PMT stands for photomultiplier tube, TCSPC is time-correlated single photon counter and CCD denotes charged-coupled device.

more concentrated short pulses with a duration of ps or 100 s of fs are generated at time intervals determined by the laser's repetition rate. The choice depends on the purpose of study.

PL studies can usually be time integrated or time resolved. For time-integrated PL, the PL signal is generally excited over a longer period with a continuous-wave laser and the wavelength-resolved photons are collected at the detector in steady state. For time-resolved PL, the photon counts are recorded as a function of time with excitation from ultrafast (ps or fs pulse width) lasers. A spectrometer is used to collect the sample's PL signal. The parallel beam is focused into the entrance slit of the spectrometer (see Fig. 3.4, time-integrated section) and directed to a grating of choice that can optimise the light scattered and provide a suitable resolution. Here, a trade-off between light intensity and resolution exists. With a higher grating number, we can get better resolution. However, this means that the number of bins into which photons are detected also increases. With the same number of photons, i.e., PL intensity, the light collected in each bin becomes less, resulting in a lower signal count despite the improved resolution.

Hence, depending on the purpose of the investigation, the researcher needs to decide if the experiment should prioritise higher resolution or stronger signal intensity. This question is especially important for valleytronic materials such as TMDs. This is because they may not always exhibit a high number of photon output under laser excitation. Therefore, despite building a well-designed PL collection system with high efficiency, the total intensity might not be sufficient to satisfy a study that requires both high intensity and resolution. Furthermore, as laser intensity increases, the TMD sample's tolerance to low-efficiency absorption also decreases. With extraneous heat generated, the sample's integrity might be compromised. Under ambient conditions, such high powers of PL excitation could also alter the original PL signature, a phenomenon that can be observed during the experiments. This could be manifested in a significant drop in PL intensity, redshift of PL central wavelength, or in some cases a broadening of the PL linewidth. Fig. 3.5 shows typical time-integrated PL spectrum of a CVD-grown monolayer WS_2 placed on Si/SiO_2 that exhibits emission across 1.9–2.1 eV (590–680 nm) at room temperature. Here the exciton peak ~2.02 eV has a linewidth (Γ_X) of ~30 meV (shown in blue shaded region).

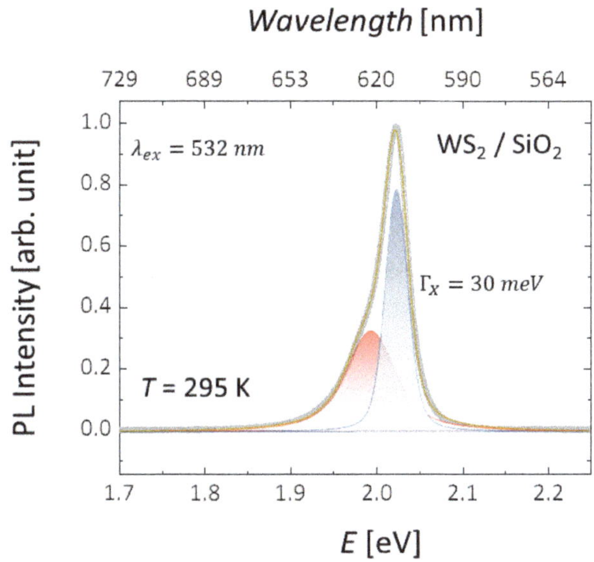

Fig. 3.5. Typical time-integrated PL spectrum of a CVD-grown monolayer WS_2 placed on Si/SiO_2 that exhibits excitonic modes of emission across 1.9–2.1 eV (590–680 nm) at room temperature. In valleytronic materials that emit across different energy ranges, confined emission local maxima can also sometimes be observed at cryogenic temperatures. Here the exciton peak ~2.02 eV has a linewidth ~30 meV (shown in blue shaded region). PL is collected under 532 nm (~2.33 eV) laser excitation with power of ~100 μW.

If a study aims to investigate the quality and emissivity of a sample in terms of their emission intensity, a choice of smaller grating numbers, while reducing the resolution of the PL spectrum, can still highlight the total emission intensity. This would allow the researcher to have a better comparison of the intensity differences of various samples. On the other hand, if a study primarily focuses on the linewidth and emission profile of a high-quality TMD sample, the choice of higher grating numbers would be needed to confirm the assessment of the transition uncertainties as far as possible. However, this must be done with caution, because a certain degree of emission intensity is still needed to ensure the data's usefulness. If the total signal count is very close to the background noise intensity, the emission profile will be very challenging to ascertain. In these cases, it is often not possible to determine the exact linewidth or emission centre. Even if it can be done, the accuracy would be very low and the large uncertainty ranges would make the results less meaningful.

PL investigations also allow us to gain further insights into the sample's uniformity. This is usually achieved by *PL mapping*. With several commercial optical systems, PL mapping can now be done automatically over an area of interest. The systems will usually divide the area of interest by a number of rows and columns set by the user and collect the PL from each of these coordinates individually. The more rows and columns used, the better the mapping resolution, at the expense of longer mapping time. It is important to balance the individual PL acquisition time such that each point of collection has just enough data, and then optimise the mapping parameters such that the entire area can be studied with reasonable accuracy.

Fig. 3.6 is an example that shows (a) the variations of linewidth of the sample and (b) the changes in central emission wavelength. If a sample

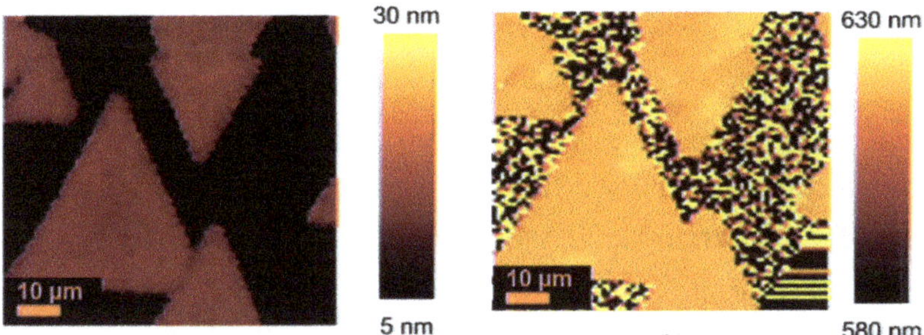

(a) (b)

Fig. 3.6. An example of the PL mapping of CVD-grown TMD samples at room temperature. (a) Mapping of the sample's PL transition linewidth. Most of the linewidth across the sample is below 30 nm and exhibits little variation across different sections. (b) Mapping of the sample's centre emission wavelength. The variation of the central wavelength is mostly between 618.8 and 619.5 nm, exhibiting very small changes across the area of interest.

has many local variations of defects, they could act as carrier-trapping centres. The Coulomb interaction of these centres with the excitons in their vicinity will affect not only central emission wavelength, but also contribute to the quantum transition uncertainty. This degree of uncertainty then adds up to increase the linewidth of the TMD transition. Therefore, large variation of linewidth across a sample could be construed as an indication that the sample has a random distribution of defects that negatively affects the uniformity of the valleytronic materials. These studies are especially useful to ascertain the sample's integrity, especially after CVD growth. With additional help from optical microscopy, these investigations would help the researchers obtain an initial understanding of the quality of the sample, see whether the growth routine produces desired outcomes and whether the samples are suitable for the building of devices and more complex optical studies such as Kerr rotation, which will be covered later in the chapter.

3.3.2 Polarisation of μPL from Valleytronic Materials

Optical polarisation is another parameter that can also be introduced to the aforementioned studies of time-integrated and time-resolved studies. For experimental realisation of polarisation-resolved studies, a set of half- or quarter-wave plates or polarisers can be inserted before the PL reaches the spectrometer for detection. As explained in the theory chapter, the K and K′ valleys of valleytronic materials have opposite orbital magnetic moments. This is one of the distinct properties of materials such as TMDs and arises from the broken inversion symmetry and the strong spin-orbit interactions [30]–[38]. The difference in orbital magnetic moment also results in distinct transitions for the K and K′ valleys. By convention, right circularly polarised light, or σ^+ photons, tend to couple directly with the K valley. Similarly, left circularly polarised light, or σ^- photons, interacts with the K′ valley preferentially.

Hence, we can make use of this valley selectivity to address either of the two valleys by optically exciting a valleytronic material with circularly polarised light. For instance, right circularly polarised light (with the appropriate laser wavelength and intensity) would in principle only create electron-hole pairs in the K valley. When the excitons recombine, the PL released would also be right circularly polarised. In this way, we can optically address each of the two valleys.

For instance, a schematic of a possible CDPL setup design is shown in Fig. 3.7, where the use of quarter-wave plates converts linearly polarised laser excitation to circularly polarised light to address either of the two different valleys. The CDPL emission is then collected by the objective, and then passes through the same quarter-wave plate to be re-converted back to linearly polarised light. The resultant PL

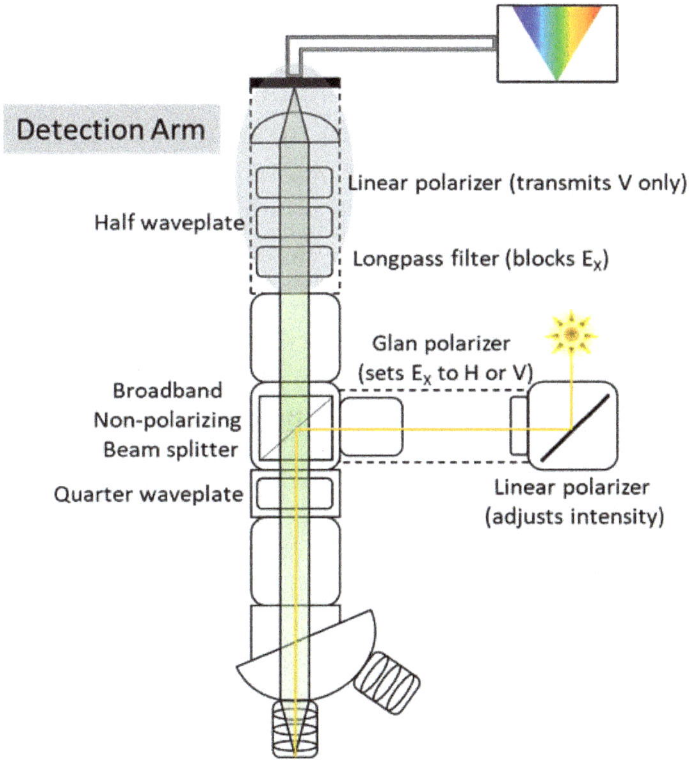

Fig. 3.7. A schematic of the excitation arm of a circular dichroic photoluminescence setup that can be built to achieve circularly polarised excitation and emission collection. The quarter-wave plate converts the excitation light, which is linearly polarised, into circularly polarised light to address the different valleys, and then re-converts the photoemission back into linearly polarised light, which can be separately detected to ascertain the degree of polarisation.

can then be separately detected as direct indications of the polarisation-resolved responses of the two valleys for the sample. One possible method to separate the polarised PL beams is via polarisation-sensitive optical block which breaks the incident beam into orthogonally linear polarisations, in this case a Glan-Thompson prism.

Having the ability to address different valleys is fundamental to the realisation of valley-selective read-outs in valleytronic devices, the details of which will be covered in the valley control sections later in this chapter, and also used frequently in conjunction with the experimental study of valley physics in Chapter 5, and of valley devices in Chapter 6. Naively, if we use 100% circularly polarised light as the source of excitation, we should expect to get 100% circular polarisation in the detected signals from the valleytronic sample. In principle, this is true. However, several mechanisms degrade the polarisation performance from ideal, causing a phenomenon called depolarisation [39]–[41], which we will discuss shortly. With a non-zero counter-circular polarisation (when excitation and detection are orthogonal to each other), there is now a need to define the degree of optical circular polarisation (DOCP) as a property and performance "metric" of the valleytronic material. DOCP is defined as follows [42], [43]:

$$\mathrm{DOCP} = \frac{|I(\sigma^{+}) - I(\sigma^{-})|}{I(\sigma^{+}) + I(\sigma^{-})} \tag{3.6}$$

where I is the intensity and is co- (when excitation and detection are both clockwise or anticlockwise) or counter-circular polarisation (when excitation is clockwise but detection anticlockwise or vice versa). As defined above, the DOCP is expressed as the ratio between the difference of opposite circular polarisation intensities and the sum of them.

As an aside, this definition is similar to that of the degree of optical linear polarisation (DOLP) which is widely used in other quantum systems. More detailed discussions on valley coherence, represented through the Bloch sphere, have already been presented in Chapter 2.

Here we provide a brief review for completeness. When the material is excited with a linearly polarised light, it can excite both the valleys due to valley degeneracy. The linearly polarised light can be considered as a coherent superposition of $(\sigma^{+})/(\sigma^{-})$ light. Here the degree of linear polarisation can be expressed as follows,

$$\mathrm{DOLP} = \frac{|I(H) - I(V)|}{I(H) + I(V)} \tag{3.7}$$

where *I(H/V)* denotes horizontally/vertically polarised light. In monolayer TMD, the neutral exciton is linearly polarised while the charged exciton is not [44], [45]. As defined above, DOCP is technically not a "percentage" *per se*, which means that it is mathematically not the percentage contribution of the intensity of one circular polarisation to the total PL intensity. The polarisation degree values should normally be indicated as a number equal or less than 1, like in these works [30], [35].

Equipped with the knowledge of DOCP and the necessary optical setup to study optical valley addressability, we now turn to look at some sample PL spectra. Fig. 3.8 is taken from a 2012 study on 2D valleytronic materials by Mak *et al.* [35]. In this investigation, the CDPL of MoS_2 monolayer and bilayer samples are studied in detail under various laser excitation wavelengths (on and off resonance). From Fig. 3.8(a), we can see that the DOCP of a monolayer MoS_2 reaches near unity under resonant excitation but falls off significantly when we moved to a bilayer MoS_2 in Fig. 3.8(d). This can be explained by the broken inversion symmetry in a monolayer and the restored inversion symmetry in a bilayer. Without broken inversion symmetry, excitons will not have preferential coupling to the K and K′ valleys and the measured DOCP drops drastically.

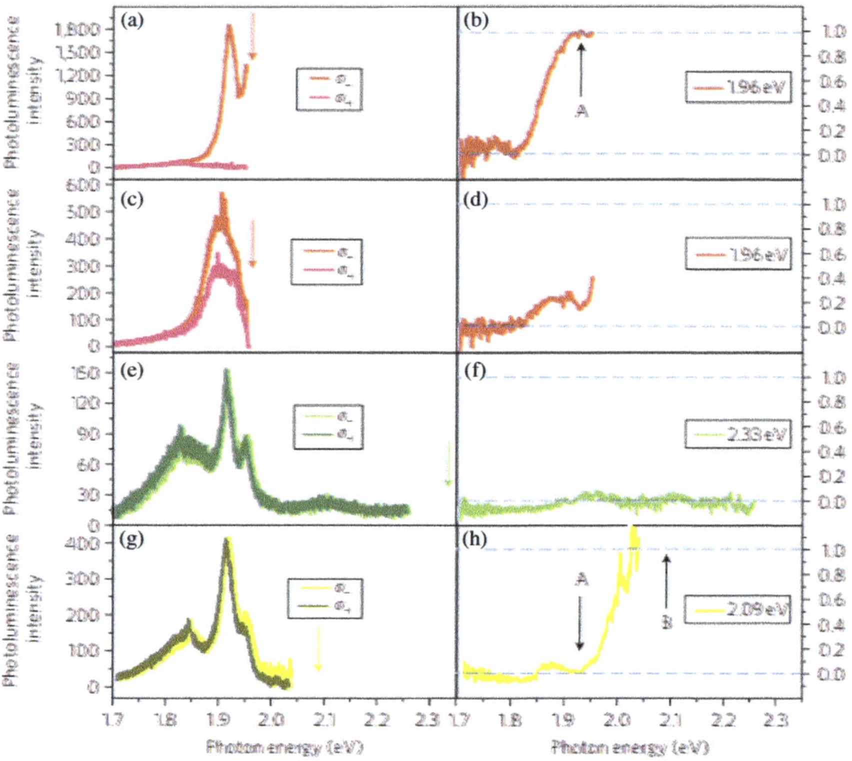

Fig. 3.8. Addressing the valley of MoS_2 by controlling the degree of optical circular polarisation. (a) and (b) show the results from monolayer MoS_2 while (c) and (d) are obtained from bilayer MoS_2, and the laser's excitation is near resonance. (e)–(h) indicate the situation where the laser's excitation is off resonance. From this result, it is observed that laser resonance excitation and monolayer TMD in unison would produce the highest degree of polarisation of near unity. Reproduced with permission from Ref. [35].

Furthermore, the source of excitation could also affect the extent of depolarisation with valleytronic materials. At resonant excitation, i.e., when the excitation wavelength matches the PL emission wavelength, the degree of circular polarisation is the highest. As shown in Fig. 3.8, the DOCP drops close to zero when the excitation is at 2.09 eV or 2.33 eV, away from the near-resonant frequency of 1.96 eV. This could be attributed to the better spatial overlap of electron and hole wavefunctions with respect to the exciton dipole transition.

Apart from reasons arising from monolayer/bilayer and on/off laser resonant excitation, there are other reasons that can cause depolarisation. One key reason is that electrons and holes exhibit Coulomb interactions even when they are in different valleys [41]. During a valley-selected laser excitation with a specific circular polarisation, this intervalley exchange interaction can create carriers in the opposite valley, which then recombine and emit photons that have a counter-circular polarisation with respect to the source of excitation. In addition, it is reasonable to suggest that there is little energy left from the excitation and PL emission to create more phonons that may perturb the distinguishability between K and K′ valleys.

Indeed, factors such as electron-phonon interactions [40], [46], [47] also could induce a greater extent of intervalley scattering, reducing the distinguishability between the K and K′ valleys under optical excitation. The interactions with phonons could be especially important in ascertaining the performance of a valleytronic material. This can be manifested in two major ways:

(1) Defects and quality of the valleytronic material: When a 2D material contains numerous defects that can act as charge-trapping sites, more phonon interaction pathways could be opened up, which then reduces the DOCP due to increased intervalley scattering caused by electron-phonon coupling. Moreover, charges trapped in these sites could also exhibit Coulomb interactions with the exciton in their vicinity, thereby potentially increasing carriers in the valley that is not being addressed optically and reducing DOCP [48].

    These carrier traps and defects are often associated with a Boltzmann-type activation energy, and the effects become more prominent with increasing temperature when more of these sites are thermally activated. This leads us to the next point of discussion.

(2) The temperature performance of the valleytronic material: As we increase the temperature, the degree of phonon interaction within the material becomes enhanced. The enhancement of phonon activities directly impacts several properties of the PL of a quantum system, including exciton central transition energy, transition linewidth uncertainty, and quantum efficiency [49]–[51]. These are observations that we can make in most quantum systems including TMDs. An example series of temperature-dependent spectra for a typical quantum system is shown in Fig. 3.9, where emission

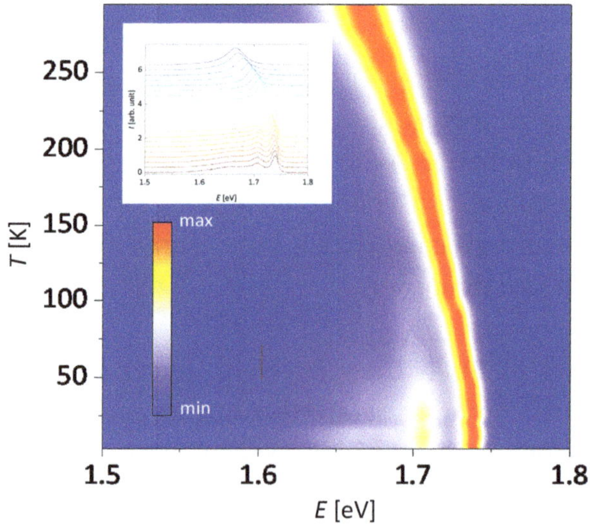

Fig. 3.9. Temperature-dependent PL spectra from 5–290 K for a monolayer WSe_2. Three key observations can be made in the redshift of the exciton transition energy, broadening of exciton transition linewidth, and decrease of the quantum efficiency. These behaviours are the case of TMDs, where these elevated phonon interactions would result in a lower DOCP likely due to the greater extent of phonon-mediated intervalley scattering. Inset shows the individual spectra at different temperatures.

energy redshift, linewidth broadening, and intensity quenching can be seen.

For the exciton transition energy or central emission wavelength, we can start with Bose-Einstein statistics to model phonons, which are bosons:

$$\bar{n}(T) = 1/(\exp(\langle\hbar\omega\rangle/k_B T) - 1) \tag{3.8}$$

where n is the phonon number and $\hbar\omega$ is the average phonon energy. Multiplying this by the phonon energy and manipulating the equation, we can obtain [52]

$$\varepsilon_g(T) = \varepsilon_g(0) - S\langle\hbar\omega\rangle(\coth(\langle\hbar\omega\rangle/2k_B T) - 1) \tag{3.9}$$

This empirical model is an alternative to the more commonly known Varshni equation [53]. Here ε_g is the bandgap energy and S is a constant that we call the Huang Rhys factor, a constant that measures the strength of electro-phonon coupling of the material. This model has been empirically shown to agree with the behaviour of quantum systems, can be used to model the temperature dependence of TMD bandgaps, and also correlates observed PL redshifts with the increase of phonon interactions.

Having shown that phonon interactions do indeed increase at elevated temperatures, we see that electron-phonon interactions can be a possible reason for the drop in DOCP.

3.3.3 Exciton and Spin Lifetime of Valleytronic Materials

Beyond time-integrated microanalysis PL, time-resolved PL is also very useful to understand spin-valley lifetime, depolarisation, and valley coherence [54]–[59]. In time-resolved photoluminescence or TRPL, an ultrafast short pulse (ps or fs) is excited in the 2D material and the photon emission from the sample is recorded at the detector as a function of time. The laser pulse is further integrated by a time-correlated single photon counter and an avalanche photodetector, as shown in Fig. 3.4. Several reports indicate that the bright excitonic lifetime (τ) in 2D materials is of the order of sub-ps [55]–[59]. However, this lifetime can be varied depending on the quality of the sample, defect bound states and external modulation [60], [61]. The charged exciton or trion and other excitonic complex states also have similar recombination lifetimes (~ten ps) to excitons [56], [62]–[65]. Polarisation-resolved TRPL gives further clues to the generation, recombination, and decay dynamics of excitons.

However, spin lifetime may differ from exciton recombination time. While the exciton lifetime refers to the recombination of the electron-hole pair, the spin lifetime is generally considered for the free charge carriers, i.e., electron or hole. Therefore, charged excitons are usually the more suitable candidate to study spin lifetime. Beyond TRPL, time-resolved Kerr rotation can extend our study to include spin dynamics, as explained in the next section. The degree of valley polarisation at steady state depends mainly on the lifetime of the excitons and intervalley scattering time according to the following relation [54], [66]:

$$P_c = P_0 \Big/ \left(1 + \frac{\tau}{\tau_s}\right) \tag{3.10}$$

where P_c is the degree of valley polarisation, P_0 is the initial polarisation and τ_s is the intervalley scattering time. High degree of valley polarisation at steady state condition is possible when the intervalley scattering time is longer compared to the exciton lifetime or exciton recombination time. Fig. 3.10(a) shows a typical decay dynamic of exciton in monolayer $MoSe_2$ capped with BN layers probed via TRPL [56]. A survey of exciton recombination and valley polarisation lifetimes for different 2D materials reported in literature [54], [56] is also provided for reference in Fig. 3.10(b). Compared to the relatively

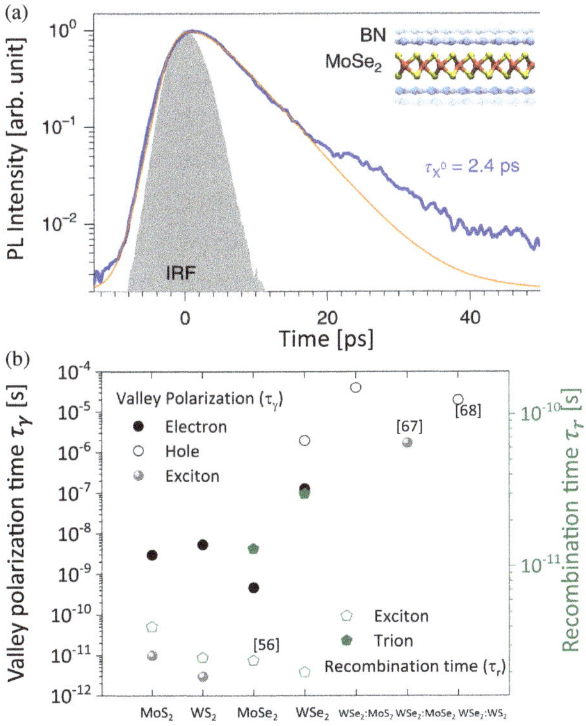

Fig. 3.10. (a) Time-resolved PL of the neutral exciton feature (X^0) in hBN-capped monolayer $MoSe_2$ with lifetime of ~ 2.4 ps. IRF is the instrument response function. (b) Exciton recombination lifetime and valley polarisation lifetimes reported for different monolayer TMDs at low temperature. Unlabelled data points are based on Ref. [54] and references therein. Labelled data points are based on Refs. [56], [67], [68]. Reproduced with permission from Refs. [54], [56].

shorter exciton recombination times of the order of pico- to nanoseconds, valley lifetimes in TMDs such as WSe_2 and WSe_2:MoS_2 and other heterostructures can reach tens of nanoseconds to hundreds of microseconds [67], [68].

3.4 Kerr Rotation Spectroscopy in 2D Materials

The Kerr effect is named after the Scottish physicist John Kerr, who discovered the effect in 1875 [69], [70]. In the Kerr effect, the optical field creates the dipole in the system that interacts with the laser pulses, and preferentially aligns them to the polarisation axis, which is further read out by another probe pulse. This is known as the optical Kerr effect and was theoretically shown by Mayer and Gires in 1964 [71] and experimentally demonstrated in 1969 by Duguay and Hansen [72].

Kerr rotation spectroscopy (KRS) is a powerful and effective tool to probe the population of spin-polarised carriers in a material system. KRS is based on the magneto-optical Kerr effect where linearly polarised light experiences a small rotation (θ_k) in the polarisation plane when interacting with either a magnetic material or a material under an external magnetic field. The monolayer TMD has a time-reversal symmetry but a broken inversion symmetry, which results in a non-zero Berry curvature. As discussed in Chapter 2, the sign-inequivalent Berry curvature acts as an effective magnetic field in the monolayer TMDs that further leads to experimental observation of the KRS even without the requirement of any external magnetic field.

A typical KRS setup consists of pump and probe laser pulses where the pump signal is circularly polarised as shown in Fig. 3.11(a). Depending

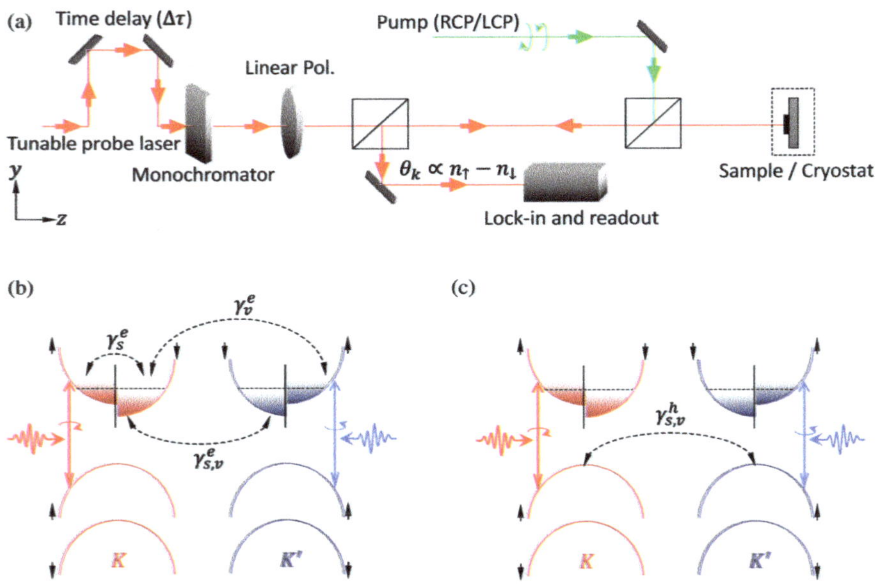

Fig. 3.11. (a) Time-resolved Kerr rotation (TRKR) setup where locked spin-valley polarised carriers are photoexcited with the right- or left-circularly polarised (RCP/LCP) pump pulse and the linearly polarised probe pulse experience the Kerr rotation (θ_k) proportional to the net polarised population imbalance. For TRKR the θ_k is recorded as a function of time delay ($\Delta\tau$, created through a physical path delay) between pump and the probe pulse. (b–c) Schematic diagram of the possible relaxation pathways for carriers (electrons in b, holes in c) in W-based TMD monolayer where γ_s^e is the rate of spin relaxation within a valley, γ_v^e is the spin-conserving and $\gamma_{s,v}^{e,h}$ is the spin-flip intervalley scattering for electrons and holes. Adapted with permission from Ref. [76].

upon the polarisation of the pump pulse, it can selectively populate the either K or K′ valley in monolayer TMD. The population imbalance between two valleys is then detected by a wavelength-tuned linearly polarised weak probe laser via KRS. Upon photoexcitation, the system generates a net spin and valley polarisation that is proportional to the spin-polarised carriers present in the system and is further proportional to the Kerr rotation (θ_k) experienced by the probe laser ($\theta_k \propto n_\uparrow - n_\downarrow$, where $n_\uparrow - n_\downarrow$ is the spin-valley polarisation) [73]–[75]. The pump pulses are modulated between right- and left-circular polarisation by a photo elastic modulator, and the pump-induced θ_k is detected by balanced photodiodes and measured via standard lock-in techniques. The probe laser is typically tuned in photon energy to the peak of the induced Kerr response, which coincides with the spectral position of the exciton or charged exciton resonances [76]–[78], as discussed in more detail below. Any long-lived polarisation of the carriers can be directly monitored using Kerr spectroscopy. This differentiates KRS from polarised PL studies, where both the electron and hole must participate (recombine) to emit a photon. Kerr effects depend only on the difference between a material's circularly polarised absorption and net imbalance of the spin-polarised carriers.

KRS further allows us to understand the transient dynamics of the spin-polarised carriers or the "spin lifetime" where the Kerr decay dynamics is probed as a function of pump-probe delay. This is known as time-resolved Kerr rotation (TRKR). A schematic diagram of a TRKR setup is shown in Fig. 3.11(a). In this setup, it is possible to selectively inject spins into a valley within a certain timescale by shining an ultrashort circularly polarised pump laser pulse. Subsequently, the optical readout of the rotation of the reflected pulse is possible by linearly polarised probe pulse as a function of time delay ($\Delta\tau$) between pump and probe. Now, by varying the delay time between those two pulses as shown in Fig. 3.11(a), the initial spin polarisation and decay of the spin ensemble can be observed.

One of the earliest reports on KRS in CVD-grown monolayer MoS_2 depicts that although the locked spin-valley polarisation can be probed via circularly polarised photoluminescence, it is limited by the ultrashort timescale of the electron-hole recombination or the exciton lifetime [73]. One of the easier approaches to study the coupled spin-valley dynamics via KRS is using electron-doped MoS_2 and WS_2 monolayers. The TRKR of the same system reveals very long electron spin lifetimes, exceeding 3 ns at 5 K (two to three orders of magnitude longer than typical exciton recombination times). The valley polarisation of the 2D material is sensitive to the wavelength of the optical excitation. Generally, the material is excited near the exciton resonance and probed over the entire desired spectral range. The long-lived resident electron (or hole) come from the charged excitons when it is probed near the trion region while it is still being excited (via pump) near exciton resonance. This can further be extended to probe the intra- and intervalley trion (singlet and triplet) states [74], and the long-lived dark states in monolayer WSe_2/WS_2 at cryogenic temperature.

The exciton in monolayer TMD can easily form a charged exciton or trion in a doped system either via external gating or photoexcitation. Depending on doping (*n* or *p* type), it creates a negatively (or positively) charged exciton by capturing an electron (or hole). The trion emits a photon via recombination of electron-hole pair and the resident electron (or hole) goes back to the conduction band (or valence band) edge. This recombination time of trion is in picoseconds, the same order as exciton, but the resident electron (or hole) is long lived compared to the recombination lifetime and can live up to nanoseconds or even microseconds [76]. The transfer of valley pseudospin from photo-generated states to resident carriers in a specific valley is responsible for the long-lived valley polarisation. Similar observations are also well reported in III-V semiconductors [79]–[81]. For 2D materials, this valley pseudospin of free carriers remains robust even at room temperature. This directly correlates with the spin coherence and finds a plethora of applications in "valleytronics" devices.

To probe the dynamics of a specific energy state, it's important to know the spectral position of the corresponding state. For example, the trion lies around 30 meV below the exciton in the energy spectrum of TMDs in Fig. 3.12(a). Hence, the pump and probe wavelengths can be selected accordingly to get the maximised θ_k. When the pump

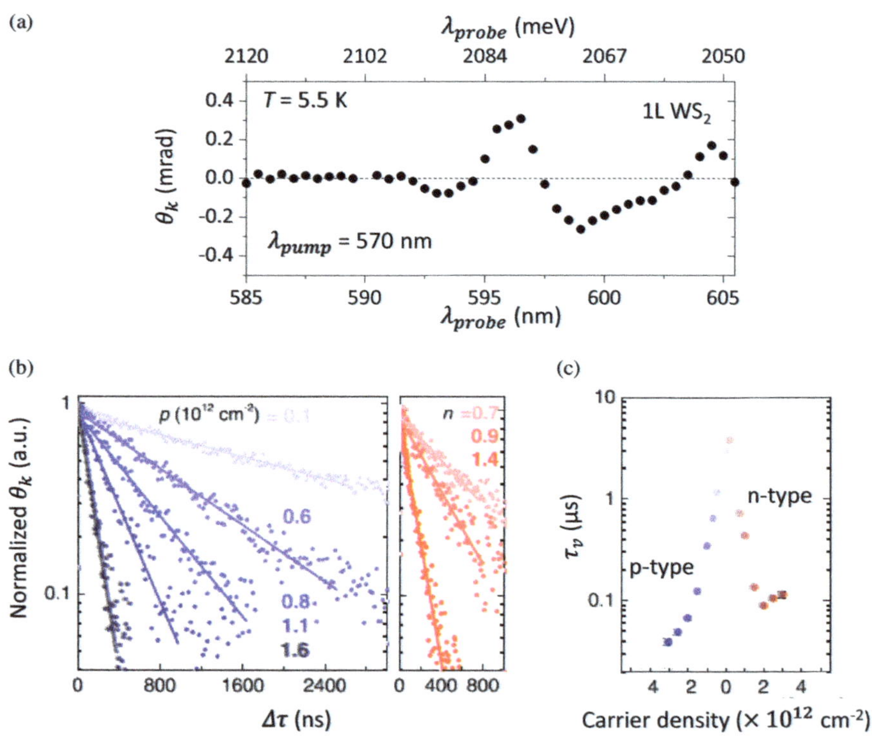

Fig. 3.12. (a) KRS for monolayer WS_2 capped with hBN layers excited with 570 nm pump pulse and probed within exciton and trion resonance region at T = 5.5 K. (b) TRKR signals from a fully encapsulated WSe_2 sample, at different electron (n) and hole (p) carrier densities. The signals reveal the relaxation of optically induced valley polarisation of resident carriers, which in monolayer TMDs is intimately linked to carrier spin polarisation due to spin-valley locking. (c) The corresponding valley relaxation time τ_v versus carrier density; τ_v is sharply peaked at low carrier density. Adapted with permission from Ref. [77].

excites the exciton, it can easily form a trion by down conversion via capturing a free carrier present in the Fermi Sea. However, the carriers might get trapped in the defect states present within the material, causing a slower decaying time dynamic. The population density of the free carriers can be efficiently tuned by an external gate voltage.

Since the θ_k measured at normal incidence is also sensitive to the net magnetisation projection along the surface normal, this technique is ideal for detecting the net spin and valley pseudospin even after carrier recombination and the temporal dynamics of KRS directly reflects the valley depolarisation. Hsu *et al.* [82] has experimentally shown that by using TRKR, a long-lived valley polarisation for positively charged trion with a lifetime approaching 1 ns at low temperatures can be observed, which is much longer than the trion recombination lifetime (~10–20 ps) in CVD-grown WSe_2 when probed near trion resonance while keeping the pump 12 meV above the probe energy. In the same experimental scheme, when the exciton is probed, the typical lifetime is in the order of tens of ps. The short valley lifetime for exciton is an indication of the strong electron-hole Coulomb exchange interaction. This further suggests that the exciton would depolarise within a timescale faster than (or comparable to) the recombination lifetime, and hence the formation of long-lived valley-polarised holes is suppressed. These numbers are further supported by other groups in similar KRS and TRKR experiments [83]–[86].

As mentioned earlier, the gate dependence of the charged exciton states allows us to modulate the population density of the carriers in the system. Hence, the spin-polarised electrons or holes can be probed depending on the gate bias. Dey *et al.* [76] showed that for a hBN-capped WSe_2, the resident electrons can show long valley polarisation ~130 ns in the *n*-type regime, while an extraordinary long-lived valley polarisation can be observed for holes (~2 μs) in the *p*-type regime (shown in Fig. 3.12(b)–(c)). The difference of lifetime for the resident electron and holes has further been argued with the difference of spin-orbit splitting conduction (Δ_c) and valence bands (Δ_v). Because of the large spin-orbit splitting for holes, the resident holes can only scatter through valleys by changing both the spin and valley whereas for electrons spin-conserving valley scattering is also possible (electron and hole relaxation pathways are shown in Fig. 3.11(b)–(c)). Li *et al.* [77] further measured the θ_k and TRKR, the low-temperature valley dynamics of resident electrons and holes in exfoliated WSe_2 monolayers as a systematic function of carrier density. While the valley relaxation time for holes (τ_v^h) is quite long, it decreases rapidly as carrier density increases. Furthermore, the decay dynamic is sensitive to the quality of the material. While the monolayer kept directly on the SiO_2 substrates enhances the probability of capturing free carriers, the carriers also get trapped to the defect states generating multi-exponential complex decay dynamics. On the other hand, the hBN-capped monolayer sample supresses the scattering channels for the resident carriers to decay and yields mostly mono-exponential decay as a function of pump-probe delay. Overall, it is well established that, by changing the Fermi level via external gating, TMDs can be tuned to allow long-lived localised spin states or allow valley lifetimes of conduction and valence band states exceeding nanoseconds across a wide range of temperatures, which are adequate timescales for both spin manipulation and valley transport [87].

Although the Berry curvature present in the system acts as an effective out-of-plane magnetic field (B_z) in the monolayer TMDs (discussed in detail in the previous chapter), an external magnetic field can also be applied on the samples. KRS produces different results depending on whether the magnetic field is in-plane or out-of-plane. The spin polarisation decays sharply even with a very small transverse magnetic field (B_y). This is known as the Hanle-Kerr effect [88]. This experiment also reveals rapid reduction of optically induced spin polarisation by very small B_y, suggesting long spin relaxation times of resident electrons. This is complementary to the traditional Hanle effect in conventional semiconductors [80], [81]. The B_y dependence of the θ_k generally follows a Lorentzian shape and the corresponding spin lifetime (τ_s) can be extracted from the curve if the corresponding *g*-factor (discussed earlier in Chapter 2 and in the next section) is known from the relation as:

$$\tau_s^{-1} = g\mu_B B_{y0} / \hbar \quad (3.11)$$

where μ_B is Bohr magneton, B_{y0} is the half width at half maximum of the Hanle peak and $\hbar$ is the reduced Plank's constant [73] as shown in Fig. 3.13(a). On the other hand, resident electrons and holes experience

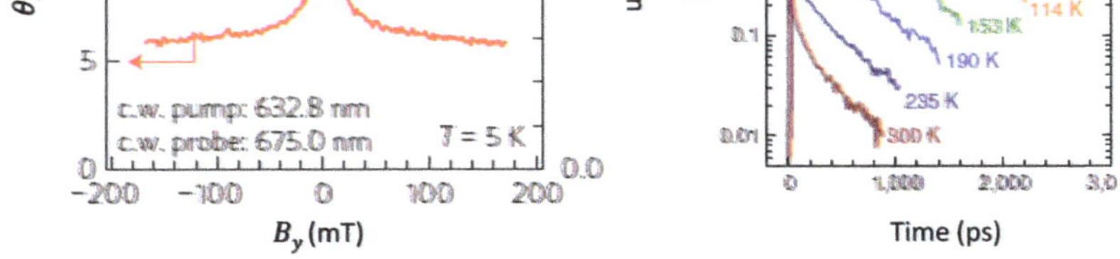

Fig. 3.13. (a) The induced KR ($\boldsymbol{\theta_k}$) versus applied magnetic field $\boldsymbol{B_y}$ from monolayer MoS_2 at 5 K. This Hanle-Kerr experiment reveals rapid reduction of optically induced spin polarisation by very small $\boldsymbol{B_y}$ (the half width at half maximum, ~ 7mT, is determined as half the width indicated by the two inward arrows at half the peak maximum), suggesting long spin relaxation times of resident electrons. The PL polarisation (dashed line) shows no appreciable change over this field range. (b) TRKR trace of monolayer WSe_2 measured at different temperatures. The depolarisation increases with increasing temperature. Adapted with permission from Refs. [73], [82].

a large effective out-of-plane magnetic field (B_{soc}) because of the strong spin-orbit coupling. Depending upon conduction and valence band splitting (Δ_c and Δ_v), the effective B_{soc} can range from tens of Tesla (for electrons) to >10^3 Tesla (for holes). Hence the electron and holes are affected differently by external magnetic field. Thus, the spin-valley locking of the holes in valence band supports the long-lived hole polarisation.

The fast intervalley scattering of the resident electrons leads to rapid fluctuation of the effective magnetic field. However, it does not cause any spin dephasing. But when an external transverse magnetic field is applied, even with a small magnitude, the total magnetic field deviates from the out-of-plane direction and rapidly fluctuates, leading to the fast spin dephasing of resident electrons [73], [76], [89]. Therefore, the spin polarisation of resident electrons in the conduction band decays too fast to show periodic spin precession signals with the remaining precession signals solely originating from localised electrons. This transverse magnetic field dependence can further be expressed in terms of the Larmor precession frequencies described by the equation as

$$\theta(t) = Ae^{-\gamma t}\cos(2\pi ft) \tag{3.12}$$

where f is the precession frequency and γ is the corresponding dephasing rate. The frequency also holds a linear relationship with the applied transverse magnetic field as

$$2\pi f = g\mu_B B_y / \hbar \tag{3.13}$$

A collection of independent spins with distributed g-factor directly correlates with the ensemble-dephasing rate.

A temperature-dependent study reveals that the spin depolarisation of the carriers increases with increasing temperature as shown in Fig. 3.13(b). This is consistent for both types of carriers. This has been attributed to the phonon-related scattering processes (e.g., Elliott-Yafet mechanisms) [90], [91] which cause the rapid relaxation of valley polarisation. The scattering processes can also be influenced by external pathways such as mid-gap states or dielectric environments. Depending on the scattering rate of electron and hole, the decay dynamic might show a sign reversal of θ_k, especially under influence of phonon interaction at elevated temperature [73], [82].

3.5 Valley Controllability

Now we will move forward the discussion towards control of spin-valley properties via external manipulation. Manipulations of the valley properties include the control of spin-polarised carrier population in different valleys with the help of, e.g., external electrical bias, magnetic field, or temperature. Successful control is measured by one of the accessibility methods mentioned in the previous section.

3.6 Electrical Control of Valley Carriers

The large spin-orbit coupling in the valence band of W-based TMD monolayer enables access to the long-lived hole polarisation spatially distributed over the monolayer induced by valley carrier imbalance through the valley Hall effect (VHE) along with the help of KRS. The Berry curvature that acts as an effective magnetic field ($B_{soc}\hat{z}^{\pm}$) and spin of carriers in the valence and conduction bands around the K and K′ points are equal in magnitude but opposite in sign. Hence if the coupled spin-valley carrier populations are different, the dielectric function of the two circularly polarised light will also differ. And the linearly polarised light also will acquire θ_k accordingly in passing through or being reflected from the sample.

Barrè *et al.* [92] used the VHE in WSe_2 to electrically control the spatial separation of valley-coupled spins both in *n*- and *p*-type regime and image the spatial distribution with the help of KRS. The Kerr rotation measurements probe the spatial distribution of the valley carrier imbalance induced by the VHE as shown in Fig. 3.14(a). The accumulation and distribution of the spin-polarised (spin-valley polarised for monolayer WSe_2) carriers can be further analysed by the spin diffusion models developed for the III-V semiconductors domain [93]. Since the valence band spin splitting exceeds the thermal energies even at room temperature, we might expect to see room temperature valleytronic devices with similar measurement schemes.

While the inversion symmetry is broken in monolayers, it is restored in the case of bilayers. The most stable configuration of bilayer TMD consists of two individual monolayers rotated by 180° to each other where the Berry curvature vanishes in 2H bilayer TMDs. The inversion symmetry, however, can be broken by applying an electric field perpendicular to the plane of the material. Thus, the VHE can be controlled by the application of an external gate voltage which creates an interlayer potential difference between two individual monolayers. In monolayers, the valley magnetisation can be imaged by KRS (as explained under valley addressability in Section 3.4) by breaking the rotational symmetry under application of, e.g., a uniaxial stress [94]. Lee *et al.* [95] showed the valley polarisation in bilayer MoS_2 induced by longitudinal electrical current subsequently imaged by KRS, shown in Fig. 3.14(b)–(e) for different biasing conditions.

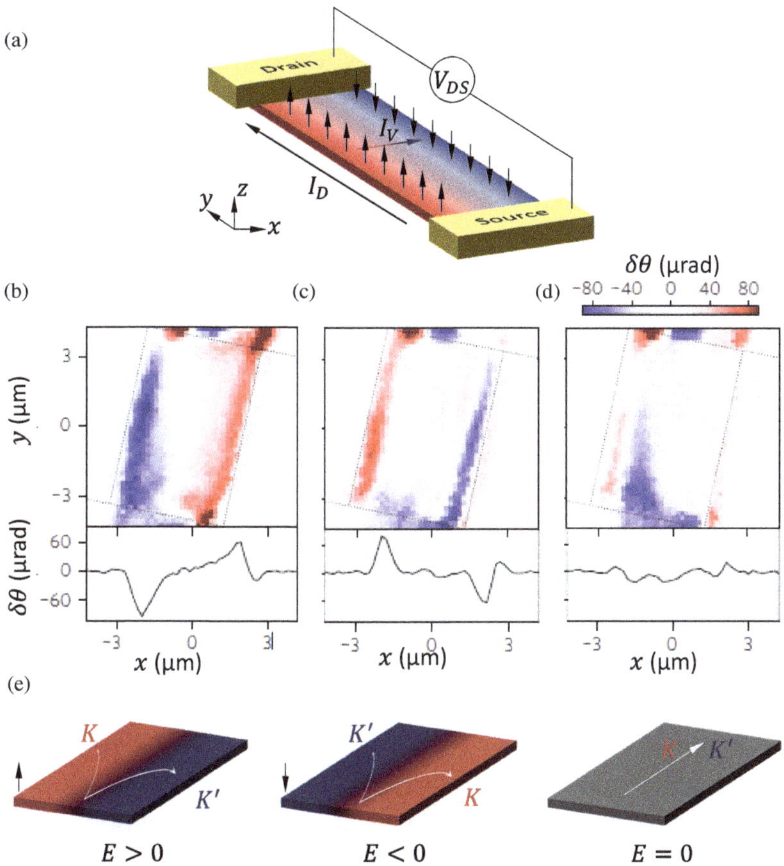

Fig. 3.14. (a) Schematics of the VHE under transverse electric field-induced spin accumulation (shown with red and blue colour shades) at the edges of the monolayer sample. This can further be probed via KRS. (b–d) Spatial map of the induced KR (θ_k) distribution for a bilayer sample under different biasing conditions (Vg = 20 V, –5 V and 4 V in (b), (c) and (d), respectively, for a bilayer MoS_2). Representative horizontal line cuts are shown in the bottom panel of the 2D colormap. (e) Illustration of the field-induced VHE where the arrow represents the direction of the vertical field. Adapted with permission from Refs. [92], [95].

Similar imaging was previously done for GaAs devices to study the spin Hall effect (SHE). However, the TMD offers to image the valley (or orbital) magnetisation through VHE instead of detecting the spin magnetisation in SHE. This out-of-plane magnetisation strongly depends on the applied gate voltage where the θ_k is maximised at the edges with opposite sign. Also, the θ_k sign flips with the reversal of the longitudinal current direction. These observations are consistent with the symmetry-dependent Berry curvature and valley Hall conductivity in bilayer MoS_2.

3.7 Magnetic Field Control — the Valley Zeeman Effect

The spin-valley coupling in TMD materials can further be controlled and manipulated by an external magnetic field. Magneto spectroscopy allows us to investigate the valley pseudospin with the help of the valley Zeeman effect. Magneto-optic spectroscopy also enables the experimental observation and analysis of the *g*-factor of the individual electron and holes as well as for the excitons. It also helps us to calculate the exciton-reduced masses, Bohr radii, and gives the opportunity to observe the optically bright and dark excitons, trions, their phonon replicas in TMD monolayers and other many-body excitonic complexes such as biexcitons or charged biexcitons [96]–[99].

Since the emergence of valleytronics, magneto-optic spectroscopy proved to be integral to the experimental realisation of the coupled spin-valley physics in 2D materials. As mentioned earlier, because of the broken inversion symmetry the K (K′) valley can couple to the right (left) hand circularly polarised light (σ^+/σ^-). The valley degeneracy can be broken by applying an external magnetic field perpendicular to the 2D plane, where the K valley energy is shifted up and K′ valley energy is shifted down or vice versa. This has been discussed in more detail in Chapter 2 and is often called the valley Zeeman splitting as depicted in Fig. 3.15(a). The conduction band and valence band individually can experience the Zeeman shift under magnetic field. However, for TMD materials the effective g-factor (g_{eff}) is defined in terms of the valley Zeeman splitting of the excitonic (e.g., exciton, trion or other excitonic complexes) resonances as

$$\Delta\varepsilon = \varepsilon(\sigma^{+}) - \varepsilon(\sigma^{-}) = g_{eff}\,\mu_B B \tag{3.14}$$

Here $\varepsilon(\sigma^{+}/\sigma^{-})$ denotes the helicity-resolved excitonic resonances, and g_{eff} is the effective g-factor for the corresponding transition energy.

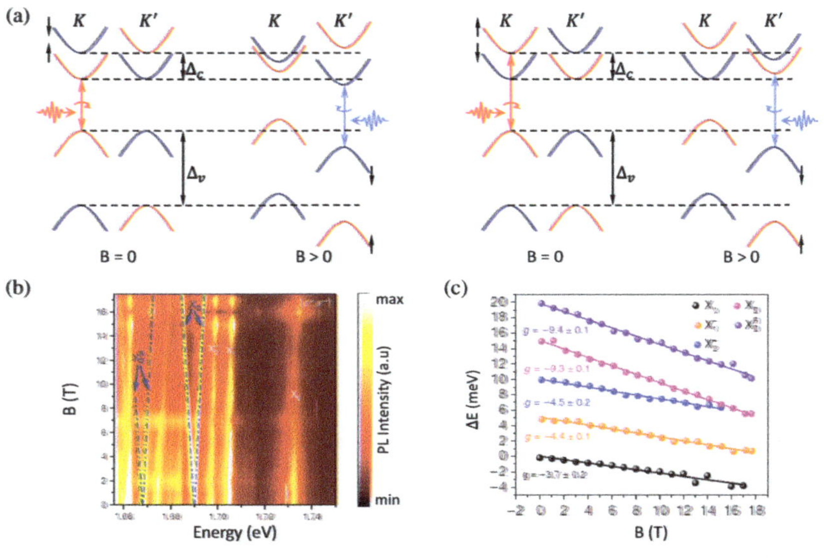

Fig. 3.15. (a) Schematic band diagram for Mo (W)-based TMD monolayer band structure with relative shift under magnetic field in left (right) panel. Depending upon the spin-allowed (-forbidden) optical transition the lowest energy state is bright (dark) for Mo (W)-based TMD monolayer. (b) 2D colour plot for spin-valley polarisation-resolved magneto PL for monolayer WSe_2 under the application of out-of-plane magnetic field. (c) Calculated g-factor for the different excitonic complexes obtained from the fitting of the PL spectra under Zeeman splitting. Adapted with permission from Refs. [97], [108].

Depending on the optically allowed or forbidden transition at the lowest energy state, the TMD monolayers can further be classified as bright or dark. Generally, Mo (W)-based TMD monolayers are optically bright (dark) where the lowest possible transition energy is spin allowed (forbidden). Now, although this spin-forbidden dark states cannot be accessed directly, there is a rich interest in the community especially due to the expectation of long-lived states present as dark states. This optically forbidden dark states can be explored via magnetic brightening (shown in Fig. 3.16) with the application of a magnetic field and probed via helicity-resolved photoluminescence, micro-reflectance or transmittance, etc. [100]–[103]. Along with the

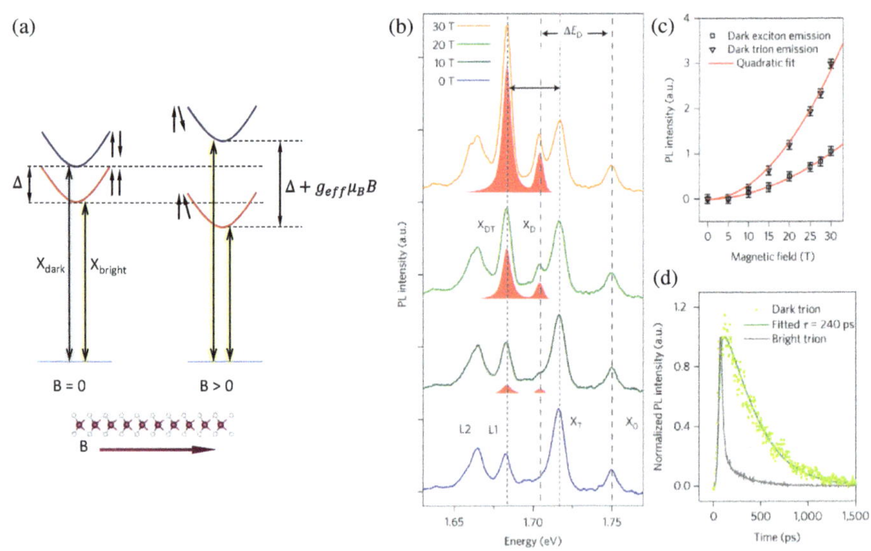

Fig. 3.16. (a) Schematic diagram for magnetic brightening of the dark excitons with an in-plane magnetic field. (b) PL spectra of monolayer WSe_2 under in-plane magnetic field at T = 30 K. The dark exciton and dark trion emerge as the magnetic field increases. (c) Magnetic field-dependent intensity evolution for the dark excitonic complexes at T = 30 K. (d) Time traces of the dark and bright trion PL intensity at T = 4 K under B = 17.5 T. Adapted with permission from Refs. [96], [108].

effective *g*-factor, an attempt was made to calculate the spin, orbital, and valley contribution to the exciton *g*-factor [104].

The dark excitons in monolayer TMD system undergo a fine structure doublet (< 1 meV) due to exchange interaction [105]–[107]. Although the dark excitons cannot directly couple to the out-of-plane polarisation of light, the grey excitons (with partial out-of-plane distribution) can. Alternatively, the dark states can be brightened with applied magnetic field. From magnetic field-driven optical spectroscopy, we can field the *g*-factor of –4 and –9 for bright and dark excitonic species, respectively (similarly, a *g*-factor of –12 has also been experimentally observed for the spin- and momentum-forbidden phonon replicas [108]) as shown in Fig. 3.15(b)–(c). Like the excitons, the charged excitons or trions also undergo similar Zeeman splitting under a magnetic field. Hence one can expect a similar feature in the experimental observation of the *g*-factor (mostly observed in W-based TMDs) of –4 and –9 for the bright and dark trions, respectively (similar observations were found for the charged biexcitons) [99].

For monolayer TMDs the electrons and holes are confined within a single layer forming only an intralayer exciton, while for bilayer the electron and hole wavefunction can distribute within two opposite layers and form an interlayer exciton. Depending on the two layers coming from same or different material, these interlayer excitons can be observed both in homo-bilayer and hetero-bilayer systems. While the *g*-factor is negative for intralayer excitons, it has a positive sign for the interlayer [103], [109]. The *g*-factor is weakly varied as a function of layer thickness in TMD systems. This is primarily because the TMD multilayer also possesses 2D-like behaviour even in the bulk limit where individual monolayers can act on their own [110].

When the valley degeneracy is lifted under a magnetic field, there is an expectation of population imbalance to the spin-polarised carriers in the system. Thus, the valley polarisation (depolarisation) can be controlled externally with magnetic fields and probed via magneto-optic spectroscopy. Thus, we can make inferences on the valley coherence effect, extract the coherence lifetime, and control the same. As mentioned in the previous section, things get more interesting if we probe the valley coherence (or spin lifetime) of the resident carriers with the help of KRS and TRKR.

3.8 Optical Field Control — the AC Stark Effect

We have seen in the earlier section about the potential of the external magnetic field to break the valley degeneracy and thereby allowing potential application of the coupled spin-valley physics in TMD systems. Similarly, the optical field can also be used to lift the valley degeneracy, generate the spin-valley selective carrier population and control the same. By applying an intense laser pulse (generally circularly polarised), we can break the time-reversal symmetry by applying circularly polarised light of the degenerate valleys in monolayer TMDs and selectively control the valley pseudospin further.

The energy shift of the two-level system under application of optical pulse (non-resonant excitation) is known as the optical Stark effect (OSE). For TMD systems, the circularly polarised pump pulse can selectively control the energy bands of the valleys (K or K′) that are subsequently explored by another probe pulse. This is termed as valley-selective OSE.

The coherent interaction between exciton energy and non-resonant polarised optical pulse can shift the exciton energy states. By ignoring the higher-lying energy states, this can be described by a simple two-level system. In this picture, the coherent interaction of the ground ($|g\rangle$) and excited states ($|e\rangle$) with the light field gives rise to so-called Floquet states ($|g\rangle \pm h$, $|e\rangle \pm h$) [111], [112] as shown in Fig. 3.17(a). These states hybridise further and form separate energy states due to the Coulomb repulsion.

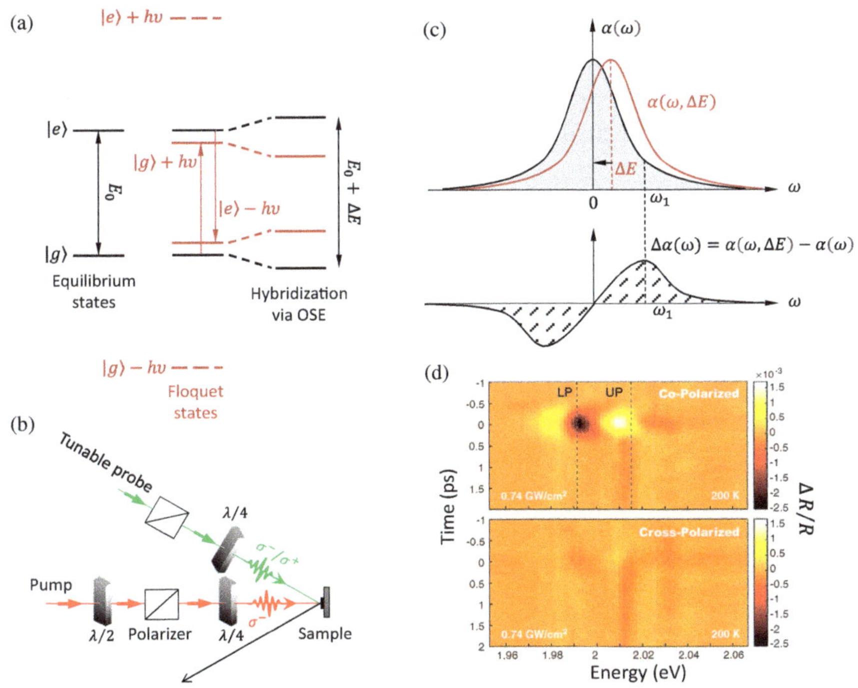

Fig. 3.17. (a) Energy level diagram for a two-level system at equilibrium (left panel) and hybridisation of the Floquet states via OSE (right panel). (b) Schematic representation of the experimental setup. (c) Top panel: OSE-induced absorption spectra before (black) and after (red) absorption. Bottom panel: The change in absorption spectra with energy shift of $\mathbf{\Delta E}$. (d) 2D colour plot of the transient differential reflectance for co- (and cross-)polarised pump probe excitation in the top (bottom) panel. The sharp signal at $\mathbf{t = 0}$ ps is attributed to the OSE of the upper and lower polaritons. Adapted with permission from Refs. [111], [115].

Because of the valley selectivity of the monolayer TMD, the K valley can be populated with circularly polarised light with a particular helicity. When the monolayer TMDs are excited with a non-resonant σ^+ pulse, the transition energy experiences a blue shift that can further be probed by another σ^+ probe pulse (schematically presented in Fig. 3.17(b). However, because of the valley-selective optical transition, the other valley does not exhibit any OSE. The observed blue shift arises from repulsion between coherently created virtual excitons, which dominates when the applied photon field is detuned from resonance by less than the exciton binding energy. Hence the change in absorbance ($\Delta\alpha$) can only be observed near "zero time" delay $\Delta\tau \rightarrow 0$ between pump and probe pulse as shown in Fig. 3.17(c).

In 2014, Sie *et al.* first experimentally showed that by applying an intense circularly polarised light the valley degeneracy can be lifted in monolayer WS_2 and that the exciton level can be selectively tuned as much as 18 meV through OSE [111]. At around the same time, Kim *et al.* also observed the valley-selective OSE in monolayer WSe_2 [113]. They have further estimated the corresponding pseudo-magnetic field (B_{eff}) for the valley excitons with optical Stark shift of $\Delta\varepsilon$ as

$$B_{eff} = \Delta\varepsilon / 2g_{eff}\mu_B \tag{3.15}$$

Cunningham *et al.* further observed the OSE in WS_2 for excitation above the bandgap and they have explained it through exciton OSE where they considered the exciton-exciton interactions along with the exciton-photon interaction [114]. By placing the monolayer WS_2 in a distributed Bragg reflector substrate, LaMountain *et al.* showed the valley-selective control of polariton energies using the OSE [115] as shown in Fig. 3.17(d), thus extending coherent valley manipulation to the hybrid light-matter regime [116] which is also sensitive to KRS. For biexcitons in monolayer $MoSe_2$, OSE control has also been experimentally demonstrated with both valley-selective red and blue detuning of the optical pump with respect to the exciton energy [117].

3.9 Temperature-Induced Tuning of Valley Polarisation

Finally, we shall discuss another possible control knob for the valley population reported in literature, i.e., temperature gradient. It has been predicted that spin and valley current can be generated perpendicular to the applied thermal gradient in the two-dimensional plane of the monolayer TMDs. When a temperature gradient is applied, an electric field develops in the opposite direction due to the Seebeck effect. Besides the parallel effect, in the presence of the temperature gradient, the holes (or electrons) experience a Lorentz-like force and thus move in the direction perpendicular to the diffusion current, which is the anomalous Nernst effect induced by the intrinsic non-vanishing Berry curvature [118]. The measured current originates thus from a single valley with a single spin component. Therefore, by lowering the Fermi level, a single spin current (down spin) carrying single valley information can be generated by a temperature gradient (ΔT), which hints at possible applications in spin caloritronics and valleytronics. This has been further shown experimentally that a spatial temperature gradient created through the laser illumination allows the exciton to flow across the gradient (presented in Fig. 3.18). By helicity-selective laser probing, this exciton diffusion can be mapped. The advantage of this method is the observation of exciton Hall effect instead of individual transport of electrons or holes induced via VHE. The Hall angles (described in Chapter 2) in this study are orders of magnitude larger than the conventional VHE reported in TMD monolayers [119].

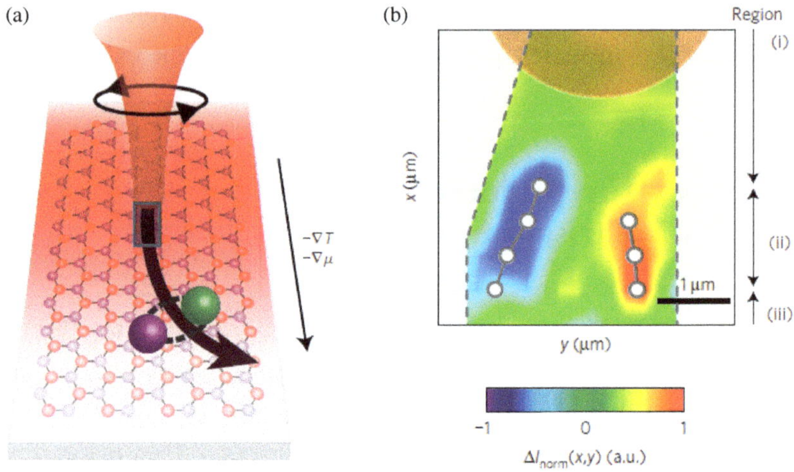

Fig. 3.18. (a) Schematic representation of the exciton Hall effect (EHE) under the illumination of the laser that creates a temperature gradient across the material. This causes the transverse diffusion of the excitons. (b) Colour plot of the normalised intensity for the difference of different helicity excitation (ΔI_{norm}) in real space for EHE. White dots correspond to the peak positions of ΔI_{norm}. Adapted with permission from Ref. [119].

3.10 Summary

In this chapter, we examined the methods to achieve valley addressability — one "initial state" characterisation technique with ARPES and SARPES, and two "final state" optical methods represented by polarisation-resolved μPL and KRS. We explained the operational principles of ARPES and summarised the reported key results obtained with this photoemission spectroscopy technique in studying the band structure of valleytronic materials. The relevant theories of PL characterisation, possible setups optimised for the study of TMDs, and discussions on optical polarisation in valleytronic materials have also been included. Another method that has been successfully employed to valley-selective addressability is KRS. We introduced the technique with salient examples of its demonstrations in reported literature. As emphasised, valley addressability is not only a prerequisite for eliciting valley-related information from the materials, but also an indispensable cornerstone for achieving valley controllability and building practical valleytronic devices. On the aspect of control, we have discussed the possible techniques to efficiently tune valley pseudo-spin in the layered materials. Often, the mechanisms for addressing and controlling the valley degree of freedom are intricately related. For example, the optical KRS technique enables us to address and control the net spin induced into the system while exciting with circularly polarised pump laser, and the TRKR allows us to look further into the decay dynamics of the spin polarisation/depolarisation of the resident carriers. We have also discussed possible control knobs to manipulate the valley carrier dynamics, e.g., the electrical bias, magnetic field (valley Zeeman effect), intense light pulses via OSE and temperature gradient (anomalous Nernst effect). All these control parameters individually or in combination can help us to understand the mechanisms for valley population/depopulation and motivate a plethora of applications for novel valleytronic devices. Some of these, e.g., electrical and optical control, might appear more feasible for device implementation. However, it remains to be seen which combination of control knobs eventually make it into mainstream valleytronics.

References

[1] M. Cardona and L. Ley, *Photoemission in Solids I: General Principles.* Springer, 1978. doi: 10.1007/3-540-08685-4.

[2] A. Damascelli, Z. Hussain, and Z.-X. Shen, 'Angle-resolved photoemission studies of the cuprate superconductors', *Rev. Mod. Phys.*, vol. 75, no. April, pp. 473–541, 2003, doi: 10.1103/RevModPhys.75.473.

[3] D. N. Basov, M. M. Fogler, A. Lanzara, F. Wang, and Y. Zhang, 'Colloquium: Graphene spectroscopy', *Rev. Mod. Phys.*, vol. 86, no. 3, pp. 959–954, 2014, doi: 10.1103/RevModPhys.86.959.
[4] M. P. Seah and W. A. Dench, 'Quantitative electron spectroscopy of surfaces: A standard data base for electron inelastic mean free paths in solids', *Surf. Interface Anal.*, vol. 1, no. 1, pp. 2–11, 1979, doi: 10.1002/sia.740010103.
[5] J. Osterwalder, 'Spin-polarized photoemission', in *Magnetism: A Synchrotron Radiation Approach*. Springer, 2006, pp. 95–120. doi: 10.1007/3-540-33242-1_5.
[6] M. Kolbe *et al.*, 'Highly efficient multichannel spin-polarization detection', *Phys. Rev. Lett.*, vol. 107, no. 20, p. 207601, 2011, doi: 10.1103/PhysRevLett.107.207601.
[7] D. Vasilyev and J. Kirschner, 'Design and performance of a spin-polarized electron energy loss spectrometer with high momentum resolution', *Rev. Sci. Instrum.*, vol. 87, no. 8, p. 083902, 2016, doi: 10.1063/1.4961471.
[8] W. Jin *et al.*, 'Direct measurement of the thickness-dependent electronic band structure of MoS_2 using angle-resolved photoemission spectroscopy', *Phys. Rev. Lett.*, vol. 111, no. 10, p. 106801, 2013, doi: 10.1103/PhysRevLett.111.106801.
[9] W. Jin *et al.*, 'Substrate interactions with suspended and supported monolayer MoS_2: Angle-resolved photoemission spectroscopy', *Phys. Rev. B*, vol. 91, no. 12, p. 121409(R), 2015, doi: 10.1103/PhysRevB.91.121409.
[10] H. Yuan *et al.*, 'Evolution of the valley position in bulk transition-metal chalcogenides and their monolayer limit', *Nano Lett.*, vol. 16, no. 8, pp. 4738–4745, 2016, doi: 10.1021/acs.nanolett.5b05107.
[11] I. Tanabe *et al.*, 'Band structure characterization of WS_2 grown by chemical vapor deposition', *Appl. Phys. Lett.*, vol. 108, no. 25, p. 252103, 2016, doi: 10.1063/1.4954278.
[12] S. Ulstrup *et al.*, 'Spatially resolved electronic properties of single-layer WS_2 on transition metal oxides', *ACS Nano*, vol. 10, no. 11, pp. 10058–10067, 2016, doi: 10.1021/acsnano.6b04914.

[13] S. Forti *et al.*, 'Electronic properties of WS_2 on epitaxial graphene on SiC(0001)', *Nanoscale*, vol. 9, pp. 16412–16419, 2017, doi: 10.1039/c7nr05495e.
[14] D. J. Trainer *et al.*, 'Inter-layer coupling induced valence band edge shift in mono- to few-layer MoS_2', *Sci. Rep.*, vol. 7, no. August 2016, p. 40599, 2017, doi: 10.1038/srep40559.
[15] M. M. Ugeda *et al.*, 'Giant bandgap renormalization and excitonic effects in a monolayer transition metal dichalcogenide semiconductor', *Nat. Mater.*, vol. 13, no. 12, pp. 1091–1095, 2014, doi: 10.1038/nmat4061.
[16] M. Dendzik *et al.*, 'Growth and electronic structure of epitaxial single-layer WS_2 on Au(111)', *Phys. Rev. B*, vol. 92, no. 24, p. 245442, 2015, doi: 10.1103/PhysRevB.92.245442.
[17] J. A. Miwa *et al.*, 'Electronic structure of epitaxial single-layer MoS_2', *Phys. Rev. Lett.*, vol. 114, no. 4, p. 046802, 2015, doi: 10.1103/PhysRevLett.114.046802.
[18] Y. Zhang *et al.*, 'Direct observation of the transition from indirect to direct bandgap in atomically thin epitaxial $MoSe_2$', *Nat. Nanotechnol.*, vol. 9, no. 2, pp. 111–115, 2014, doi: 10.1038/nnano.2013.277.
[19] N. Alidoust *et al.*, 'Observation of monolayer valence band spin-orbit effect and induced quantum well states in MoX_2', *Nat. Commun.*, vol. 5, no. 1, p. 4673, 2014, doi: 10.1038/ncomms5673.
[20] Y. Zhang *et al.*, 'Electronic structure, surface doping, and optical response in epitaxial WSe_2 thin films', *Nano Lett.*, vol. 16, no. 4, pp. 2485–2491, 2016, doi: 10.1021/acs.nanolett.6b00059.
[21] F. Bussolotti, J. Yang, H. Kawai, J. Y. Chee, and K. E. J. Goh, 'Influence of many-body effects on hole quasiparticle dynamics in a WS_2 monolayer', *Phys. Rev. B*, vol. 103, no. 4, 2021, doi: 10.1103/PhysRevB.103.045412.
[22] F. Bussolotti, H. Kawai, S. L. Wong, and K. E. J. Goh, 'Protected hole valley states in single-layer MoS_2', *Phys. Rev. B*, vol. 99, p. 045134, 2019, doi: 10.1103/PhysRevB.99.045134.
[23] K. Sugawara, T. Sato, Y. Tanaka, S. Souma, and T. Takahashi, 'Spin- and valley-coupled electronic states in monolayer WSe_2 on bilayer graphene', *Appl. Phys. Lett.*, vol. 107, no. 7, p. 071601, 2015, doi: 10.1063/1.4928658.

[24] S.-K. Mo *et al.*, 'Spin-resolved photoemission study of epitaxially grown $MoSe_2$ and WSe_2 thin films', *J. Phys. Condens. Matter*, vol. 28, no. 45, p. 454001, 2016, doi: 10.1088/0953-8984/28/45/454001.

[25] J. M. Riley *et al.*, 'Direct observation of spin-polarized bulk bands in an inversion-symmetric semiconductor', *Nat. Phys.*, vol. 10, no. 11, pp. 835–839, 2014, doi: 10.1038/nphys3105.

[26] M. Gehlmann *et al.*, 'Quasi 2D electronic states with high spin-polarization in centrosymmetric MoS_2 bulk crystals', *Sci. Rep.*, vol. 6, no. June 2015, p. 26197, 2016, doi: 10.1038/srep26197.

[27] E. Razzoli *et al.*, 'Selective probing of hidden spin-polarized states in inversion-symmetric bulk MoS_2', *Phys. Rev. Lett.*, vol. 118, no. 8, p. 086402, 2017, doi: 10.1103/PhysRevLett.118.086402.

[28] X. Zhang, Q. Liu, J. W. Luo, A. J. Freeman, and A. Zunger, 'Hidden spin polarization in inversion-symmetric bulk crystals', *Nat. Phys.*, vol. 10, no. 5, pp. 387–393, 2014, doi: 10.1038/nphys2933.

[29] K. E. J. Goh, F. Bussolotti, C. S. Lau, D. Kotekar-Patil, Z. E. Ooi, and J. Y. Chee, 'Toward valley-coupled spin qubits', *Adv. Quantum Technol.*, vol. 3, no. 6, p. 1900123, 2020, doi: 10.1002/qute.201900123.

[30] H. Zeng, J. Dai, W. Yao, D. Xiao, and X. Cui, 'Valley polarization in MoS_2 by optical pumping', *Nat. Nanotechnol.*, vol. 7, no. 8, pp. 490–493, 2012, doi: 10.1038/nnano.2012.95.

[31] R. Suzuki *et al.*, 'Valley-dependent spin polarization in bulk MoS_2 with broken inversion symmetry', *Nat. Nanotechnol.*, vol. 9, no. 8, pp. 611–617, 2014, doi: 10.1038/nnano.2014.148.

[32] B. Zhu, H. Zeng, J. Dai, Z. Gong, and X. Cui, 'Anomalously robust valley polarization and valley coherence in bilayer WS_2', *Proc. Natl. Acad. Sci.*, vol. 111, no. 32, pp. 11606–11611, 2014, doi: 10.1073/pnas.1406960111.

[33] G. Kioseoglou, A. T. Hanbicki, M. Currie, A. L. Friedman, D. Gunlycke, and B. T. Jonker, 'Valley polarization and intervalley scattering in monolayer MoS_2', *Appl. Phys. Lett.*, vol. 101, no. 22, p. 221907, 2012, doi: 10.1063/1.4768299.

[34] H. Su, C. Wei, A. Deng, D. Deng, C. Yang, and J. F. Dai, 'Anomalous enhancement of valley polarization in multilayer WS_2 at room temperature', *Nanoscale*, vol. 9, no. 16, pp. 5148–5154, 2017, doi: 10.1039/c7nr00554g.

[35] K. F. Mak, K. He, J. Shan, and T. F. Heinz, 'Control of valley polarization in monolayer MoS_2 by optical helicity', *Nat. Nanotechnol.*, vol. 7, no. 8, pp. 494–498, 2012, doi: 10.1038/nnano.2012.96.
[36] P. K. Nayak, F.-C. Lin, C.-H. Yeh, J.-S. Huang, and P.-W. Chiu, 'Robust room temperature valley polarization in monolayer and bilayer WS_2', *Nanoscale*, vol. 8, no. 11, pp. 6035–6042, 2016, doi: 10.1039/C5NR08395H.
[37] G. Sallen *et al.*, 'Robust optical emission polarization in MoS_2 monolayers through selective valley excitation', *Phys. Rev. B*, vol. 86, no. 8, p. 081301, 2012, doi: 10.1103/PhysRevB.86.081301.
[38] D. Xiao, G.-B. Liu, W. Feng, X. Xu, and W. Yao, 'Coupled spin and valley physics in monolayers of MoS_2 and other group-VI dichalcogenides', *Phys. Rev. Lett.*, vol. 108, no. 19, p. 196802, 2012, doi: 10.1103/PhysRevLett.108.196802.
[39] T. Yan, X. Qiao, P. Tan, and X. Zhang, 'Valley depolarization in monolayer WSe_2', *Sci. Rep.*, vol. 5, pp. 1–7, 2015, doi: 10.1038/srep15625.
[40] V. Chellappan, A. L. C. Pang, S. Sarkar, Z. E. Ooi, and K. E. J. Goh, 'Effect of phonons on valley depolarization in monolayer WSe_2', *Electron. Mater. Lett.*, vol. 14, no. July, p. 766, 2018, doi: 10.1007/s13391-018-0086-2.
[41] T. Yu and M. W. Wu, 'Valley depolarization due to intervalley and intravalley electron-hole exchange interactions in monolayer MoS_2', *Phys. Rev. B*, vol. 89, no. 20, p. 205303, 2014, doi: 10.1103/PhysRevB.89.205303.
[42] S. Wu, C. Huang, G. Aivazian, J. S. Ross, D. H. Cobden, and X. Xu, 'Vapor–solid growth of high optical quality MoS_2 monolayers with near-unity valley polarization', *ACS Nano*, vol. 7, no. 3, pp. 2768–2772, 2013, doi: 10.1021/nn4002038.
[43] T. Cao *et al.*, 'Valley-selective circular dichroism of monolayer molybdenum disulphide', *Nat. Commun.*, vol. 3, p. 887, 2012, doi: 10.1038/ncomms1882.
[44] G. Wang *et al.*, 'Control of exciton valley coherence in transition metal dichalcogenide monolayers', *Phys. Rev. Lett.*, vol. 117, no. 18, p. 187401, 2016, doi: 10.1103/PhysRevLett.117.187401.

[45] A. M. Jones *et al.*, 'Optical generation of excitonic valley coherence in monolayer WSe_2', *Nat. Nanotechnol.*, vol. 8, no. 9, pp. 634–638, 2013, doi: 10.1038/nnano.2013.151.

[46] E. del Corro *et al.*, 'Atypical exciton-phonon interactions in WS_2 and WSe_2 monolayers revealed by resonance raman spectroscopy', *Nano Lett.*, vol. 16, no. 4, pp. 2363–2368, 2016, doi: 10.1021/acs.nanolett.5b05096.

[47] N. F. Hinsche *et al.*, 'Spin-dependent electron-phonon coupling in the valence band of single-layer WS_2', *Phys. Rev. B*, vol. 96, no. 12, p. 121402(R), 2017, doi: 10.1103/PhysRevB.96.121402.

[48] G. Moody *et al.*, 'Microsecond valley lifetime of defect-bound excitons in monolayer WSe_2', *Phys. Rev. Lett.*, vol. 121, no. 5, 2018, doi: 10.1103/PhysRevLett.121.057403.

[49] M. Selig *et al.*, 'Excitonic linewidth and coherence lifetime in monolayer transition metal dichalcogenides', *Nat. Commun.*, vol. 7, p. 13279, 2016, doi: 10.1038/ncomms13279.

[50] G. Moody *et al.*, 'Intrinsic homogeneous linewidth and broadening mechanisms of excitons in monolayer transition metal dichalcogenides', *Nat. Commun.*, vol. 6, no. 1, p. 8315, 2015, doi: 10.1038/ncomms9315.

[51] S. Das, G. Gupta, and K. Majumdar, 'Layer degree of freedom for excitons in transition metal dichalcogenides', *Phys. Rev. B*, vol. 99, no. 16, p. 165411, 2019, doi: 10.1103/PhysRevB.99.165411.

[52] K. P. O'Donnell and X. Chen, 'Temperature dependence of semiconductor band gaps', *Appl. Phys. Lett.*, vol. 58, no. 25, pp. 2924–2926, 1991, doi: 10.1063/1.104723.

[53] Y. P. Varshni, 'Temperature dependence of the energy gap in semiconductors', *Physica*, vol. 34, no. 1, pp. 149–154, 1967, doi: 10.1016/0031-8914(67)90062-6.

[54] F. Bussolotti *et al.*, 'Roadmap on finding chiral valleys: Screening 2D materials for valleytronics', *Nano Futures*, vol. 2, no. 3, p. 032001, 2018, doi: 10.1088/2399-1984/aac9d7.

[55] T. Korn, S. Heydrich, M. Hirmer, J. Schmutzler, and C. Schller, 'Low-temperature photocarrier dynamics in monolayer MoS_2', *Appl. Phys. Lett.*, vol. 99, no. 10, p. 102109, 2011, doi: 10.1063/1.3636402.

[56] E. Lorchat *et al.*, 'Filtering the photoluminescence spectra of atomically thin semiconductors with graphene', *Nat. Nanotechnol.*, vol. 15, no. 4, pp. 283–288, 2020, doi: 10.1038/s41565-020-0644-2.

[57] Y. Ye *et al.*, 'Electrical generation and control of the valley carriers in a monolayer transition metal dichalcogenide', *Nat. Nanotechnol.*, vol. 11, no. 7, pp. 598–602, 2016, doi: 10.1038/nnano.2016.49.

[58] D. Lagarde *et al.*, 'Carrier and polarization dynamics in monolayer MoS_2', *Phys. Rev. Lett.*, vol. 112, no. 4, p. 047401, 2014, doi: 10.1103/PhysRevLett.112.047401.

[59] T. Yan, X. Qiao, X. Liu, P. Tan, and X. Zhang, ' Photoluminescence properties and exciton dynamics in monolayer WSe_2', *Appl. Phys. Lett.*, vol. 105, no. 10, p. 101901, Sep. 2014, doi: 10.1063/1.4895471.

[60] S. Chatterjee, G. Gupta, S. Das, K. Watanabe, T. Taniguchi, and K. Majumdar, 'Trion-trion annihilation in monolayer WS_2, *Phys. Rev. B*, vol. 105, no. 12, pp. L121409, 2022, doi: 10.1103/PhysRevB.105.L121409.

[61] M. Dandu, G. Gupta, P. Dasika, K. Watanabe, T. Taniguchi, and K. Majumdar, 'Electrically tunable localized versus delocalized intralayer Moiré excitons and trions in a twisted MoS_2 bilayer', *ACS Nano*, vol. 16, no. 6, pp. 8983–8992, 2022, doi: 10.1021/acsnano.2c00145.

[62] P. Nagler *et al.*, 'Zeeman splitting and inverted polarization of biexciton emission in monolayer WS_2', *Phys. Rev. Lett.*, vol. 121, no. 5, p. 57402, 2018, doi: 10.1103/PhysRevLett.121.057402.

[63] S. Das, G. Gupta, S. Chatterjee, K. Watanabe, T. Taniguchi, and K. Majumdar, 'Highly nonlinear biexcitonic photocurrent from ultrafast interlayer charge transfer', *ACS Nano*, vol. 16, no. 6, pp. 9728–9735, 2022, doi: 10.1021/acsnano.2c03397.

[64] Y. You, X. X. Zhang, T. C. Berkelbach, M. S. Hybertsen, D. R. Reichman, and T. F. Heinz, 'Observation of biexcitons in monolayer WSe_2', *Nat. Phys.*, vol. 11, no. 6, pp. 477–481, 2015, doi: 10.1038/nphys3324.

[65] S. Chatterjee, S. Das, G. Gupta, K. Watanabe, T. Taniguchi, and K. Majumdar, 'Probing biexciton in monolayer WS_2 through controlled many-body interaction', *2D Mater.*, vol. 9, no. 1, p. 015023, 2022, doi: 10.1088/2053-1583/ac3b1c.

[66] G. Kioseoglou, A. T. Hanbicki, M. Currie, A. L. Friedman, and B. T. Jonker, 'Optical polarization and intervalley scattering in single layers of MoS_2 and $MoSe_2$', *Sci. Rep.*, vol. 6, p. 25041, 2016, doi: 10.1038/srep25041.

[67] C. Jiang *et al.*, 'Microsecond dark-exciton valley polarization memory in two-dimensional heterostructures', *Nat. Commun.*, vol. 9, no. 1, p. 753, 2018, doi: 10.1038/s41467-018-03174-3.

[68] C. Jin *et al.*, 'Imaging of pure spin-valley diffusion current in WS_2-WSe_2 heterostructures', *Science*, vol. 360, no. 6391, pp. 893–896, 2018, doi: 10.1126/science.aao3503.

[69] J. Kerr, 'A new relation between electricity and light: Dielectrified media birefringent', *Philos. Mag.*, vol. 50, no. 332, pp. 337–348, 1875, doi: 10.1080/14786447508641302.

[70] J. Kerr, 'A new relation between electricity and light: Dielectrified media birefringent (Second paper)', *Philos. Mag.*, vol. 50, no. 333, pp. 446–458, 1875, doi: 10.1080/14786447508641319.

[71] G. Mayer and F. Gires, *Physique moleculaire-action dune onde lumineuse intense sur lindice de refraction des liquides*, vol. 258. GAUTHIER-VILLARS/EDITIONS ELSEVIER 23 RUE LINOIS, 75015 PARIS, FRANCE, 1964.

[72] M. A. Duguay and J. W. Hansen, 'An ultrafast light gate', *Appl. Phys. Lett.*, vol. 15, no. 6, pp. 192–194, 1969, doi: 10.1063/1.1652962.

[73] L. Yang *et al.*, 'Long-lived nanosecond spin relaxation and spin coherence of electrons in monolayer MoS_2 and WS_2', *Nat. Phys.*, vol. 11, no. 10, pp. 830–834, 2015, doi: 10.1038/nphys3419.

[74] G. Plechinger *et al.*, 'Trion fine structure and coupled spin-valley dynamics in monolayer tungsten disulfide', *Nat. Commun.*, vol. 7, p. 12715, 2016, doi: 10.1038/ncomms12715.

[75] G. Plechinger, T. Korn, and J. M. Lupton, 'Valley-polarized exciton dynamics in exfoliated monolayer WSe_2', *J. Phys. Chem. C*, vol. 121, no. 11, pp. 6409–6413, 2017, doi: 10.1021/acs.jpcc.7b01468.

[76] P. Dey *et al.*, 'Gate-controlled spin-valley locking of resident carriers in WSe_2 monolayers', *Phys. Rev. Lett.*, vol. 119, no. 13, p. 137401, 2017, doi: 10.1103/PhysRevLett.119.137401.

[77] J. Li, M. Goryca, K. Yumigeta, H. Li, S. Tongay, and S. A. Crooker, 'Valley relaxation of resident electrons and holes in a monolayer semiconductor: Dependence on carrier density and the role of substrate-induced disorder', *Phys. Rev. Mater.*, vol. 5, no. 4, p. 044001, 2021, doi: 10.1103/PhysRevMaterials.5.044001.

[78] M. Ersfeld *et al.*, 'Unveiling valley lifetimes of free charge carriers in monolayer WSe_2', *Nano Lett.*, vol. 20, no. 5, pp. 3147–3154, 2020, doi: 10.1021/acs.nanolett.9b05138.

[79] S. A. Crooker, D. D. Awschalom, J. J. Baumberg, F. Flack, and N. Samarth, 'Optical spin resonance and transverse spin relaxation in magnetic semiconductor quantum wells', *Phys. Rev. B*, vol. 56, no. 12, p. 7574, 1997, doi: 10.1103/PhysRevB.56.7574.

[80] Y. K. Kato, R. C. Myers, A. C. Gossard, and D. D. Awschalom, 'Observation of the spin Hall effect in semiconductors', *Science*, vol. 306, no. 5703, pp. 1910–1913, 2004, doi: 10.1126/science.1105514.

[81] N. P. Stern, D. W. Steuerman, S. MacK, A. C. Gossard, and D. D. Awschalom, 'Drift and diffusion of spins generated by the spin Hall effect', *Appl. Phys. Lett.*, vol. 91, no. 6, p. 062109, 2007, doi: 10.1063/1.2768633.

[82] W. T. Hsu *et al.*, 'Optically initialized robust valley-polarized holes in monolayer WSe_2', *Nat. Commun.*, vol. 6, p. 8963, 2015, doi: 10.1038/ncomms9963.

[83] M. Kempf, A. Schubert, R. Schwartz, and T. Korn, 'Two-color Kerr microscopy of two-dimensional materials with sub-picosecond time resolution', *Rev. Sci. Instrum.*, vol. 92, no. 11, p. 113904, 2021, doi: 10.1063/5.0058110.

[84] L. Yang, W. Chen, K. M. McCreary, B. T. Jonker, J. Lou, and S. A. Crooker, 'Spin coherence and dephasing of localized electrons in monolayer MoS_2', *Nano Lett.*, vol. 15, no. 12, pp. 8250–8254, 2015, doi: 10.1021/acs.nanolett.5b03771.

[85] M. Jiang *et al.*, 'Coherent spin dynamics of localized electrons in monolayer MoS_2', *J. Phys. Chem. Lett.*, vol. 13, no. 11, pp. 2661–2667, 2022, doi: 10.1021/acs.jpclett.2c00329.

[86] M. Ersfeld *et al.*, 'Unveiling valley lifetimes of free charge carriers in monolayer WSe_2', *Nano Lett.*, vol. 20, no. 5, pp. 3147–3154, 2020, doi: 10.1021/acs.nanolett.9b05138.
[87] M. Ersfeld *et al.*, 'Spin states protected from intrinsic electron-phonon coupling reaching 100 ns lifetime at room temperature in $MoSe_2$', *Nano Lett.*, vol. 19, no. 6, pp. 4083–4090, 2019, doi: 10.1021/acs.nanolett.9b01485.
[88] M. Furis *et al.*, 'Local Hanle-effect studies of spin drift and diffusion in n:GaAs epilayers and spin-transport devices', *New J. Phys.*, vol. 9, p. 347, 2007, doi: 10.1088/1367-2630/9/9/347.
[89] C. Robert *et al.*, 'Spin/valley pumping of resident electrons in WSe_2 and WS2 monolayers', *Nat. Commun.*, vol. 12, p. 5455, 2021, doi: 10.1038/s41467-021-25747-5.
[90] L. Wang and M. W. Wu, 'Electron spin relaxation due to D'yakonov-Perel' and Elliot-Yafet mechanisms in monolayer MoS_2: Role of intravalley and intervalley processes', *Phys. Rev. B*, vol. 89, no. 11, p. 115302, 2014, doi: 10.1103/PhysRevB.89.115302.
[91] H. Ochoa, F. Finocchiaro, F. Guinea, and V. I. Fal'Ko, 'Spin-valley relaxation and quantum transport regimes in two-dimensional transition-metal dichalcogenides', *Phys. Rev. B*, vol. 90, no. 23, p. 235429, 2014, doi: 10.1103/PhysRevB.90.235429.
[92] E. Barré *et al.*, 'Spatial separation of carrier spin by the valley hall effect in monolayer WSe_2 transistors', *Nano Lett.*, vol. 19, no. 2, pp. 770–774, 2019, doi: 10.1021/acs.nanolett.8b03838.
[93] S. Matsuzaka, Y. Ohno, and H. Ohno, 'Electron density dependence of the spin Hall effect in GaAs probed by scanning Kerr rotation microscopy', *Phys. Rev. B*, vol. 80, no. 24, p. 241305(R), 2009, doi: 10.1103/PhysRevB.80.241305.
[94] J. Lee, Z. Wang, H. Xie, K. F. Mak, and J. Shan, 'Valley magnetoelectricity in single-layer MoS_2', *Nat. Mater.*, vol. 16, no. 9, pp. 887–891, 2017, doi: 10.1038/nmat4931.
[95] J. Lee, K. F. Mak, and J. Shan, 'Electrical control of the valley Hall effect in bilayer MoS_2 transistors', *Nat. Nanotechnol.*, vol. 11, no. 5, pp. 421–425, 2016, doi: 10.1038/nnano.2015.337.

[96] X. X. Zhang *et al.*, 'Magnetic brightening and control of dark excitons in monolayer WSe_2', *Nat. Nanotechnol.*, vol. 12, no. 9, pp. 883–888, 2017, doi: 10.1038/nnano.2017.105.
[97] Z. Li *et al.*, 'Emerging photoluminescence from the dark-exciton phonon replica in monolayer WSe_2', *Nat. Commun.*, vol. 10, no. 1, p. 2469, 2019, doi: 10.1038/s41467-019-10477-6.
[98] E. Liu, J. van Baren, T. Taniguchi, K. Watanabe, Y. C. Chang, and C. H. Lui, 'Valley-selective chiral phonon replicas of dark excitons and trions in monolayer WSe_2', *Phys. Rev. Res.*, vol. 1, no. 3, p. 032007, 2019, doi: 10.1103/PhysRevResearch.1.032007.
[99] M. Zinkiewicz *et al.*, 'Excitonic complexes in n-doped WS_2 monolayer', *Nano Lett.*, vol. 21, no. 6, pp. 2519–2525, 2021, doi: 10.1021/acs.nanolett.0c05021.
[100] Y. Li *et al.*, 'Valley splitting and polarization by the Zeeman effect in monolayer $MoSe_2$', *Phys. Rev. Lett.*, vol. 113, no. 26, p. 266804, 2014, doi: 10.1103/PhysRevLett.113.266804.
[101] A. A. Mitioglu *et al.*, 'Optical investigation of monolayer and bulk tungsten diselenide (WSe_2) in high magnetic fields', *Nano Lett.*, vol. 15, no. 7, pp. 4387–4392, 2015, doi: 10.1021/acs.nanolett.5b00626.
[102] A. Arora *et al.*, 'Valley Zeeman splitting and valley polarization of neutral and charged excitons in monolayer $MoTe_2$ at high magnetic fields', *Nano Lett.*, vol. 16, no. 6, pp. 3624–3629, 2016, doi: 10.1021/acs.nanolett.6b00748.
[103] A. Arora *et al.*, 'Interlayer excitons in a bulk van der Waals semiconductor', *Nat. Commun.*, vol. 8, no. 1, p. 639, 2017, doi: 10.1038/s41467-017-00691-5.
[104] M. Koperski *et al.*, 'Orbital, spin and valley contributions to Zeeman splitting of excitonic resonances in $MoSe_2$, WSe_2 and WS_2 Monolayers', *2D Mater.*, vol. 6, no. 1, p. 015001, 2019, doi: 10.1088/2053-1583/aae14b.
[105] M. R. Molas *et al.*, 'Probing and manipulating valley coherence of dark excitons in monolayer WSe_2', *Phys. Rev. Lett.*, vol. 123, no. 9, p. 096803, 2019, doi: 10.1103/PhysRevLett.123.096803.
[106] C. Robert *et al.*, 'Fine structure and lifetime of dark excitons in transition metal dichalcogenide monolayers', *Phys. Rev. B*, vol. 96, no. 15, p. 155423, 2017, doi: 10.1103/PhysRevB.96.155423.

[107] G. Wang *et al.*, 'In-plane propagation of light in transition metal dichalcogenide monolayers: optical selection rules', *Phys. Rev. Lett.*, vol. 119, no. 4, p. 047401, 2017, doi: 10.1103/PhysRevLett.119.047401.

[108] A. Arora, 'Magneto-optics of layered two-dimensional semiconductors and heterostructures: Progress and prospects', *J. Appl. Phys.*, vol. 129, no. 12., p. 120902, 2021, doi: 10.1063/5.0042683.

[109] A. Arora *et al.*, 'Zeeman spectroscopy of excitons and hybridization of electronic states in few-layer WSe_2, $MoSe_2$ and $MoTe_2$', *2D Mater.*, vol. 6, no. 1, p. 015010, 2019, doi: 10.1088/2053-1583/aae7e5.

[110] J. Huang, T. B. Hoang, and M. H. Mikkelsen, 'Probing the origin of excitonic states in monolayer WSe_2', *Sci. Rep.*, vol. 6, p. 22414, 2016, doi: 10.1038/srep22414.

[111] E. J. Sie, J. W. McLver, Y. H. Lee, L. Fu, J. Kong, and N. Gedik, 'Valley-selective optical Stark effect in monolayer WS_2', *Nat. Mater.*, vol. 14, no. 3, pp. 290–294, 2015, doi: 10.1038/nmat4156.

[112] P. D. Cunningham, A. T. Hanbicki, T. L. Reinecke, K. M. McCreary, and B. T. Jonker, 'Resonant optical Stark effect in monolayer WS_2', *Nat. Commun.*, vol. 10, no. 1, p. 5539, 2019, doi: 10.1038/s41467-019-13501-x.

[113] J. Kim *et al.*, 'Ultrafast generation of pseudo-magnetic field for valley excitons in WSe_2 monolayers', *Science*, vol. 346, no. 6214, pp. 1205–1208, 2014, doi: 10.1126/science.1258122.

[114] C. Bao, P. Tang, D. Sun, and S. Zhou, 'Light-induced emergent phenomena in 2D materials and topological materials', *Nat. Rev. Phys.*, vol. 4, p. 33–48, 2022, doi: 10.1038/s42254-021-00388-1.

[115] T. LaMountain *et al.*, 'Valley-selective optical Stark effect of exciton-polaritons in a monolayer semiconductor', *Nat. Commun.*, vol. 12, no. 1, p. 4530, 2021, doi: 10.1038/s41467-021-24764-8.

[116] Y. Xiong, M. S. Rudner, and J. C. W. Song, 'Multistable excitonic Stark effect', *Phys. Rev. Res.*, vol. 4, no. 2, p. 023168, 2022, doi: 10.1103/physrevresearch.4.023168.

[117] C. K. Yong *et al.*, 'Biexcitonic optical Stark effects in monolayer molybdenum diselenide', *Nat. Phys.*, vol. 14, no. 11, pp. 1092–1096, 2018, doi: 10.1038/s41567-018-0216-7.

[118] X. Q. Yu, Z. G. Zhu, G. Su, and A. P. Jauho, 'Thermally driven pure spin and valley currents via the anomalous Nernst effect in monolayer group-VI dichalcogenides', *Phys. Rev. Lett.*, vol. 115, no. 24, p. 246601, 2015, doi: 10.1103/PhysRevLett.115.246601.

[119] M. Onga, Y. Zhang, T. Ideue, and Y. Iwasa, 'Exciton Hall effect in monolayer MoS_2', *Nat. Mater.*, vol. 16, no. 12, pp. 1193–1197, 2017, doi: 10.1038/nmat4996.

https://doi.org/10.1142/9789811229107_0004

Chapter 4

Material Preparation and Development for Valleytronics

Zhuojun Jiang, Fabio Bussolotti, Wei Fu, Calvin Pei Yu Wong, and Kuan Eng Johnson Goh

4.1 Introduction

The ability to obtain high-quality atomically thin 2D transition metal dichalcogenide (TMD) crystals is central to the development of valleytronic devices towards applications because the observation of valley-based physics is predicated upon the preservation of the valley-differentiated band structure, which can be sensitive to the material's crystalline quality. However, the crystal quality of 2D TMD flakes can vary drastically depending on the preparation and growth method. Typical preparation methods such as mechanical exfoliation, chemical vapour deposition growth, or liquid exfoliation can result in flakes that have drastically different properties. Thus, it is important to choose a suitable method to prepare materials for fabricating valleytronic devices. When choosing a method for producing these 2D TMD layers, some important factors to consider include:

1. Precise thickness control,
2. High material quality (e.g., measured in terms of low defect densities and large single crystal domains),
3. Ease of integration with other materials, such as other layered materials, forming 2D heterostructures,
4. Ease of doping/functionalisation (e.g., for tuning material properties or enabling adhesion of other bulk materials such as metals or dielectrics), and
5. Ability to scale up.

Fig. 4.1 shows two general approaches — top-down exfoliation method and bottom-up growth method for obtaining 2D TMD layers. The "top-down" strategy involves the physical (mechanical) or chemical exfoliation of monolayer or few-layer flakes from the bulk layered materials which are typically grown using traditional chemical vapour transport bulk crystal growth methods. Tape exfoliation or other exfoliation techniques such as ultrasonication, ion intercalation and electrochemical methods are subsequently used to separate the crystalline layers into their 2D monolayer or few-layer forms and then deposited on a substrate. While top-down methods can give very good control of the material quality (dependent on the starting material), the main drawback of this technique is its poor ability to scale up. The "bottom-up" strategy involves the synthesis of 2D TMD layers directly on a suitable growth substrate using vapour-phase methods, such as chemical vapour deposition (CVD), atomic layer deposition (ALD), molecular beam epitaxy (MBE) and physical vapour deposition (PVD). As the bottom-up method involves crystal growth on a suitable substrate, it has greater potential for scaling up,

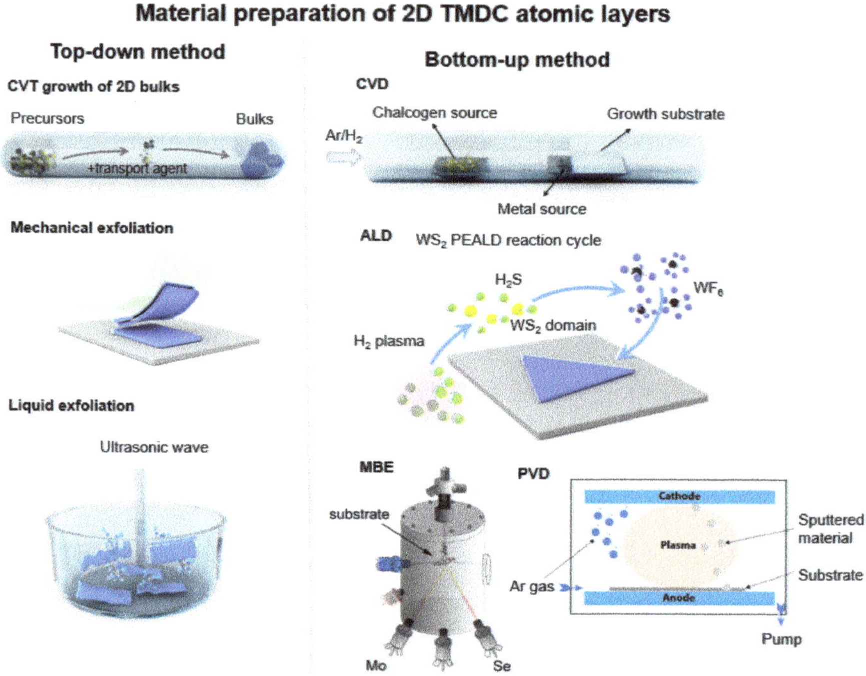

Fig. 4.1. Schematic presentation of various material preparation techniques for fabrication of atomically thin 2D TMD layers.

but control over other parameters such as crystal quality or doping are challenging. Among these growth methods, the CVD process is the most widely used method to synthesise 2D TMD layers on a suitable substrate such as SiO_2 or sapphire wafers, and this process will be discussed further in Section 4.3 while the other methods will be briefly covered here.

ALD is a gaseous deposition technique which relies on a sequence of precisely controlled self-limiting surface reactions using dosed gas-phase precursors. The surface of the growth substrate is first functionalised with an appropriate chemical group that serves to anchor the first gas precursor on the surface in a self-teminating surface reaction. Then, the first gas precursor is purged from the reaction chamber and the second precursor is introduced into the chamber to react with the first precursor which is now the top-most layer anchored to the substrate. Of utmost importance to the ALD technique is the availability of the self-limiting surface reaction kinetics to enable layer-by-layer precise control of the growth thickness, with good reproducibility and scalability. When the gaseous precursors are pulsed onto a functionalised surface, it adsorbs and reacts with all available surface groups, and due to its self-limiting reaction kinetics, excess precursors are pumped away before the next precursor is dosed, and the cycle repeats. Typically, ALD produces highly conformal films but they tend to have a high nucleation density, resulting in amorphous materials (or small grain sizes) which may have a high defect density, due to the low reaction temperatures required to maintain the self-limiting kinetics [1], [2].

MBE is another gas-phase method widely used in the growth of 2D TMDs. It offers excellent control of the growth precursor and condition via ultra-high vacuum (UHV) condition and precise evaporation of molecular precursors using effusion cells. These effusion cells produce dilute beams of high-purity precursors which have very little chance of interacting in the gas phase and rely on the surface diffusion and reaction of the precursors on a suitable growth substrate. Because MBE allows precise control of the growth mechanism and precursor flux in a UHV environment, it is possible to grow crystals with very high quality, but thus far, large single crystals of 2D TMDs grown by MBE remain elusive, though some works have shown the possibility of producing large-area uniform polycrystalline films with small domains [3], [4].

PVD techniques which rely on sputter deposition of precursors onto a suitable substrate have also been used to grow scalable 2D material film on various substrates [5]. In a typical sputtering process, bulk 2D materials or elemental precursors are placed in a target holder facing upwards and growth substrates are placed above it facing downwards. High electrical bias (or radiofrequency excitation) is applied to the target holder in the presence of an inert argon gas atmosphere, which causes a plasma of ionised Ar^+ gas to be accelerated towards the precursor targets. The bombardment of the heavy Ar^+ ions with the precursor target results in the ejection of precursor molecules or fragments onto the substrates, seeding the growth of a 2D polycrystalline film with small domains. While PVD can produce good stochiometric films when optimised, the harsh conditions of the plasma and heavy-particle bombardment during the process can result in 2D materials with a high defect density [6], [7].

As the crystal quality of the 2D material is extremely important to the preservation of valley-coupled spins for valleytronics, the next sections will assess two most commonly used methods of 2D material preparation: mechanical exfoliation and CVD. These methods are available in laboratories worldwide and have gained popularity due to the ease of use and simple equipment required. The mechanical exfoliation method from bulk layered materials is typically relied upon to provide a high-quality materials platform for studying fundamental physical properties, while the CVD method offers significant opportunities for scalable and large-area growth of high-quality 2D materials [8]. In the next section, we first discuss the mechanical exfoliation of bulk crystals to form 2D materials and the subsequent stacking and transfer techniques needed to form 2D heterostructures. Then, we explore relevant works on the CVD growth of 2D materials, focusing on its precursors, growth substrates, and the growth of 2D heterostructures.

4.2 Mechanical Exfoliation from Bulk Crystals and Stacking Techniques

4.2.1 Introduction to Mechanical Exfoliation

Following the revolutionary discovery and isolation of the first 2D graphene layers using the 'scotch tape method' by Novoselov and Geim in 2004 [9], [10], mechanical exfoliation from bulk crystals have been used to successfully isolate many other types of 2D materials ranging from insulators such as hexagonal boron nitride [11], to semiconducting TMDs with common examples such as MoS_2, WS_2, $MoSe_2$, WSe_2 [10], [12]–[16], topological insulators WTe_2 [17], [18] and even superconductors such as $NbSe_2$ [19], [20].

While large-area CVD-grown samples have already been attempted since the early days of 2D materials, the tape exfoliation method remains the dominant approach to obtain 2D materials of the highest quality due to the ability to obtain extremely high-quality bulk starting materials and the ease of exfoliation, requiring only a very simple equipment — the optical microscope. This technique of using a sticky tape to prepare 2D materials (see Fig. 4.2) has now matured into a subtopic of many different techniques, where different types of adhesive tapes, polymers, substrate materials and preparation methods have been developed to increase the yield and to minimise contamination, or even stacked into novel structures. The next few subsections will describe some exfoliation and stacking techniques based upon the now famous scotch tape method for obtaining 2D materials.

4.2.1.1 *Thinning of Bulk Flakes on the Tape*

This method can be simply described as the interfacial peeling of monolayer or few layers from bulk van der Waals layered materials and transfer of the exfoliated materials onto targeted substrates [9]. Fig. 4.2 shows a typical mechanical exfoliation process. A piece of adhesive tape is used to peel off a thin flake of crystal from the bulk material, and subsequently thinned further by pressing the initially exfoliated crystal flake against more clean areas of tape until the flakes on the tape show a lustreless and dull appearance [21], [22]. To help the crystal conform to the tape, a gentle and uniform force is applied to the tape surface during the peeling and pressing process. Excessive force should be avoided as it tends to fragment the exfoliated flakes into smaller flakes, making the subsequent fabrication steps difficult. Scotch tape and blue semiconductor wafer dicing tape are the most commonly used adhesive tapes for exfoliation [23]. Scotch tape uses a stronger acrylic adhesive which could leave residue contaminations on the surface of the 2D material [24] or the substrate wafer. Hence, while this method of preparing 2D materials is commonly referred to as the "Scotch tape" method, Scotch tape is not commonly used in present research laboratories. Instead, most laboratories now use the semiconductor wafer dicing tape instead, commonly referred to as the "blue tape".

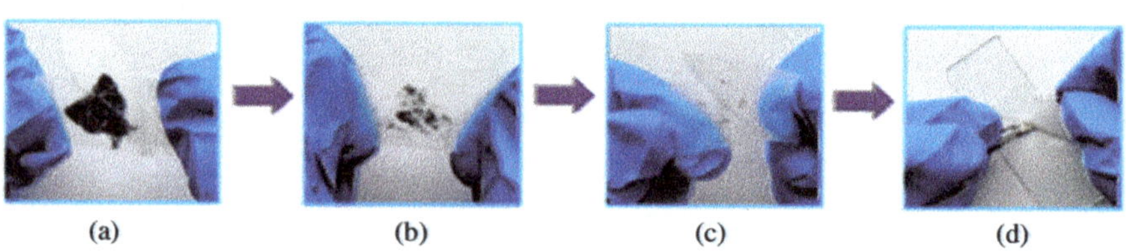

Fig. 4.2. Process of the 'scotch tape' exfoliation method [24], showing (a) a piece of adhesive tape is used to peel off a thin flake from the bulk layered crystal, (b) subsequent thinning by further pressing the initially exfoliated crystal flake against more clean areas of tape, (c) the repetition of step (b) until the flakes on the tape show a lustreless and dull appearance, and (d) the final thinned flake pressed gently onto a clean target substrate.

4.2.1.2 *Transfer of Thinned Flakes to Substrate*

After a few cycles of crystal thinning on the tape, the exfoliated thin flakes on the tape are then pressed gently onto a clean target substrate. Van der Waals force between the clean substrate and the 2D materials will hold some mono- or few-layer flakes on the substrate by cleaving them out from the thicker flakes on the tape, when the tape is slowly peeled away [12]. With some good luck and effort searching for the flakes under an optical microscope, one can obtain some few tens of micron-sized monolayer 2D flakes on the substrate, usually after a few tries. However, the size and thickness distribution and quality of the exfoliated flake can be rather inconsistent and the success rate of exfoliation remains highly skill dependent [12], [24], [25]. To an untrained eye, identifying the monolayer flakes among all the flakes on the substrate can be quite challenging. Fortunately, the optical contrast on a suitable substrate such as silicon wafer with 90 nm or 280 nm SiO_2 is usually sufficient to allow differentiation between the layers [26], [27]. While the process of exfoliation and searching for flakes can be an arduous process, it is still the go-to method to locate and obtain monolayer 2D materials, and can be used to stack and fabricate more complicated van der Waals heterostructures [28] that have shown novel physics such as superconductivity in twisted bilayer graphene devices [18], [20].

To fabricate van der Waals heterostructures by the transfer and stacking of different layers of van der Waals layered materials, many transfer or assembly methods have been developed over the last decade. In the next subsection, we will describe a few types of dry transfer methods and the generic experimental setup required for fabrication of such 2D van der Waals heterostructures.

4.2.2 Setup for Dry Transfer of 2D Materials

The simplest and most commonly used setup for 2D materials transfer is shown in Fig. 4.3 — a long working distance microscope equipped with a xyz-micromanipulator used to hold a mask or glass slide and a temperature-controlled heating stage [25], [28], [29]. This setup is relatively simple and low-cost, and can be assembled in a do-it-yourself (DIY) manner [28]. Other methods include modification of an old mask aligner or probe station into a 2D material transfer setup. The detailed instructions to DIY build a transfer station, with the list of required parts as well as the sourcing information, can be found in publications [29], [30].

During dry transfer, the top flake is carried by a transfer holder which is normally a glass slide, while the bottom flake is placed on the sample stage. The two flakes are then brought close to each other by adjusting the micromanipulator and sample stage along the x-, y-, and z-directions. The ability to tilt and rotate the sample holder or the glass slide can be helpful when control of the alignment and angle with micron-level precision is required, yielding the desired heterostructure.

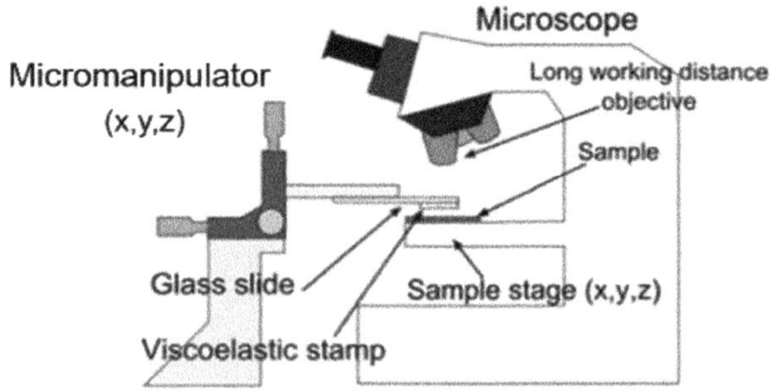

Fig. 4.3. Schematic of a typical transfer stacking station used for the fabrication of 2D heterostructures. Sample heating is provided for the sample stage to help with film adhesion if necessary. Reproduced with permission from Ref. [28].

4.2.2.1 *Polymethyl-Methacrylate Supported Transfer Method*

In this method, a spin-coated polymethyl-methacrylate (PMMA) film is used as the supporting layer for the exfoliation, transfer and stacking process [25]. Firstly, a polymer stack is prepared by spin coating a layer of water-soluble polymer and another layer of PMMA onto a piece of silicon substrate. The 2D flakes are then mechanically exfoliated onto this polymer stack by Scotch tape. The substrate is subsequently floated on a deionised (DI) water bath, with the water-soluble layer partially submerged. The water-soluble layer is slowly dissolved by the ingress of water from the sample edge and the hydrophobic PMMA layer with the flake on top of it is released from the silicon substrate and floated on the water surface. This floating layer of PMMA is then gently scooped out using a glass slide and dried over a hot plate. After drying, the glass slide is flipped over and transfer of the flake onto the desired location can be performed using a transfer system with micron-level precision such as the one shown in Fig. 4.3.

By looking through the microscope, the operator can monitor the alignment of the PMMA-held 2D flake to a feature (typically another 2D material flake but can also be a marker) on the receiving substrate. The manipulator is carefully adjusted till the PMMA-held 2D flake is in contact with the target on the receiving substrate. Then the substrate is heated to over 100°C to increase the adhesion between the PMMA film and target substrate and to avoid trapping water vapour in between the two flakes. Once transferred, the PMMA layer is washed off by a suitable solvent, such as acetone [28], [31], [32].

There are a few things to take note during this process: firstly, even though DI water is used in this method, it is still considered a dry transfer method because water or solvent does not directly contact the side of the flake which contacts the targeted structure as the PMMA layer acts as a hydrophobic polymer to block the ingress of water [25]. Hence, it is essential for the operator to carefully manage the water ingress to dissolve the water-soluble layer and release of the PMMA from the silicon substrate, by avoiding full submersion of the PMMA. Secondly, the thickness of the PMMA layer should be tuned for optical identification of monolayer flakes under the microscope, else identification of monolayers can be difficult [31]. Lastly, the water-soluble film as a sacrificial layer can be chosen from polyvinyl alcohol [22], [33], [34], aquaSave [31], and polydimethylglutarimide (PMGI) [35] (dissolving PMGI requires use of water-based solvents such as Microposit MF-319).

4.2.2.2 *Polydimethylsiloxane Stamping Method*

In contrast to the previous method, the polydimethylsiloxane (PDMS) transfer method does not involve the use of any aqueous solvent, which enables it to be used inside a glovebox for processing of air- and water-sensitive materials. PDMS is a commercially available viscoelastic polymer material that can be purchased in different sizes, thickness and tackiness. By utilising the viscoelastic property of PDMS, a strong adhesion to the 2D flakes can be obtained by using the sticky rubbery property of PDMS when the flakes (or tape) are pulled at a high detaching speed, whereas PDMS behaves as a viscous liquid with low adhesion when the flakes are peeled off very slowly [28], [36].

Fig. 4.4 shows schematically the PDMS stamping method, which starts by placing the PDMS stamp on the glass slide, and the Scotch tape with the exfoliated flakes will be directly pasted onto the PDMS stamp. By ripping the scotch tape which carries the bulk flakes off

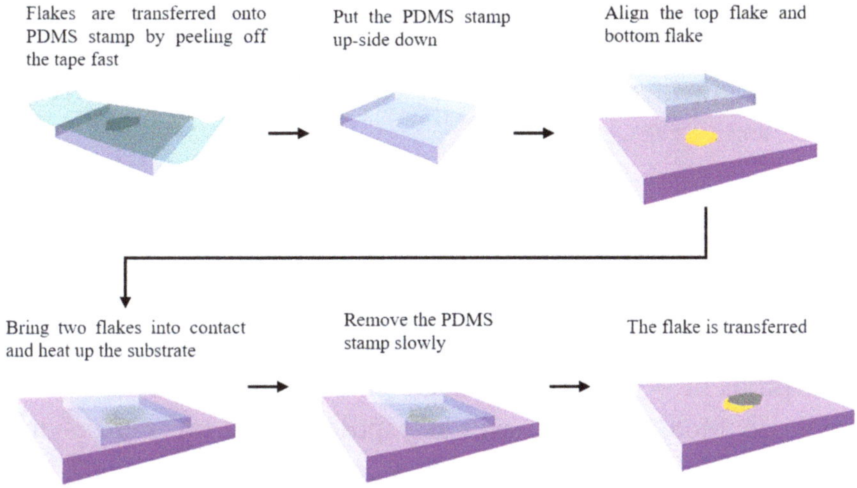

Fig. 4.4. Schematic transfer process for PDMS stamping method. Reproduced with permission from Ref. [28].

the PDMS quickly, some flakes will be transferred onto the PDMS stamp which is in its rubbery state. Like the PMMA transfer method, the glass slide with the PDMS stamp and the exfoliated flakes will be clamped upside down on the micromanipulator and brought closer to the target substrate which has the bottom flake prepared. Once the two flakes are aligned and placed in contact, heating up the stamp to 80°C and slowly lifting up the glass slide along with PDMS will cause the top flake to be released from the stamp and left on the bottom flake [28], [37]. Heating up the PDMS reduces the viscosity of the PDMS stamp and also helps to release the strain caused by pressing it down onto the substrate, thus reducing the probability of forming cracks or wrinkles on the transferred flakes.

While this method appears easy and effective in assembling 2D material heterostructures since it is completely solvent free and does not require spin coating or other pre-preparation, this method may leave

PDMS oligomer residues on the surface and interface of the heterostructure which can be difficult to remove [37]. This drawback led to the development of the ultra-clean pick-up method, which is described in the next section.

4.2.2.3 *Pick-up Method*

The two methods introduced in the previous sections have one common problem — at least one side of the channel material (such as graphene, MoS_2, WS_2, etc.) will be exposed to the transfer polymer, which may leave polymeric residues on the 2D material [37]. Even though post-transfer annealing [38] and pre-transfer UV-ozone treatment [37] have been reported to be effective post-contamination cleaning and prevention methods, a novel method based on a 'pick-up' methodology has been developed to prepare ultra-clean 2D material heterostructure devices with extremely high channel mobility [39]–[41]. This technique relies on the presence of van der Waals forces between the layers of 2D materials and is used as the adhesion medium to attach to the 2D material and to detach the target flake from the original substrate (such as SiO_2 wafers). We shall use the fabrication of hBN/graphene/hBN heterostructure as an example to expatiate the pick-up method shown schematically in Fig. 4.5. A polypropylene carbonate (PPC) layer [38], polycarbonate (PC) [41], or thermoplastic polymer layer [42] is used to facilitate this method. The PPC layer is spin-coated onto a PDMS stamp, and this polymer stack is then affixed onto a glass slide, which is again mounted facing downwards on a micromanipulator arm. Flakes of hBN and graphene are exfoliated in advance onto the Si/SiO_2 substrate, and the thickness and location of the target flakes identified via optical microscopy. A typical desired thickness of the top hBN is 10–30 nm [31]. The glass

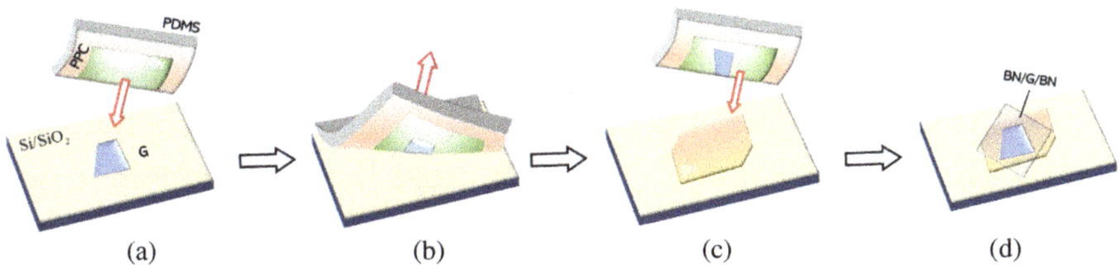

Fig. 4.5. Schematic of the pick-up method. (a) A PDMS/PPC stamp with thick hexagonal boron nitride (hBN) crystal is brought close to the target graphene. (b) The graphene is picked up by the hBN and polymer stamp. (c) The hBN/graphene is brought into contact with bottom hBN. (d) The PDMS/PPC is slowly removed when the whole stack is heated up to 110°C and the hBN/graphene/hBN stack is transferred to a new substrate. Reproduced with permission from Ref. [39].

slide with the PPC stamp is brought down into contact with the first hBN flake and the stage is heated up so that the substrate and the PPC stamp equilibrate at ~40°C. At this temperature, the PPC stamp is hard and tacky, so retracting the glass slide makes it possible for the PPC stamp to pick-up the hBN from the SiO_2 substrate. This few-layer hBN flake will then be further used as the stamp to pick-up monolayer or few-layer graphene like in the previous step, this time using the van der Waals forces between the 2D layers as the adhesive medium. Here, graphene would adhere more strongly to hBN than to SiO_2 and will be lifted from the substrate. Thereafter, the hBN/graphene stack can be released onto the bottom hBN when they are in contact by heating the whole stack to 110°C, which softens and decreases the tackiness of the PPC and by slowly removing the glass slide away from the surface, the 2D heterostructure stack will be left on the target substrate [38], [39], [43]. During the release step, a higher temperature of 110°C is required as at this temperature the polymer softens and behaves as a viscous liquid, helping hBN and graphene to contact conformally, avoiding bubbles and facilitating the removing of the PPC from the heterostructure.

In summary, while many dry transfer methods have been developed to align and stack 2D materials to fabricate heterostructures, the pick-up method is currently the cleanest method to fabricate 2D heterostructures with the highest quality. This is because the pick-up method does not expose the active 2D material to solvents or polymers, which may leave residues on the surface of the active material. The van der Waals forces between 2D materials also have a self-cleaning effect where weakly bound adsorbates can be pushed out of the 2D material stack when the assembled heterostructure is vacuum annealed. Although exfoliation and stacking using the pick-up technique is time consuming, arduous and difficult to scale up, this method is still the go-to method for the fabrication of devices of the highest quality to demonstrate novel physics phenomena.

In the next section, we will discuss a scalable method to grow 2D TMD materials which are currently used for the fabrication of devices that are less stringent on material quality. It is important to note that while grown material is presently known to be of poorer quality, this field is rapidly maturing and we anticipate that soon the material quality of grown flakes could be comparable to or even better than the exfoliated flakes in the near future.

4.3 Chemical Vapour Deposition Growth of Transition Metal Dichalcogenides

4.3.1 Introduction to Chemical Vapour Deposition

Compared to other growth techniques briefly discussed in Section 4.1, CVD benefits from having a more straightforward and simple instrumentation that is easier to set up in laboratories and allows a wide range of freedom to tune the structure and electronic properties

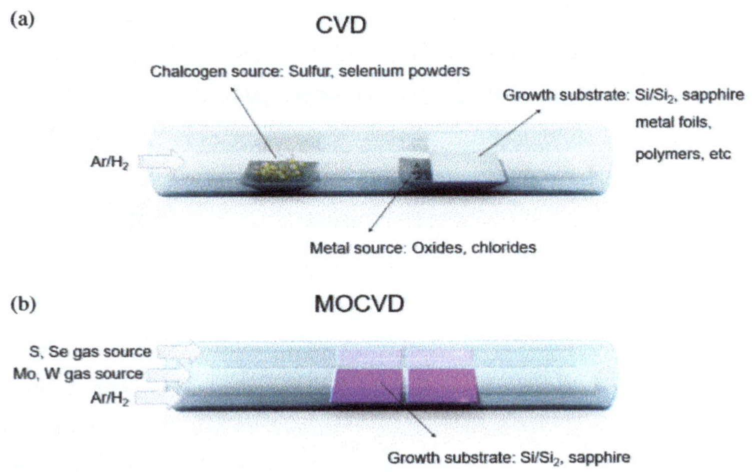

Fig. 4.6. (a) Representative setup for CVD growth of TMD materials by using solid metal (oxides, salt) and solid chalcogens (S, Se in the example) as source materials. (b) Representation setup for MOCVD growth of TMD with metal-containing gas-phase source precursors. The tubes are typically made of quartz and form the enclosures for controlled CVD growth. Heat is applied to various regions of the tube via external thermal furnaces or induction coils that heat up graphite susceptors inside the tube. Typically, the growth substrate is in the high-temperature zone of the quartz tube which may reach temperatures above 1000°C.

of the targeted TMDs [44]. In essence, the aim of CVD growth is to create low-defect-density, large-area 2D materials with quality comparable to or better than their exfoliated counterpart, combined with a precise control over layer numbers, domain size, and morphology. Fig. 4.6(a) shows a typical CVD setup using two different solid sources (chalcogen X and metal M) for the growth of MX_2-TMD. In this configuration, the metal source is loaded in the high-temperature zone of a furnace while a chalcogen source is placed in the low-temperature zone. The process relies on the sublimation of the solid precursor into vapour species, followed by their transport via a carrier gas (usually Ar/H_2 mixture) towards a suitable substrate located in the high-temperature region. The chalcogens and metal precursors react and deposit into the targeted material on the substrate's surface. As the solid precursor powder consists of particles with varying sizes, it is usually challenging to grow a uniform film of TMD materials as these particles vaporise at different rates. A possible solution to this problem is to use thin film deposition or spin coating to pre-deposit a uniform film of metal source, then introduce the chalcogen vapour by gas source or evaporating their powder to promote uniform growth of 2D materials. Utilising all-gas-phase precursor in a method called metal-organic CVD (MOCVD) is another approach. This method allows precise and independent control of the precursor concentration reaching the growth substrate's surface, as shown in Fig. 4.6(b). The volatile metal-organic and chalcogen compound sources are introduced into the reaction chamber with well-controlled ratios in an inert carrier gas using precise mass flow controllers. The reactive precursor molecules are decomposed in the hot zone of the growth chamber where the desired material can be deposited onto the target substrate.

In the past decade, these CVD-based strategies were implemented with varied degrees of success for the large-scale growth of various TMD materials [45]–[53]. Several important factors such as precursor types and the local concentration of precursors on the substrate regions [46], [49], [50], [53], [54], flow rate of carrier gas, seed and growth promoters [55], as well as the nanoscale morphology of substrates [51], [52] are critical degrees of freedom that lead to changes in the morphology, domain size, uniformity, and electrical performance of the grown TMD materials. These will be discussed in detail in the following sections.

4.3.2 Role of Precursors and Seeds in CVD

Vapour-phase reactions of TMD growth precursors by thermal evaporation of metal oxide powder (e.g., MoO_3 or WO_3) and elemental chalcogen powder (e.g., S or Se powder) as shown in Fig. 4.6(a) has been widely used in the synthesis of TMD thin films. The growth reaction mechanism is complex, not well understood and can result in different shapes or edges of the as-grown TMDs as well as different coverages on the selected supporting substrates. A simplified mechanism for growing MoS_2 monolayers can be the following sulfuration pathway [55]:

$$2\ MoO_3 + 7\ S \rightarrow 2\ MoS_2 + 3\ SO_2$$

During the reaction, local precursor concentration can vary along the length and width of the growth substrate and greatly influence the shape of the grown MoS_2, with hexagonal or triangular flakes grown in sulfur-deficient or sulfur-rich conditions, respectively [56], [57] as shown in Fig. 4.7(a),(b). It was also suggested that three-point star-shaped MoS_2 crystals such as those seen in Fig. 4.7(c) are obtained under low-temperature or sulfur-rich growth conditions which can originate from the significant mismatch between the growth rates of Mo zigzag and S zigzag edge terminations. For example, in Ref. [58], the growth rate along the Mo edge is determined to be three times faster than that of the S edge as represented schematically in Fig. 4.7(d). This results in the growth of dendritic-like crystals which can be observed in a non-equilibrium growth condition, such as when high carrier gas flow rate is employed, leading to a greater precursor flux and an increased growth rate [58]. In addition, incomplete sulfurisation of MoO_3 may result in the presence of residual molybdenum oxides (MoO_{3-x}) [59] and molybdenum oxysulphides ($MoO_{3-x}S_y$; $0 \le x \le 3$; $0 \le y \le 2$) [60] which can affect the quality of the as-grown MoS_2. Metal halides such as $MoCl_5$ have also shown some success as a metal precursor for the growth of monolayer MoS_2 films [61], with the as-grown material showing good continuity and uniformity over large lateral scale (~5 cm). But it has poor electrical transport properties (electron mobility = 0.003–0.03 $cm^2\ V^{-1}\ s^{-1}$) likely due to a more defective film, and may not be ideal for the fabrication of electrical devices.

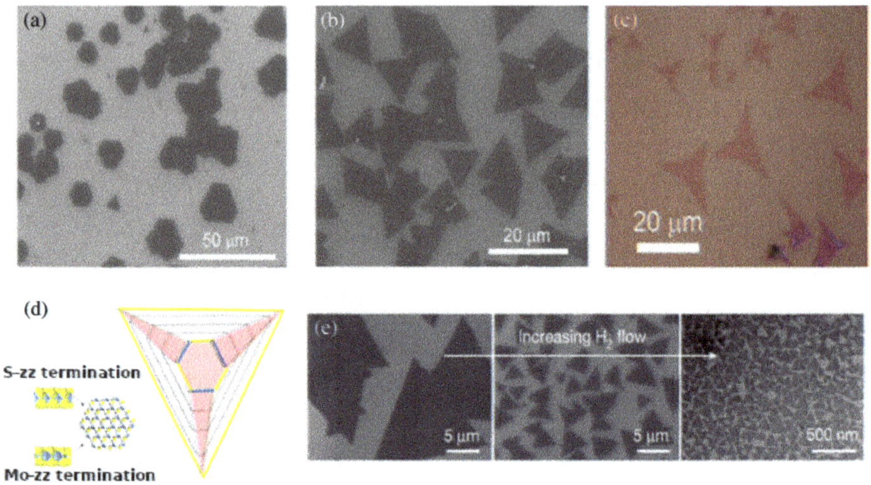

Fig. 4.7. (a)–(c) Evolution of domain shape of CVD-grown MoS_2 monolayer on SiO_2 in (a) low, (b) ideal and (c) high S concentration conditions. Figures adapted with permission from Ref. [56]. (d) Schematic illustration of different types of atomic zigzag (zz) termination in MoS_2 structure, i.e., S-zz (yellow line) and Mo-zz (blue line) with formation mechanism of the three-point star domain in high S concentration condition. Figure adapted with permission from Ref. [58]. (e) Domain size of MoS_2 monolayer crystals becoming smaller with increasing hydrogen flow rate during a MOCVD growth process on SiO_2 substrate. Figure adapted with permission from Ref. [62].

While solid precursors are easily available and have shown great success in growing micron-sized flakes, it is difficult to control the local precursor concentrations along the length of the growth substrate. As a result, gas-phase precursor growth methods have been developed to improve the uniformity and controllability of the morphology of the grown TMD layers. In this context, Kang *et al.* reported uniform growth of high-quality MoS_2 and WS_2 monolayer films on inch-scale quartz wafers by using the MOCVD method using $(C_2H_5)_2S$ as a gas-phase source for sulfur, and $Mo(CO)_6$ or $W(CO)_6$ as a gas-phase source for Mo or W [62]. The MOCVD growth was found to follow a layer-by-layer growth mechanism, which facilitated the growth of uniform monolayer MoS_2 and WS_2 film with a lateral grain size as large as 10 μm as shown in Fig. 4.7(e) when an optimal amount of hydrogen was added in the carrier gas [62]. MoS_2 transistors fabricated on the resulting 2D TMD films showed reasonably high electron mobility of 30 $cm^2\ V^{-1}\ s^{-1}$ at room temperature and 114 $cm^2\ V^{-1}\ s^{-1}$ at 90 K. While this work successfully demonstrated wafer-scale growth of high-quality TMD films by MOCVD method, the single crystal domains are small, and more efforts are still needed to increase the grain size of the films to enable high-performance device applications on single crystal flakes.

The use of seeding promoters on the supporting substrate prior to the growth process has also shown some success to increase the lateral growth area of TMD materials produced by CVD with solid metal and chalcogen precursors [63], [64]. For example Lee *et al.* [64] reported that by introducing graphene oxide, perylene-3,4,9,10-tetracarboxylic acid tetrapotassium salt (PTAS), or perylene-3,4,9,10-tetracarboxylic dianhydride coating on SiO_2 substrate, large-area (up to several mm of lateral size) MoS_2 layers can be grown at a relatively low temperature (650°C) in an atmospheric pressure CVD system. The density (or concentration) of the seed on the supporting substrate directly controls the TMD nucleation density and, consequently, their domain sizes before merging into a continuous polycrystalline film. The presence of seeding promoters on the surface may also alter and improve the MoS_2 surface adhesive force and thus reduce the nucleation's free energy barrier [63]. These two factors together facilitate the large-area layered growth of MoS_2 and lead to a decrease in the required growth temperature, making the growth conditions milder, allowing processes that have a lower thermal budget. The use of PTAS as a seed promoter is not only limited to MoS_2; it has also shown some success when used for the CVD synthesis of other TMD materials including WS_2 and $MoSe_2$ [47], [61].

4.3.3 Large-Scale Epitaxial Growth — Substrate Effect

Although the substrate material was initially thought to be arbitrary and played little role in the CVD growth of TMD materials (unlike MBE, where the lattice geometry of the substrate is of utmost importance), substrates were found to play an extremely important role in TMD growth — the lattice geometry of the substrate functions as the epitaxial template and is highly sensitive to the nanoscale morphology, where the terminating atoms, lattice constant, and symmetry of the growth substrates play an important role during CVD growth. To realise high-quality van der Waals epitaxy-aligned growth of TMDs, an atomically flat and chemically inert surface with hexagonal in-plane lattice is required.

The *c*-plane sapphire substrate was quickly found to be suitable for the synthesis of epitaxially grown TMD due to its close symmetry and compatible lattice constant with semiconducting TMD materials [65]. Dumcenco *et al.* [51] and Lim *et al.* [46] observed three basic alignment directions of MoS_2 domains on the sapphire surface with relative rotation angles of 0°, ±60° and ±30°, shown in Fig. 4.8(a). Similar results were also reported for WSe_2 and WS_2, which are commonly attributed to the preferred nucleation of the TMD along the periodic atomic step-terraces of the sapphire surface, thus resulting in the aligned growth [66].

Besides *c*-plane sapphire, other substrates with a hexagonal lattice symmetry and lattice constant match show similar functionality towards the growth of highly in-plane oriented TMD flakes. Unsurprisingly, Yu *et al.* [67] reported the epitaxial growth of monolayer MoS_2 on hBN. Only relative rotation angles of 0° and ±60° between MoS_2 and hBN basal plane were observed as shown in Fig. 4.8(b). Furthermore,

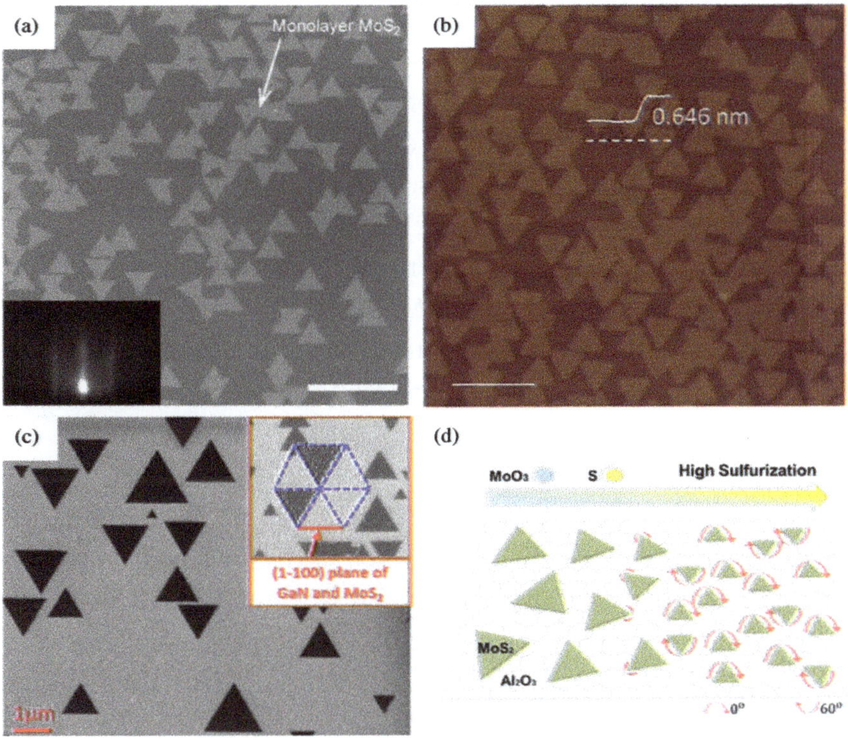

Fig. 4.8. (a) Optical image of monolayer MoS_2 flakes grown on atomically smooth sapphire exhibiting relative rotation angles of 0°, ±60° and ±30°; inset depicts the reflection high-energy electron diffraction pattern. Figure adapted with permission from Ref. [51]. (b) Atomic force microscopy image of monolayer MoS_2 domains grown on a smooth hBN surface; the height of domain was 0.646 nm, only 0° and 60° domains were observed on the hBN surface. Figure adapted with permission from Ref. [67]. (c) Scanning electron microscopy image of MoS_2 monolayer triangles on GaN epitaxial crystal. The scale bar is 1 μm and is aligned with the (1-100) plane of GaN. (Inset) Orientation of the (1100) plane of hexagonal crystals of GaN and MoS_2 is shown. Figure adapted with permission from Ref. [68]. (d) Schematic illustration of vapour concentration ratio-dependent growth direction. Figure adapted with permission from Ref. [72].

when the domains with same orientation stitched, they formed single crystals, while domains with the opposite orientation stitched and formed mirror twin boundaries. Hence, while highly desired, it is still a challenge to have all the domains align in the exact same direction. Other hexagonal substrates have also been shown to work; Ruzmetov *et al.* [68] utilised a semiconducting substrate GaN for orienting the growth of monolayer MoS_2 shown in Fig. 4.8(c) favoured by the close lattice match (<1% difference) with MoS_2 crystal. Other substrates, for example, graphene [69], [70] and mica [71] have also demonstrated a similar ability for controlling the growth direction of TMD monolayer during the CVD process.

Due to the complexity of the CVD process, random misaligned growth is frequently obtained even when lattice-matched supporting substrates are used. To fully understand the growth and nucleation mechanism, researchers have therefore investigated the impact of other factors concurring with substrate symmetry in establishing epitaxial growth condition during CVD. According to Aljarb *et al.* [72] and further confirmed by Suenaga *et al.* [73], the ratio of precursor concentration plays a key role in determining the monolayer MoS_2 orientation on a *c*-sapphire substrate. In particular, high S/MoO_3 ratio is needed in the early stage of growth to form small seeds that can align easily to the substrate lattice structure, while the ratio should be decreased to enlarge the size of the monolayer at the next stage to promote lateral growth as presented in Fig. 4.8(d). When compared with randomly oriented grains, the aligned MoS_2 domains showed weaker photoluminescence signal which is attributed to the strong interfacial coupling of the aligned flake to the substrate when compared to the randomly oriented ones.

While single crystals can be formed via growth originating from a single nucleation point (see next section), the multinucleation approach toward large single crystals might be more practical for scalable device fabrication where a nearly ideal domain alignment can be obtained through substrate engineering to reduce grain boundaries and defect formation. Future works may therefore focus on novel substrates for better controlling the growth direction of TMD. Substrates such as single crystals (different faces), amorphous metals (inert to chalcogen precursors), oxides, and nitrides with engineered surface morphologies offer many options to be explored.

4.3.4 Large-Scale Growth of TMD Single Crystal

While millimetre-sized TMD single crystals can be achieved by using CVD methods, merging the various TMD domains into continuous film, especially in the absence of domain alignment, may cause the formation of grain boundaries and defects in the as-grown film, with detrimental effects on its electronic properties. One approach to reduce grain boundaries and the related defects is reducing the nucleation density during the film growth which can result in the formation of large-sized TMD crystals.

Recently, a simple but effective approach [74] was introduced by using a solution-processed precursor deposition for growing large-scale and highly crystalline MoS_2 monolayers. The normally high nucleation density due to the uncontrolled precursor vapour pressure during CVD process was reduced by introducing an extremely small amount of MoO_3 (~0.01 mg) to suppress the supersaturation state of the vapour reactants. This method resulted in large MoS_2 single crystal growth (~0.5 mm) on SiO_2 substrate as shown in Fig. 4.9(a). In a different investigation [46], nickel oxide (NiO) foam was added to the CVD reactor as a chemical barrier. The nucleation density was lowered by the NiO barrier's reaction with MoO_3, which decreased the concentration of precursors reaching the substrate, obtaining MoS_2 grains with sizes of up to 170 μm and continuous monolayers at the centimetre length scale.

Ren *et al.* [75] reported the synthesis of millimetre-sized WS_2 single crystals on Au foils. They used density functional theory calculations to explain that the energy barrier for the sulfurisation of WO_3 by sulfur atoms diffusing on Au is lower than the direct sulfurisation of WO_3 in a sulfur vapour atmosphere, which leads to a large-area synthesis of WS_2 on Au. Moreover, the authors also synthesised millimetre-size WSe_2 domains on reusable Au substrates at a very high growth rate. Fig. 4.9(b) shows the typical image of an as-grown WSe_2 domain obtained with a growth time of only 10 s.

Chen *et al.* [76] reported the largest single crystal TMD domain to date, using molten glass as the substrate. The molten glass formed a

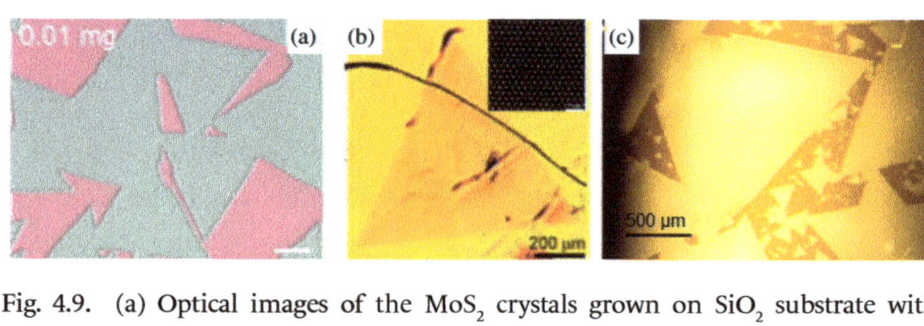

Fig. 4.9. (a) Optical images of the MoS_2 crystals grown on SiO_2 substrate with 0.01 mg of MoO_3 precursor. Scale bar is 100 μm. Figure adapted with permission from Ref. [74]. (b) An optical image of a typical WSe_2 single crystal triangle CVD deposited on gold foil. Inset: transmission electron microscopy image. Figure adapted with permission from Ref. [75]. (c) An optical image of synthesized WS_2 triangles on molten glass substrate. Figure adapted with permission from Ref. [76].

homogeneous and quasi-atomic smooth surface at elevated temperatures, reducing the nucleation sites for growing large $MoSe_2$ crystals monolayer up to 2.5 mm in lateral size as shown in Fig. 4.9(c).

Halide salts (e.g., NaCl and KCl) have also been shown to play a critical role in synthesising large-area TMD single crystals. The halide salts, when introduced on the supporting substrate, can react with Mo or W precursors, forming active gaseous oxyhalide species (e.g., $MoOCl_4$ and $WOCl_4$). Large TMD crystals including WS_2 (200 μm), WSe_2 (140 μm), $MoTe_2$ (1 mm), and WTe_2 (350 μm) have been achieved based on this halide salt-assisted growth method [77], [78].

Strategies for wafer-scale growth of 2D single crystal have also been reported recently. Wang and co-workers [79] reported the growth of wafer-scale 2D TMD single crystals on 2-inch-size sapphires. The miscut orientation towards the *a*-axis of *c*-plane sapphire creates a set of unique step edges, which break formation energy degeneracy for the antiparallel domains and result in a >99% unidirectional alignment of domains for single crystal film growth, as schematically shown in Fig. 4.10(a). Moreover, Liu *et al.* [80] reported another growth strategy of epitaxial growth of wafer-scale single-crystal WS_2 monolayers on vicinal *a*-plane sapphire shown in Fig. 4.10(b). The cooperative interaction of sapphire plane atomic structure and step edges to WS_2 breaks the symmetry of the antiparallel orientations and eventually results in unidirectional alignment of nearly all the WS_2 islands.

The report of a single crystal 2D TMD up to 2-inch wafer size synthesised through CVD [80] presents a promising outlook. While progress in the field is evolving rapidly, the TMD lateral sizes remain well below those of their CVD-grown graphene counterpart. More efforts

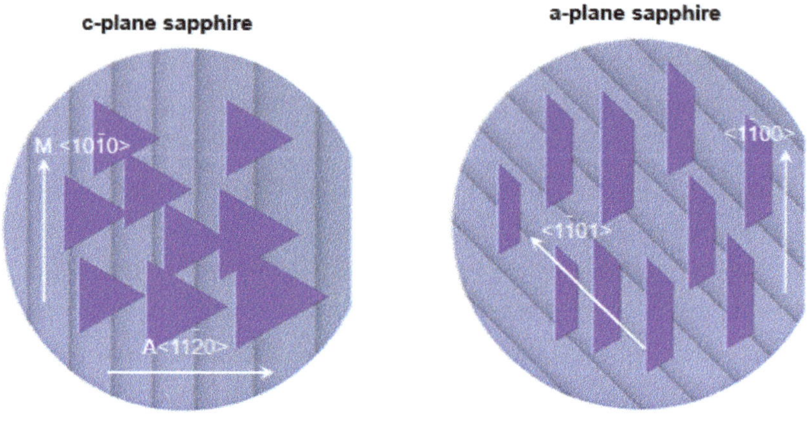

(a) (b)

Fig. 4.10. Strategies for wafer-scale 2D single crystal growth exploiting atomic steps on different crystal facets of the sapphire substrate. (a) Schematic of domain alignment on *c*-plane sapphire (0001) wafer; (b) Schematic of the domain orientation on vicinal *a*-plane sapphire.

are therefore needed to develop novel approaches for further increasing the growth rate or decreasing the nucleation density to attain at least 6-inch wafer-scale size of 2D TMD single crystals.

4.3.5 2D Heterostructures

One-dimensional (1D) contacts [81], tunnelling FETs [82], and p-n junction [83] for next-generation electronics and optoelectronics are just a few examples of the novel devices that can be built using TMD heterostructures in a variety of combinations and energy band alignments. While a physical pick-up-and-transfer method can be used to create heterostructures [82], [84], CVD offers a different approach through the direct growth of lattice-aligned TMDs layers for both vertical heterojunctions with clean interfaces [85], [86] and seamless in-plane lateral heterostructures with atomically sharp interfaces [87], [88].

Recently, Gong *et al.* [85] reported on the CVD preparation of vertical WS_2/MoS_2 heterostructures. Using a single growth process, MoS_2 and WS_2 monolayers were obtained successively due to the distinct evaporation temperature of MoO_3 and WO_3 powders. Fig. 4.11(a) presents a typical optical image of the synthesised vertical WS_2/MoS_2 heterostructure, showing bilayer triangles with a clear optical contrast of the second WS_2 layer grown on top of the MoS_2. Photoluminescence

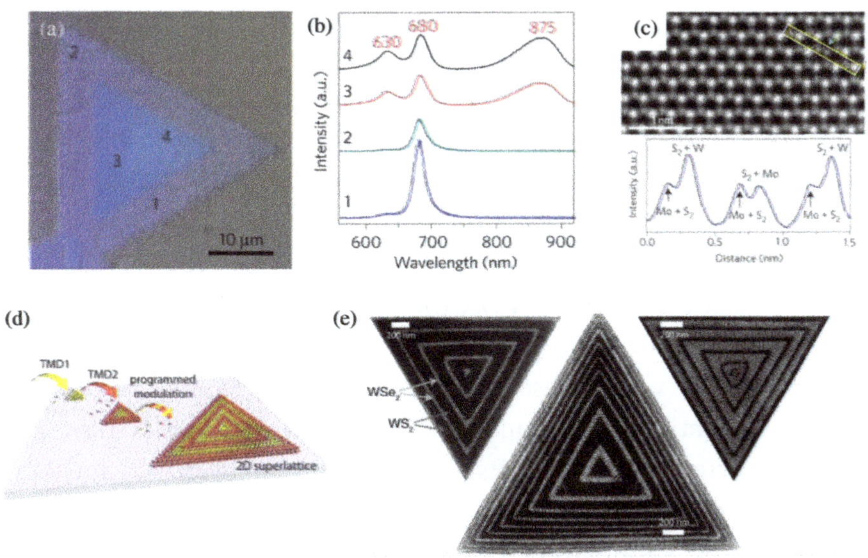

Fig. 4.11. (a) Optical microscopy image of an as-grown WS_2/MoS_2 vertical heterostructure. (b) Photoluminescence data from different regions. (c) A scanning transmission electron microscopy image of the stacked WS_2/MoS_2 heterostructure and the corresponding intensity profile. Figures adapted with permission from Ref. [85]. (d) Schematic view of the modulated growth of a TMD lateral superlattice. (e) Scanning electron microscopy images of the WS_2-WSe_2 monolayer superlattices. Scale bars: 200 nm. Figures adapted with permission from Ref. [90].

data supports the presence of strong interlayer coupling marked by the presence of the additional emission line (850 nm) not observed in the individual layers as shown in Fig. 4.11(b). The coupling reflects the epitaxial growth at the interface between TMD layers in the stacked WS_2/MoS_2 heterostructure resulting in a 2H-stacking structure shown in Fig. 4.11(c). Besides WS_2/MoS_2 heterostructures, high-quality vertical WSe_2/SnS_2 heterostructures were also synthesised using a similar CVD-based method [86].

By using a seed precursor-assisted method, Ling *et al.* demonstrated that a variety of large-area 2D heterostructures (e.g., MoS_2/WS_2, MoS_2/graphene, MoS_2/hBN) can be fabricated regardless of the lattice mismatch between the materials. The as-grown heterostructures shows an in-plane "parallel stitched" configuration at the junction [89]. Complex multi-heterostructures (e.g., WS_2-$MoSe_2$-WSe_2) and superlattices (e.g., WS_2-WSe_2-WS_2-WSe_2-WS_2) were also realised through edge-epitaxy growth [90], [91]. As schematically presented in Fig. 4.11(d), the spatial distribution of lateral WS_2-WSe_2 superlattices can be switched in a MOCVD experimental setup by simply changing the precursors during the growth process. The concentration of each precursor was precisely controlled, which finally induced the direct modulation of the supercell dimension. Fig. 4.11(e) shows that dark and bright regions corresponding to monolayer WS_2 and WSe_2 are clearly observed in the scanning electron micrograph, with the as-grown WSe_2 monolayers sharing the same crystal orientation and a seamless interface with atomically sharp bonding.

4.4 Summary

In summary, we have discussed the more commonly used processes and strategies for the preparation of van der Waals layered 2D materials. An overview of 2D materials preparation metrics of various fabrication techniques is summarised in Table 4.1. Mechanical exfoliation enables the isolation of monolayer or few-layer 2D flakes with high quality, inherently from the bulk crystals. While the stacking transfer technique can facilitate the fabrication of heterolayers, it is challenging to scale this process up to wafer scale as it is difficult to control the thickness, domain size and orientation of the layers during exfoliation. But it remains an important method for most 2D materials-related discoveries by producing various 2D material stacks with the highest quality for fundamental physics studies. While the chemical exfoliation method shows good potential to produce monolayer or few-layer 2D flakes dispersed in solutions in a large-scale process, its usage is limited due to the broad distribution of layer thickness, small flake sizes, low monolayer yield, and poor quality. Therefore, this method is not suitable for device fabrication.

Table 4.1. Comparison of various 2D materials preparation techniques. The main techniques are classified as "top-down" and "bottom-up" as discussed in depth in this chapter.

Preparation \ Property		Material Quality		Engineering Control		
		Defect density	**Domain size**	**Thickness control**	**Scale-up**	**Heterostacking**
Top-down	Mechanical exfoliation	Low	~10 μm Ref. [92]	Low	Low	High
	Liquid exfoliation	High	45 nm–5 μm Refs. [93], [94]	Medium	High	Low
Bottom-up	CVD	Low to medium	1–1000 μm Refs. [75], [76]	Medium	High	Medium
	ALD	Low to medium	100 nm Refs. [1], [2]	High	High	High
	MBE	Low	10–100 nm (epitaxial) Refs. [3], [4]	High	High	Medium
	PVD	High	<10–100 nm Refs. [5], [6]	High	High	Low

CVD has proven to be a cost-effective and relatively controllable method for producing large-area 2D materials of sufficiently high quality for devices in a scalable manner. Currently, single crystal TMDs produced by this method tend to be small and cannot fulfil the requirement for meaningful device engineering. But this field is progressing rapidly, and specific examples of breakthrough have already been demonstrated. For example, the strategy of exploiting the atomic surface of epitaxial substrates to align all domains in a single direction provides a promising opportunity for growing wafer-scale TMD single crystals. From a theoretical standpoint, it is possible to optimise the epitaxial substrate to break the symmetry of the antiparallel orientations in the growth to result in unidirectional alignment of all the growth domains, which can eventually stitch seamlessly to grow grain boundary-free single-crystal TMD films. However, it is necessary to gain more in-depth understanding of the epitaxial mechanism, and the surface growth kinetics need to be further investigated, to obtain both large-area consistency and wafer-to-wafer reproducibility of growth.

References

[1] B. Groven *et al.*, "Two-dimensional crystal grain size tuning in WS2 atomic layer deposition: an insight in the nucleation mechanism," *Chem. Mater.*, vol. 30, no. 21, pp. 7648–7663, 2018, doi: 10.1021/acs.chemmater.8b02924.

[2] W. Hao, C. Marichy, and C. Journet, "Atomic layer deposition of stable 2D materials," *2D Mater.*, vol. 6, no. 1, p. 012001, 2018, doi: 10.1088/2053-1583/aad94f.

[3] W. Fu *et al.*, "Room temperature commensurate charge density wave on epitaxially grown bilayer 2H-tantalum sulfide on hexagonal boron nitride," *ACS Nano*, vol. 14, no. 4, pp. 3917–3926, 2020, doi: 10.1021/acsnano.0c00303.

[4] X. Zhao *et al.*, "Engineering covalently bonded 2D layered materials by self-intercalation," *Nature*, vol. 581, no. 7807, pp. 171–177, 2020, doi: 10.1038/s41586-020-2241-9.
[5] C. Muratore *et al.*, "Continuous ultra-thin MoS_2 films grown by low-temperature physical vapor deposition," *Appl. Phys. Lett.*, vol. 104, no. 26, p. 261604, 2014, doi: 10.1063/1.4885391.
[6] C. Muratore, A. A. Voevodin, and N. R. Glavin, "Physical vapor deposition of 2D van der Waals materials: a review," *Thin Solid Films*, vol. 688, p. 137500, 2019, doi: 10.1016/j.tsf.2019.137500.
[7] L. Li, R. Long, and O. V. Prezhdo, "Why chemical vapor deposition grown MoS_2 samples outperform physical vapor deposition samples: time-domain ab initio analysis," *Nano Lett.*, vol. 18, no. 6, pp. 4008–4014, 2018, doi: 10.1021/acs.nanolett.8b01501.
[8] S. H. Choi *et al.*, "Large-scale synthesis of graphene and other 2D materials towards industrialization," *Nat. Commun.*, vol. 13, no. 1, pp. 1–5, 2022, doi: 10.1038/s41467-022-29182-y.
[9] K. S. Novoselov *et al.*, "Electric field in atomically thin carbon films," *Science*, vol. 306, no. 5696, pp. 666–669, 2004, doi: 10.1126/science.1102896.
[10] K. S. Novoselov *et al.*, "Two-dimensional atomic crystals," *Proc. Natl. Acad. Sci.*, vol. 102, no. 30, pp. 10451–10453, 2005, doi: 10.1073/pnas.0502848102.
[11] K. Zhang, Y. Feng, F. Wang, Z. Yang, and J. Wang, "Two dimensional hexagonal boron nitride (2D-hBN): Synthesis, properties and applications," *J. Mater. Chem. C*, vol. 5, no. 46, pp. 11992–12022, 2017, doi: 10.1039/c7tc04300g.
[12] K. S. Novoselov and A. H. Castro Neto, "Two-dimensional crystals-based heterostructures: Materials with tailored properties," *Phys. Scr.*, no. T146, 2012, doi: 10.1088/0031-8949/2012/T146/014006.
[13] S. Z. Butler *et al.*, "Progress, challenges, and opportunities in two-dimensional materials beyond graphene," *ACS Nano*, vol. 7, no. 4, pp. 2898–2926, 2013, doi: 10.1021/nn400280c.
[14] K. F. Mak, C. Lee, J. Hone, J. Shan, and T. F. Heinz, "Atomically thin MoS_2: A new direct-gap semiconductor," *Phys. Rev. Lett.*, vol. 105, no. 13, pp. 2–5, 2010, doi: 10.1103/PhysRevLett.105.136805.

[15] G. Plechinger, T. Korn, and J. M. Lupton, "Valley-polarized exciton dynamics in exfoliated monolayer WSe_2," *J. Phys. Chem. C*, vol. 121, no. 11, pp. 6409–6413, 2017, doi: 10.1021/acs.jpcc.7b01468.

[16] C. R. Zhu *et al.*, "Exciton valley dynamics probed by Kerr rotation in WSe_2 monolayers," *Phys. Rev. B*, vol. 90, no. 16, p. 161302, 2014, doi: 10.1103/PhysRevB.90.161302.

[17] S.-Y. Xu *et al.*, "Electrically switchable Berry curvature dipole in the monolayer topological insulator WTe_2," *Nat. Phys.*, vol. 14, no. 9, pp. 900–906, 2018, doi: 10.1038/s41567-018-0189-6.

[18] V. Fatemi *et al.*, "Electrically tunable low-density superconductivity in a monolayer topological insulator," *Science*, vol. 362, no. 6417, pp. 926–929, 2018, doi: 10.1126/science.aar4642.

[19] S. C. de la Barrera *et al.*, "Tuning Ising superconductivity with layer and spin–orbit coupling in two-dimensional transition-metal dichalcogenides," *Nat. Commun.*, vol. 9, no. 1, p. 1427, 2018, doi: 10.1038/s41467-018-03888-4.

[20] Y. Cao *et al.*, "Quality heterostructures from two-dimensional crystals unstable in air by their assembly in inert atmosphere," *Nano Lett.*, vol. 15, no. 8, pp. 4914–4921, 2015, doi: 10.1021/acs.nanolett.5b00648.

[21] W. Li *et al.*, "Gas sensors based on mechanically exfoliated MoS_2 nanosheets for room-temperature NO_2 detection," *Sensors (Basel)*, vol. 19, no. 9, pp. 1–12, 2019, doi: 10.3390/s19092123.

[22] X. Zhang *et al.*, "Advanced tape-exfoliated method for preparing large-area 2D monolayers: A review," *2D Mater.*, vol. 8, no. 3, 2021, doi: 10.1088/2053-1583/ac016f.

[23] M. C. F. Da Costa, H. B. Ribeiro, F. Kessler, E. A. T. De Souza, and G. J. M. Fechine, "Micromechanical exfoliation of two-dimensional materials by a polymeric stamp," *Mater. Res. Express*, vol. 3, no. 2, p. 025303, 2016, doi: 10.1088/2053-1591/3/2/025303.

[24] F. Liu, "Mechanical exfoliation of large area 2D materials from vdW crystals," *Prog. Surf. Sci.*, vol. 96, no. 2, p. 100626, 2021, doi: 10.1016/j.progsurf.2021.100626.

[25] S. Fan, Q. A. Vu, M. D. Tran, S. Adhikari, and Y. H. Lee, "Transfer assembly for two-dimensional van der Waals heterostructures," *2D Mater.*, vol. 7, no. 2, p. 022005, 2020, doi: 10.1088/2053-1583/ab7629.

[26] P. Blake *et al.*, "Making graphene visible," *Appl. Phys. Lett.*, vol. 91, no. 6, p. 063124, 2007, doi: 10.1063/1.2768624.

[27] X. Zhang, H. Kawai, J. Yang, and K. E. J. Goh, "Detecting MoS_2 and $MoSe_2$ with optical contrast simulation," *Prog. Nat. Sci. Mater. Int.*, vol. 29, pp. 667–671, 2019, doi: 10.1016/j.pnsc.2019.11.003.

[28] A. Castellanos-Gomez *et al.*, "Deterministic transfer of two-dimensional materials by all-dry viscoelastic stamping," *2D Mater.*, vol. 1, no. 1, 2014, doi: 10.1088/2053-1583/1/1/011002.

[29] K. Buapan, R. Somphonsane, T. Chiawchan, and H. Ramamoorthy, "Versatile, low-cost, and portable 2D material transfer setup with a facile and highly efficient DIY inert-atmosphere glove compartment option," *ACS Omega*, vol. 6, no. 28, pp. 17952–17964, 2021, doi: 10.1021/acsomega.1c01582.

[30] S. G. Martanov *et al.*, "Making van der Waals heterostructures assembly accessible to everyone," *Nanomaterials*, vol. 10, no. 11, pp. 1–12, 2020, doi: 10.3390/nano10112305.

[31] C. R. Dean *et al.*, "Boron nitride substrates for high-quality graphene electronics," *Nat. Nanotechnol.*, vol. 5, no. 10, pp. 722–726, 2010, doi: 10.1038/nnano.2010.172.

[32] A. Reina *et al.*, "Large area, few-layer graphene films on arbitrary substrates by chemical vapor deposition," *Nano Lett.*, vol. 9, no. 1, pp. 30–35, 2009, doi: 10.1021/nl801827v.

[33] C. R. Dean *et al.*, "Multicomponent fractional quantum Hall effect in graphene," *Nat. Phys.*, vol. 7, no. 9, pp. 693–696, 2011, doi: 10.1038/nphys2007.

[34] S. Bertolazzi, J. Brivio, and A. Kis, "Stretching and breaking of ultrathin MoS_2," *ACS Nano*, vol. 5, no. 12, pp. 9703–9709, 2011, doi: 10.1021/nn203879f.

[35] A. V. Kretinin *et al.*, "Electronic properties of graphene encapsulated with different two-dimensional atomic crystals," *Nano Lett.*, vol. 14, no. 6, pp. 3270–3276, 2014, doi: 10.1021/nl5006542.

[36] M. A. Meitl *et al.*, "Transfer printing by kinetic control of adhesion to an elastomeric stamp," *Nat. Mater.*, vol. 5, no. 1, pp. 33–38, 2006, doi: 10.1038/nmat1532.

[37] A. Jain *et al.*, "Minimizing residues and strain in 2D materials transferred from PDMS," *Nanotechnology*, vol. 29, no. 26, p. 265203, 2018, doi: 10.1088/1361-6528/aabd90.

[38] F. Pizzocchero *et al.*, "The hot pick-up technique for batch assembly of van der Waals heterostructures," *Nat. Commun.*, vol. 7, no. May, p. 11894, 2016, doi: 10.1038/ncomms11894.

[39] L. Wang *et al.*, "One-dimensional electrical contact to a two-dimensional material," *Science*, vol. 342, no. 6158, pp. 614–617, 2013, doi: 10.1126/science.1244358.

[40] N. Lindvall, A. Kalabukhov, and A. Yurgens, "Cleaning graphene using atomic force microscope," *J. Appl. Phys.*, vol. 111, no. 6, p. 064904, 2012, doi: 10.1063/1.3695451.

[41] P. J. Zomer, M. H. D. Guimarães, J. C. Brant, N. Tombros, and B. J. Van Wees, "Fast pick up technique for high quality heterostructures of bilayer graphene and hexagonal boron nitride," *Appl. Phys. Lett.*, vol. 105, no. 1, p. 013101, 2014, doi: 10.1063/1.4886096.

[42] D. G. Purdie, N. M. Pugno, T. Taniguchi, K. Watanabe, A. C. Ferrari, and A. Lombardo, "Cleaning interfaces in layered materials heterostructures," *Nat. Commun.*, vol. 9, no. 1, pp. 1–12, 2018, doi: 10.1038/s41467-018-07558-3.

[43] J. I. J. Wang *et al.*, "Electronic transport of encapsulated graphene and WSe_2 devices fabricated by pick-up of prepatterned hBN," *Nano Lett.*, vol. 15, no. 3, pp. 1898–1903, 2015, doi: 10.1021/nl504750f.

[44] H. Li, Y. Li, A. Aljarb, Y. Shi, and L.-J. Li, "Epitaxial growth of two-dimensional layered transition-metal dichalcogenides: growth mechanism, controllability, and scalability," *Chem. Rev.*, vol. 118, no. 13, pp. 6134–6150, 2018, doi: 10.1021/acs.chemrev.7b00212.

[45] B. Liu, M. Fathi, L. Chen, A. Abbas, Y. Ma, and C. Zhou, "Chemical vapor deposition growth of monolayer WSe_2 with tunable device characteristics and growth mechanism study," *ACS Nano*, vol. 9, no. 6, pp. 6119–6127, 2015, doi: 10.1021/acsnano.5b01301.

[46] Y. F. Lim *et al.*, "Modification of vapor phase concentrations in MoS_2: Growth using a NiO foam barrier," *ACS Nano*, vol. 12, no. 2, pp. 1339–1349, 2018, doi: 10.1021/acsnano.7b07682.

[47] K. M. McCreary, A. T. Hanbicki, G. G. Jernigan, J. C. Culbertson, and B. T. Jonker, "Synthesis of large-area WS_2 monolayers with exceptional photoluminescence," *Sci. Rep.*, vol. 6, p. 19159, 2016, doi: 10.1038/srep19159.
[48] X. Wang *et al.*, "Chemical vapor deposition growth of crystalline monolayer $MoSe_2$," *ACS Nano*, vol. 8, no. 5, pp. 5125–5131, 2014, doi: 10.1021/nn501175k.
[49] Y. Zhan, Z. Liu, S. Najmaei, P. M. Ajayan, and J. Lou, "Large-area vapor-phase growth and characterization of MoS_2 atomic layers on a SiO_2 substrate," *Small*, vol. 8, no. 7, pp. 966–971, 2012, doi: 10.1002/smll.201102654.
[50] S. Ganorkar, J. Kim, Y.-H. Kim, and S.-I. Kim, "Effect of precursor on growth and morphology of MoS_2 monolayer and multilayer," *J. Phys. Chem. Solids*, vol. 87, pp. 32–37, 2015, doi: 10.1016/j.jpcs.2015.07.016.
[51] D. Dumcenco *et al.*, "Large-area epitaxial monolayer MoS_2," *ACS Nano*, vol. 9, p. 4611, 2015, doi: 10.1021/acsnano.5b01281.
[52] K. K. Liu *et al.*, "Growth of large-area and highly crystalline MoS_2 thin layers on insulating substrates," *Nano Lett.*, vol. 12, p. 1538, 2012, doi: 10.1021/nl2043612.
[53] A. M. Van Der Zande *et al.*, "Grains and grain boundaries in highly crystalline monolayer molybdenum disulphide," *Nat. Mater.*, vol. 12, no. 6, pp. 554–561, 2013, doi: 10.1038/nmat3633.
[54] D. Xiao, G. Bin Liu, W. Feng, X. Xu, and W. Yao, "Coupled spin and valley physics in monolayers of MoS_2 and other group-VI dichalcogenides," *Phys. Rev. Lett.*, vol. 108, no. 19, p. 196802, 2012, doi: 10.1103/PhysRevLett.108.196802.
[55] J. V. Pondick, J. M. Woods, J. Xing, Y. Zhou, and J. J. Cha, "Stepwise sulfurization from MoO_3 to MoS_2 via chemical vapor deposition," *ACS Appl. Nano Mater.*, vol. 1, no. 10, pp. 5655–5661, 2018, doi: 10.1021/acsanm.8b01266.
[56] S. Najmaei *et al.*, "Vapour phase growth and grain boundary structure of molybdenum disulphide atomic layers," *Nat. Mater.*, vol. 12, p. 754, 2013, doi: 10.1038/nmat3673.
[57] V. Senthilkumar, L. C. Tam, Y. S. Kim, Y. Sim, M.-J. Seong, and J. I. Jang, "Direct vapor phase growth process and robust photoluminescence

properties of large area MoS_2 layers," *Nano Res.*, vol. 7, p. 1759, 2014, doi: 10.1007/s12274-014-0535-7.

[58] S. S. Wang *et al.*, "Shape evolution of monolayer MoS_2 crystals grown by chemical vapor deposition," *Chem. Mater.*, vol. 26, p. 6371, 2014, doi: 10.1021/cm5025662.

[59] X. L. Li and Y. D. Li, "Formation of MoS_2 inorganic fullerenes (IFs) by the reaction of MoO_3 nanobelts and S," *Chem. Eur. J.*, vol. 9, no. 12, pp. 2726–2731, 2003, 10.1002/chem.200204635.

[60] T. Weber, J. C. Muijsers, J. Van Wolput, C. P. J. Verhagen, and J. W. Niemantsverdriet, "Basic reaction steps in the sulfidation of crystalline MoO_3 to MoS_2, as studied by X-ray photoelectron and infrared emission spectroscopy," *J. Phys. Chem.*, vol. 100, no. 33, pp. 14144–14150, 1996, doi: 10.1021/jp961204y.

[61] Y. Yu, C. Li, Y. Liu, L. Su, Y. Zhang, and L. Cao, "Controlled scalable synthesis of uniform, high-quality monolayer and few-layer MoS_2 films," *Sci. Rep.*, vol. 3, pp. 1–6, 2013, doi: 10.1038/srep01866.

[62] K. Kang *et al.*, "High-mobility three-atom-thick semiconducting films with wafer-scale homogeneity," *Nature*, vol. 520, p. 656, 2015, doi: 10.1038/nature14417.

[63] X. Ling *et al.*, "Role of the seeding promoter in MoS_2 growth by chemical vapor deposition," *Nano Lett.*, vol. 14, no. 2, pp. 464–472, 2014, doi: 10.1021/nl4033704.

[64] Y. Lee *et al.*, "Synthesis of large-area MoS_2 atomic layers with chemical vapor deposition," *Adv. Mater.*, vol. 24, no. 17, pp. 2320–2325, 2012, doi: 10.1002/adma.201104798.

[65] Z. Q. Xu *et al.*, "Synthesis and transfer of large-area monolayer WS_2 crystals: Moving toward the recyclable use of sapphire substrates," *ACS Nano*, vol. 9, p. 6178, 2015, doi: 10.1021/acsnano.5b01480.

[66] L. Chen, B. L. Liu, M. Y. Ge, Y. Q. Ma, A. N. Abbas, and C. W. Zhou, "Step-edge-guided nucleation and growth of aligned WSe_2 on sapphire via a layer-over-layer growth mode," *ACS Nano*, vol. 9, p. 8368, 2015, doi: 10.1021/acsnano.5b03043.

[67] H. Yu *et al.*, "Precisely aligned monolayer MoS_2 epitaxially grown on h-BN basal plane," *Small*, vol. 13, no. 7, p. 1603005, 2017, doi: 10.1002/smll.201603005.

[68] D. Ruzmetov *et al.*, "Vertical 2D/3D semiconductor heterostructures based on epitaxial molybdenum disulfide and gallium nitride," *ACS Nano*, vol. 10, no. 3, pp. 3580–3588, 2016, doi: 10.1021/acsnano.5b08008.
[69] K. M. McCreary *et al.*, "Large-area synthesis of continuous and uniform MoS_2 monolayer films on graphene," *Adv. Funct. Mater.*, vol. 24, no. 41, pp. 6449–6454, 2014, doi: 10.1002/adfm.201401511.
[70] G. Piccinini *et al.*, "Deterministic direct growth of WS_2 on CVD graphene arrays," *2D Mater.*, vol. 7, no. 1, p. 14002, 2019, doi: 10.1088/2053-1583/ab49f0.
[71] Q. Ji *et al.*, "Epitaxial monolayer MoS_2 on Mica with novel photoluminescence," *Nano Lett.*, vol. 13, p. 3870, 2013, doi: 10.1021/nl401938t.
[72] A. Aljarb *et al.*, "Substrate lattice-guided seed formation controls the orientation of 2D transition-metal dichalcogenides," *ACS Nano*, vol. 11, no. 9, pp. 9215–9222, 2017, doi: 10.1021/acsnano.7b04323.
[73] K. Suenaga *et al.*, "Surface-mediated aligned growth of monolayer MoS_2 and in-plane heterostructures with graphene on sapphire," *ACS Nano*, vol. 12, no. 10, pp. 10032–10044, 2018, doi: 10.1021/acsnano.8b04612.
[74] J. Lee *et al.*, "Thermodynamically stable synthesis of large-scale and highly crystalline transition metal dichalcogenide monolayers and their unipolar n-n heterojunction devices," *Adv. Mater.*, vol. 29, no. 33, p. 1702206, 2017, doi: 10.1002/adma.201702206.
[75] Y. Gao *et al.*, "Large-area synthesis of high-quality and uniform monolayer WS_2 on reusable Au foils," *Nat. Commun.*, vol. 6, p. 8569, 2015, doi: 10.1038/ncomms9569.
[76] J. Chen *et al.*, "Chemical vapor deposition of large-size monolayer $MoSe_2$ crystals on molten glass," *J. Am. Chem. Soc.*, vol. 139, p. 1073, 2017, doi: 10.1021/jacs.6b12156.
[77] S. Li *et al.*, "Halide-assisted atmospheric pressure growth of large WSe_2 and WS_2 monolayer crystals," *Appl. Mater. Today*, vol. 1, p. 60, 2015, doi: 10.1016/j.apmt.2015.09.001.
[78] K. Chen *et al.*, "A simple method for synthesis of high-quality millimeter-scale 1T′ transition-metal telluride and near-field

nanooptical properties," *Adv. Mater.*, vol. 29, no. 38, p. 1700704, 2017, doi: 10.1002/adma.201700704.

[79] T. Li *et al.*, "Epitaxial growth of wafer-scale molybdenum disulfide semiconductor single crystals on sapphire," *Nat. Nanotechnol.*, vol. 16, no. 11, pp. 1201–1207, 2021, doi: 10.1038/s41565-021-00963-8.

[80] J. Wang *et al.*, "Dual-coupling-guided epitaxial growth of wafer-scale single-crystal WS_2 monolayer on vicinal a-plane sapphire," *Nat. Nanotechnol.*, vol. 17, no. 1, pp. 33–38, 2022, doi: 10.1038/s41565-021-01004-0.

[81] L. Wang *et al.*, "One-dimensional electrical contact to a two-dimensional material," *Science*, vol. 342, no. 6158, pp. 614–617, 2013, doi: 10.1126/science.1244358.

[82] T. Roy *et al.*, "Dual-gated MoS_2/WSe_2 van der Waals tunnel diodes and transistors," *ACS Nano*, vol. 9, p. 2071, 2015, doi: 10.1021/nn507278b.

[83] J. S. Ross *et al.*, "Interlayer exciton optoelectronics in a 2D heterostructure p–n junction," *Nano Lett.*, vol. 17, no. 2, pp. 638–643, 2017, doi: 10.1021/acs.nanolett.6b03398.

[84] A. F. Rigosi, H. M. Hill, Y. Li, A. Chernikov, and T. F. Heinz, "Probing interlayer interactions in transition metal dichalcogenide heterostructures by optical spectroscopy: MoS_2/WS_2 and $MoSe_2$/WSe_2," *Nano Lett.*, vol. 15, no. 8, pp. 5033–5038, 2015, doi: 10.1021/acs.nanolett.5b01055.

[85] Y. Gong *et al.*, "Vertical and in-plane heterostructures from WS_2/MoS_2 monolayers," *Nat. Mater.*, vol. 13, p. 1135, 2014, doi: 10.1038/nmat4091.

[86] T. Yang *et al.*, "Van der Waals epitaxial growth and optoelectronics of large-scale WSe_2/SnS_2 vertical bilayer p–n junctions," *Nat. Commun.*, vol. 8, no. 1, pp. 1–9, 2017, doi: 10.1038/s41467-017-02093-z.

[87] X.-Q. Zhang, C.-H. Lin, Y.-W. Tseng, K.-H. Huang, and Y.-H. Lee, "Synthesis of lateral heterostructures of semiconducting atomic layers," *Nano Lett.*, vol. 15, no. 1, pp. 410–415, 2015, doi: 10.1021/nl503744f.

[88] H. Qi *et al.*, "Production methods of van der Waals heterostructures based on transition metal dichalcogenides," *Crystals*, vol. 8, no. 1, p. 35, 2018, doi: 10.3390/cryst8010035.

[89] X. Ling *et al.*, "Parallel stitching of 2D materials," *Adv. Mater.*, vol. 28, p. 2322, 2016, doi: 10.1002/adma.201505070.

[90] S. Xie *et al.*, "Coherent, atomically thin transition-metal dichalcogenide superlattices with engineered strain," *Science*, vol. 359, no. 6380, pp. 1131–1136, 2018, doi: 10.1126/science.aao5360.
[91] Z. Zhang, P. Chen, X. Duan, K. Zang, J. Luo, and X. Duan, "Robust epitaxial growth of two-dimensional heterostructures, multiheterostructures, and superlattices," *Science*, vol. 357, p. 788, 2017, doi: 10.1126/science.aan6814.
[92] H. Li, J. Wu, Z. Yin, and H. Zhang, "Preparation and applications of mechanically exfoliated single-layer and multilayer MoS_2 and WSe_2 nanosheets," *Acc. Chem. Res.*, vol. 47, no. 4, pp. 1067–1075, 2014, doi: 10.1021/ar4002312.
[93] V. Nicolosi, M. Chhowalla, M. G. Kanatzidis, M. S. Strano, and J. N. Coleman, "Liquid exfoliation of layered materials," *Science*, vol. 340, no. 6139, p. 1226419, 2013, doi: 10.1126/science.1226419.
[94] C. X. Huo, Z. Yan, X. F. Song, and H. B. Zeng, "2D materials via liquid exfoliation: a review on fabrication and applications," *Sci. Bull.*, vol. 60, no. 23, pp. 1994-2008, 2015, doi: 10.1007/s11434-015-0936-3.

https://doi.org/10.1142/9789811229107_0005

Chapter 5

Valley Current Generation and Detection

Qian Ling Kee, Calvin Pei Yu Wong, Ivan Verzhbitskiy, and Kuan Eng Johnson Goh

5.1 Introduction

As discussed in Chapter 2, the valley degree of freedom (DOF) arises from the presence of two inequivalent valley states in the band structure of the material, and it is the associated non-zero Berry curvature which ultimately provides the valley contrast that allows for optical selectivity via helicity and valley Hall effect (VHE) in the absence of an external magnetic field. Although most of the current work focuses on transition metal dichalcogenides (TMDs) due to their more favourable properties for valleytronics, the revival of valleytronics based on 2D materials first started with graphene.

Graphene's unit cell consists of two inequivalent carbon atoms by virtue of two different lattice vectors that result in contrasting valleys in the graphene band structure, namely the K and K′ valleys, which are time-reversed images of each other. Therefore, the six Dirac points in the first Brillouin zone of graphene (K and K′) have valleys that are momentum inequivalent but energy degenerate. Thus, while graphene was first recognised for its potential for valleytronics [1], [2], it was non-trivial to engineer an energy gap between the valleys in graphene. To break this energy degeneracy, one may introduce inversion symmetry breaking in the crystal lattices. For graphene, this can be realised by utilising a substrate effect, where graphene/hBN Moiré superlattices are used to induce local Berry curvature potentials [3], or by applying an out-of-plane electric field to bilayer graphene [4], [5]. These inversion symmetry-breaking fields also result in bandgap opening, which provides an important knob to tune the Berry curvature of the material. Specifically, it was found that VHE can be tuned in bilayer graphene due to the inverse relationship between Berry curvature and bandgap [6].

On the other hand, monolayer TMDs have an intrinsic broken inversion symmetry as shown in Fig. 5.1(a) and (b). The typical unit cell of bulk 2H-MoS_2 shown in Fig. 5.1(a) has an inversion centre between the layers but no inversion centre within the MoS_2 layer itself due to its trigonal prismatic geometry. The intrinsically broken inversion symmetry removes the need for external mechanisms to break inversion symmetry, which makes monolayer TMDs ideal hosts to study and develop valleytronics [7]–[10]. Furthermore, unlike graphene, TMDs have large spin-orbit coupling which results in spin splitting (see Fig. 5.1(c)) at the band edges, and this can be significant (up to ~150 to 400 meV) in the valence band [7], allowing for not only valley selectivity but also energy selectivity. In other words, at the valence and conduction band edges of TMDs, the contrasting spin states will have different energies due to spin-valley coupling. This is advantageous because the electron spin state is

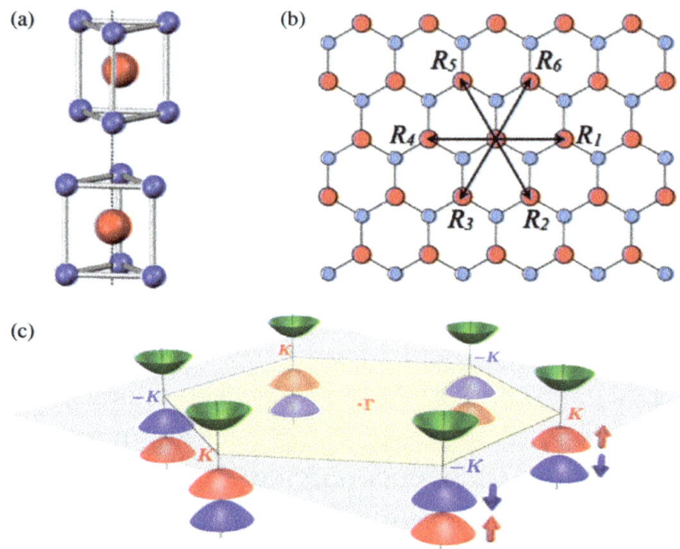

Fig. 5.1. (a) The unit cell of bulk 2H-MoS_2, which has the inversion centre located between the layers but no inversion centre within the layer itself. (b) Top view of the MoS_2 monolayer. R_i are the vectors connecting an Mo atom to its nearest neighbours. (c) Schematic drawing of the band structure at the band edges located at the K points showing significant spin splitting at the valence band edge due to strong spin-orbit interactions in MoS_2. Reproduced with permission from Ref. [7].

locked to the valleys, allowing valley control by spin manipulation or vice versa. Furthermore, the scattering of spins in monolayer TMDs is suppressed by spin-valley locking since a spin flip also requires a change in momentum. Due to the large separation in momentum space between the K and K′ valleys, this does not happen under low-energy perturbations.

Due to the presence of spin-valley locking in TMD materials, it is possible to address the valley state of carriers in these materials by controlling the spin state of carriers. This makes it possible to manipulate the valley pseudospin by manipulating the spins. Indeed, we saw in Chapter 3 that this is the key principle behind the optical technique of using circularly polarised laser excitation to selectively excite carriers in a particular valley. The resulting photoluminescence then contains information on the valley scattering and decoherence. In Sections 5.3 and 5.4, a brief review of optical techniques will be provided, along with the key literature demonstrating optical excitation/generation of the valley current. Hereinafter, we shall refer to optical excitation as optical generation.

In Sections 5.5 and 5.6, electrical generation methods will be discussed. From the perspective of engineering practical valleytronics devices, electrical methods of control and read-out are highly desirable for direct interfacing with many existing electronic technologies. It is therefore not surprising that significant efforts went into the study and development of generation and detection of valley-coupled spin currents in 2D materials.

Before that, we shall begin in Section 5.2 with a refresher of the VHE, an important phenomenon that undergirds much of the valley current generation and detection methods.

5.2 Valley Hall Effect

The VHE generates valley-coupled spin currents with the opposite spins separating in a transverse direction to an in-plane electric field, as presented in greater depth in Section 2.3. An electron moving in a crystal lattice can be treated as a Bloch wave that propagates through the crystal field. When an in-plane uniform electric field is applied in a broken inversion symmetry system, the anomalous velocity of the electron is expressed as:

$$\boldsymbol{v}(\boldsymbol{k}) = \frac{\partial \varepsilon(\boldsymbol{k})}{\hbar \partial \boldsymbol{k}} - \frac{e}{\hbar} \boldsymbol{E} \times \boldsymbol{\Omega}(\boldsymbol{k}) \tag{5.1}$$

where $\varepsilon(\boldsymbol{k})$ is the energy dispersion relation, e is the elementary charge, $\boldsymbol{E}$ is the external electric field and $\boldsymbol{\Omega}(\boldsymbol{k})$ is the Berry curvature of the Bloch electron. The Berry curvature stems from spin-orbit interaction, and it is analogous to the magnetic field but in momentum space. Similarly, the orbital magnetic moment is analogous to the Bohr magneton of the electron's spin.

For field-transverse currents to exist, we need a system that lacks either time-reversal or inversion symmetry. In the model system of graphene, there exists inequivalent K and K′ valleys depicted in Fig. 5.2(a). The K and K′ valleys are related by time-reversal symmetry resulting in valleys that are degenerate in energy but non-degenerate in spin. By engineering inversion symmetry breaking in graphene, e.g., via a substrate potential [3], and maintaining time-reversal symmetry, one acquires a Berry curvature with opposite signs in the K and K′ valleys, and consequently an anomalous velocity in opposite directions for electrons in the different valleys according to Eq. (5.1). The orbital magnetic moments associated with carriers in the K and K′ valleys will also be opposite as shown by the red peaks with opposite polarities in Fig. 5.2(b). The electrons in different valleys thus migrate in opposite directions, generating valley current in Fig. 5.2(c). This phenomenon is known as the VHE. As a result, electrons accumulate along the edges of the device and the VHE is measurable via a voltage difference across the two edges presented in Fig. 5.2(d).

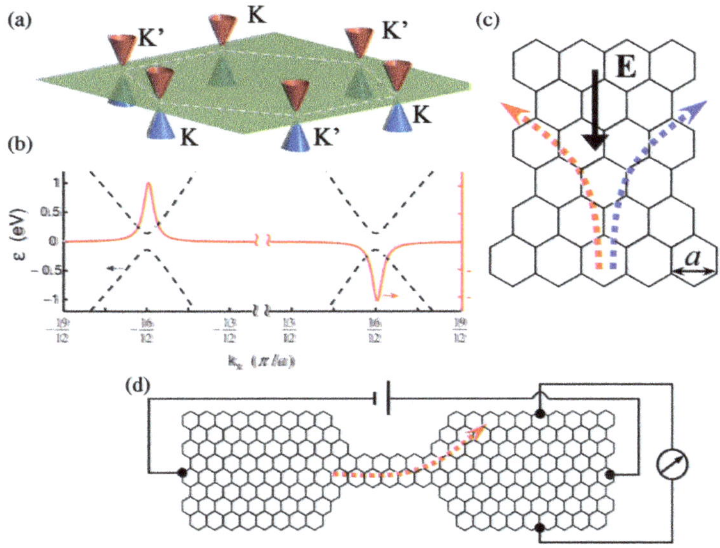

Fig. 5.2. The characteristics of valley degree of freedom in graphene with broken inversion symmetry. (a) Energy dispersion in the first Brillouin zone showing two inequivalent valleys K and K′ (labelled on the edges of the hexagon) with a slight bandgap opening. (b) Opposite orbital magnetic moments of the conduction bands at the K and K′ valleys arising from the broken inversion symmetry. (c) The valley Hall effect as a result of an in-plane electric field. (d) Transverse voltage observed as a result of valley currents. Reproduced with permission from Ref. [1].

Conversely, in an inversion-symmetric system, operations under both inversion symmetry and time-reversal symmetry results in a vanishing Berry curvature (i.e., $\mathbf{\Omega}$ goes to zero), and the electrons will not acquire an anomalous velocity, and no valley current will be generated.

The VHE is fundamental to valleytronic devices because it is topologically protected, and it results in a measurable valley current or voltage. Since the VHE exists at zero external magnetic fields, it is regarded as a type of anomalous Hall effect. In the next few sections, we will discuss the optical and electrical methods of generating and detecting valley currents, including (i) optical generation and optical detection, (ii) optical generation and electrical detection, (iii) electrical generation and optical detection, and (iv) all-electrical generation and detection.

5.3 Optical Generation and Optical Detection

Valley polarisation can be generated via optical techniques through the selective excitation of spins (and therefore valleys) using circularly polarised light [11] due to the presence of valley-dependent optical selection rules in systems with broken inversion symmetry. The key result of the valley-dependent optical selection rule is valley-selective circular dichroism, whereby the two inequivalent valleys absorb left- and right-handed polarised photons differently. Specifically, the right-circularly polarised light couples to the K valley and the left-circularly polarised light couples with the K′ valley. This has been observed in various monolayer TMDs where circularly polarised laser light is used to preferentially inject excitons into only one valley, causing a population difference in the K and K′ valley excitons [8]–[10]. This difference in valley population can then be measured optically in the form of a circular dichroic photoluminescence where the left σ^- and right σ^+ circularly polarised components of the total photoluminescence (PL) intensity will give the valley polarisation η of the material, as discussed in greater detail in Chapter 3.

$$\eta = \frac{I(\sigma^+) - I(\sigma^-)}{I(\sigma^+) + I(\sigma^-)} \tag{5.2}$$

where I (σ^+) and I (σ^-) represent the PL intensity of the left and right circularly polarised components, respectively. The circular polarisation of PL indicates the effectiveness of chiral optical selectivity. In other words, the greater the intensity of PL that has a circularly polarised component of the same helicity as the excitation laser, the greater the valley polarisation. Indeed for monolayer TMDs, valley polarisations of up to 50–100% have been demonstrated [8]–[10]. In addition to the valley polarisation, a time-dependent PL measurement can also provide the valley lifetime and recombination time of the valley-polarised carriers.

While in the valley-polarised state after optical excitation, the imbalance of charge carriers in the spin-coupled valleys of a TMD system can drive a spin-valley diffusion current [12], [13]. To determine the spin-valley diffusion length, we may employ a pump-probe spectroscopy strategy where the pump and probe are separated by a distance on the TMD sample. The working principle of this technique is shown in Fig. 5.3(a) using a monolayer WSe_2 device as an example. While there are no source and drain electrodes across the TMD sample, a side few-layer graphene (FLG) contact to the TMD and a back FLG gate under the hBN layer allows the Fermi level of the TMD to be tuned. The circularly polarised pump laser selectively polarises carriers in one valley, creating an imbalance in spin carrier population. This spin-valley polarisation then diffuses away from the pumped spot and its decay as a function of distance *d* can be detected by a weak linearly polarised probe laser. By analysing the circular polarisation of the PL emitted from the probed area, one can obtain a measure of the valley polarisation there and infer whether the valley polarisation has been quenched. For this experiment to be practically meaningful, *d* should be significantly larger than the laser beam spot.

Indeed, such experiments involving both optical generation and detection of spin-valley polarised diffusion currents have been performed in both TMD monolayers and heterostructures. In 2018, Jin *et al.* used the experimental scheme in Fig. 5.3(c)–(e) for valley current generation in hole-doped WSe_2-WS_2 heterostructures [12]. In this

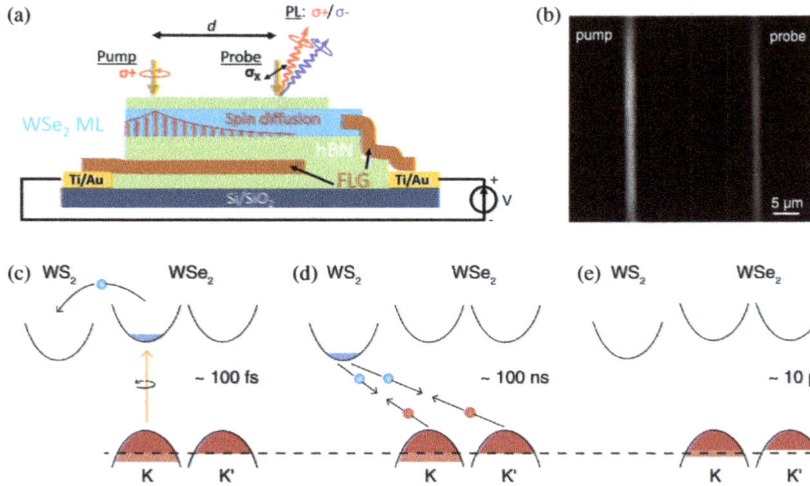

Fig. 5.3. (a) Working principle of pump-probe spectroscopy shown with monolayer WSe_2 device as an example. A side FLG contact to the monolayer WSe_2 and a back FLG gate under the hBN layer allow the Fermi level of the monolayer WSe_2 to be tuned. The spin-valley current generated by the circularly polarised pump diffuses over the distance *d* and is detected by a circular dichroic PL probe. (b)–(e) Pure spin-valley diffusion current in hole-doped WSe_2-WS_2 heterostructures. (b) Elliptically shaped pump and probe beam profiles, each with a ~1.5 μm half width used respectively for generating and detecting spin-valley currents in the heterostructures. (c) Excitons in the K valley of WSe_2 are selectively excited by circularly polarised pump light and efficiently converted into excess holes within ~100 fs through the ultrafast interlayer charge transfer process. (d) Electrons in WS_2 recombine with K valley and K′ valley holes in WSe_2 with almost equal probabilities, resulting in an excess of K valley holes and a deficiency of K′ valley holes in WSe_2. (e) This valley-polarised hole population persists for tens of microseconds. Panel (a) is reproduced with permission from Ref. [15], (b) and (c) from Ref. [12].

scheme, excitons in the K valley of WSe_2 were first selectively excited by circularly polarised pump light and efficiently converted into excess holes within ~100 fs through the ultrafast interlayer charge transfer process. The electrons in WS_2 then recombine with K valley and K′ valley holes in WSe_2 with almost equal probabilities, resulting in an excess of K valley holes and a deficiency of K′ valley holes in

WSe_2. They determined that the valley-polarised hole population subsequently persists for ~10 μs. In the WSe_2-WS_2 heterostructure studied by Jin *et al.*, the valley current is reported to live for more than 20 μs and propagate over 20 μm (Fig. 5.3(b)), and both the lifetime and diffusion length can be controlled by the gate voltage.

It is not surprising that this experiment was first demonstrated for a TMD heterostructure since the valley lifetimes are generally longer in heterostructures than in monolayers because the electron-hole pairs are physically separated by a van der Waals gap in heterostructures [13]. Consequently, valley diffusion lengths are also longer in heterostructures which makes it easier to conduct optical pump-probe spectroscopy measurements that typically require the diffusion lengths to be significantly larger than the typical micron-sized beam spots.

Nonetheless, such experiments have also recently been reported in 2022 with demonstrations of diffusion lengths in the order of microns in TMD monolayers. In particular, Cadiz *et al.* reported valley exciton diffusion lengths in the order of sub-microns at 6 K [14], and Ren *et al.* demonstrated up to 25% valley polarisation is maintained for 20 μm in monolayer WSe_2 at low temperatures (5 K) [15] using the device schematically shown in Fig. 5.3(a).

5.4 Optical Generation and Electrical Detection

Valley currents generated via optical excitation may also be detected electrically. During a PL experiment without any electrical bias on the TMD monolayer, photoexcited carriers in the form of electron-hole pairs (called excitons) are generated and some may subsequently recombine to give the PL signal while others may be subjected to thermal diffusion or become charged species/complexes. When an in-plane electric field is applied between the source and drain electrodes of the device, exciton dissociation can occur (typically at the electrodes), and the dissociated charges collectively form a photocurrent. When a circularly polarised laser light is shone on a valleytronic material, only carriers in one of the K or K′ valleys will be excited, and the photocurrent is valley polarised, i.e., the photocurrent contains only information from one valley. This is the most direct and most used valley current generation source.

Indeed, this was used in many of the earlier papers in the field of valleytronics, such as the one by Mak *et al.* published in 2014 [16]. They reported the observation of VHE in MoS_2 where the valley current was optically generated using a circularly polarised laser light source and electrically detected through a transverse voltage. Fig. 5.4 shows their experiment, where monolayer MoS_2 was illuminated with circularly polarised laser light which selectively excites the K valley, generating valley-polarised excitons. By applying an in-plane electric field between the source-drain electrodes, the VHE results in the accumulation of electrons and holes on the opposite edges of the device transverse to the electric field. This was then detected as a potential difference across the contacts A and B as shown in Fig. 5.4(c). Inversion-symmetric bilayer MoS_2 devices did not show a measurable VHE; the valley Hall resistance is nearly zero in bilayer MoS_2 (Fig. 5.4(d)), confirming that broken inversion symmetry plays a key role in creating valley polarisation.

5.5 Electrical Generation and Optical Detection

More recent works in valleytronics involve generating valley current using electrical techniques and detecting them optically. In 2016, Lee *et al.* demonstrated the electrical generation of valley current in

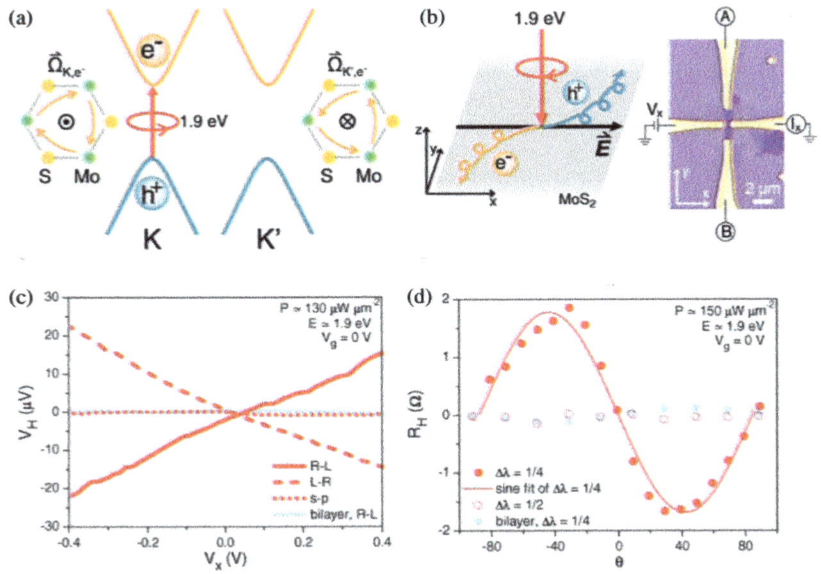

Fig. 5.4. Monolayer MoS_2 Hall bar device. (a) Schematics of the valley-dependent optical selection rules and the electrons at the K and K′ valleys that possess opposite Berry curvatures **Ω**. The orange arrows represent the clockwise and anticlockwise hopping motions of the K and K′ electrons. (b) Schematic of a photoinduced valley Hall effect driven by a net valley polarisation (left) and an image of the Hall bar device (right). (c) The source-drain bias (V_x) dependence of the Hall voltage (V_H) for the monolayer device under R-L (red solid line), L-R (red dashed line), and half-wave s-p modulations (red dotted line). Results from the bilayer device under R-L modulation (blue solid line) are also shown. (d) The valley Hall resistance of the monolayer device as a function of the incidence angle θ under quarter-wave ($\Delta\lambda$ = 1/4, solid red circles) and half-wave ($\Delta\lambda$ = 1/2, open red circles) modulations. Results for the bilayer device under quarter-wave modulation are also shown (blue circles). Reproduced with permission from Ref. [16].

bilayer MoS_2 transistor, and optical detection of the VHE using Kerr rotation (KR) microscopy [17]. The technique of KR spectroscopy is described in more detail in Chapter 3.4. While bilayer MoS_2 is inversion symmetric, its inversion symmetry can be broken using an out-of-plane electric field applied by a gate bias (Fig. 5.5). Due to the VHE,

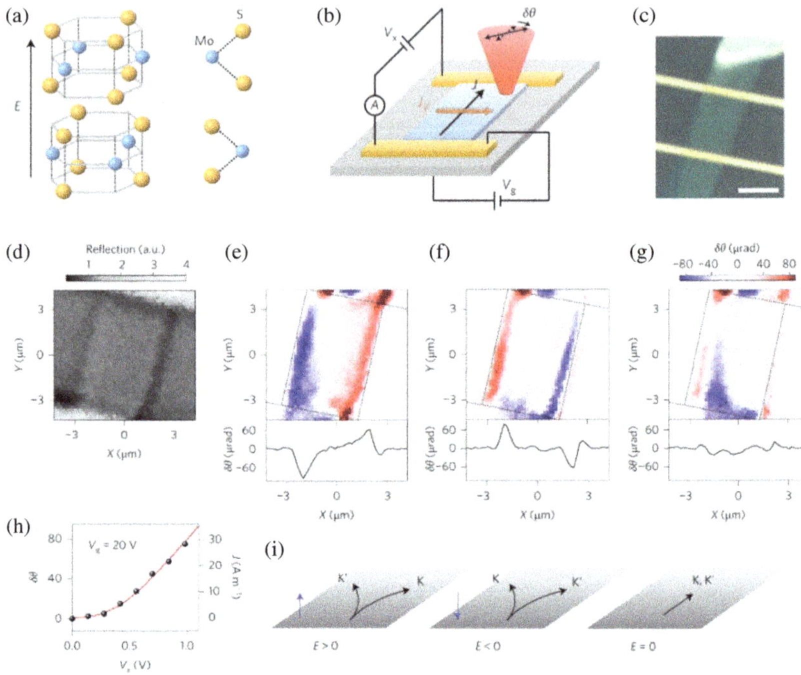

Fig. 5.5. VHE in bilayer MoS_2 device. (a) Bilayer MoS_2 crystal structure. Vertical electric field E breaks the inversion symmetry of the bilayer. (b) Schematics of a MoS_2 field-effect transistor with bias voltage (V_x) applied on the source-drain electrodes and gate voltage (V_g) applied through the Si/SiO_2 substrate. Orange arrow: valley current density J_V, black arrow: longitudinal electrical current density J. (c) Optical image of the device. Scale bar: 5 μm. (d) Reflection image of the device. The signal is normalised to the reflectivity of the MoS_2 channel. (e–g) Spatial 2D map and selected horizontal linecut of the Kerr rotation angle $\delta\theta$ under V_g = (e) 20 V, (f) –5 V and (g) 4 V. All measurements were done under the same V_x (= 2.5 V peak-to-peak). Black dashed lines show the boundary of the channel determined from the reflection image. (h) Bias voltage (V_x) dependence of $\delta\theta$ on one edge of the channel (black symbols, left y axis) and the longitudinal current density J (red line, right y axis) under V_g = 20 V. (i) Illustration of the field-controlled VHE in bilayer MoS_2. Blue arrows show the direction of the vertical E field. Reproduced with permission from Ref. [17].

a valley current density J_V in the transverse direction is induced by a longitudinal electrical field (Fig. 5.5(b)).

To image the VHE using KR microscopy, a linearly polarised light is focused onto the sample at normal incidence. Due to the interaction between the incident light and the magnetic material under study, the polarisation of the reflected beam will be rotated by an amount proportional to the net magnetisation of the sample in the out-of-plane direction. When VHE is present, the measured valley magnetisation shows that net valley polarisations of opposite sign are accumulated at the two edges of the device channel. Since the magnitude of inversion symmetry breaking is tuned by the gate bias, the measured valley magnetisation signal is highly dependent on the gate voltage (Fig. 5.5(e)–(g)). In addition, the KR angle and longitudinal current density J were found to follow the same dependence on V_x as shown in Fig. 5.5(h), indicating that the observed KR is driven by the longitudinal current. This confirms that the valley current was generated through the VHE and KR imaging allowed direct visualisation of the spin-up and spin-down currents at opposite edges of the bilayer MoS_2 channel. Furthermore, the observed VHE can be field controlled, where a change in sign of the out-of-plane gate electric field leads to a change in direction of the valley current, and the valley current vanishes when the gate bias is removed (inversion symmetry restored), as summarised in Fig. 5.5(i).

Another method for valley current generation involves the injection of spin-polarised charge carriers since spin-valley locking in TMD materials allows spin and valley degrees of freedom to be addressed interchangeably. This was demonstrated by Sanchez *et al.* in 2016 via spin injection from a ferromagnetic electrode into a WSe_2/MoS_2 heterojunction [18]. In this device, monolayer WSe_2 is contacted using a ferromagnetic electrode (permalloy) and MoS_2 is transferred on top of the WSe_2 channel, forming a heterojunction diode (Fig. 5.6(a)). By applying a positive bias voltage to the permalloy electrode, spin-polarised holes are injected from the permalloy electrode and electrons are injected from the MoS_2 side (Fig. 5.6(c)). The electrons and holes recombine in the junction, resulting in light emission. As the

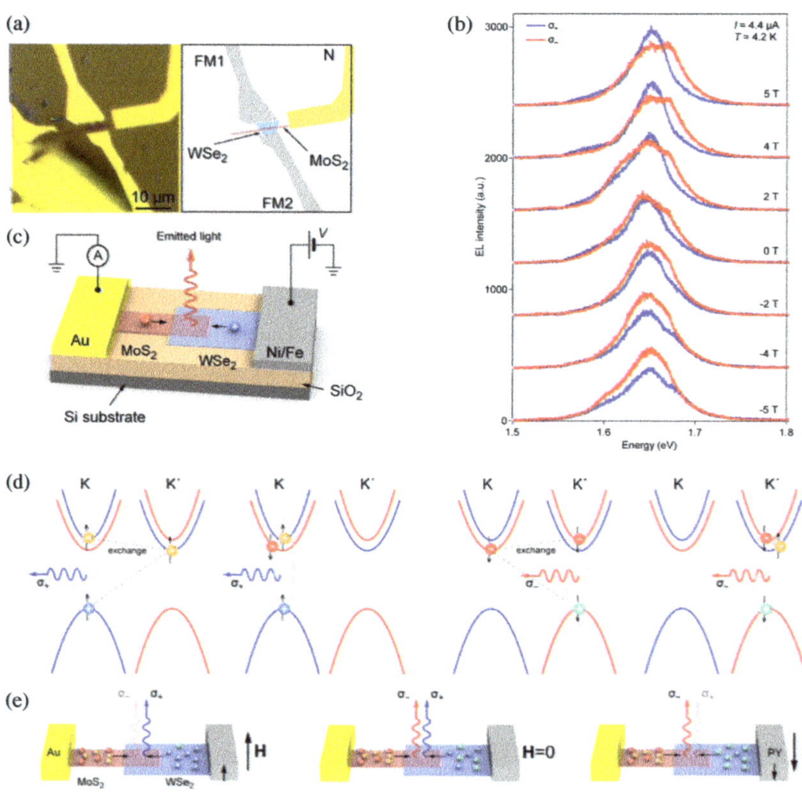

Fig. 5.6. Light generation in monolayer WSe_2/MoS_2 heterojunction diodes. (a) Optical image and schematic of the device. (b) Electroluminescence spectra from a spin/valley LED for different values of magnetic fields, acquired for σ^+ and σ^- polarisations. (c) Schematic drawing of the device. (d) Configurations of X^- charged excitons in WSe_2 that can emit light. Blue and red lines denote spin-up and spin-down polarised conductance and valence band edges in the K and K′ valleys. (e) Effect of external magnetic field on σ^+ and σ^- light. Reproduced with permission from Ref. [18].

electron-hole pair originate from the same valley, the electroluminescence is circularly polarised. Figure 5.6(d) shows the mechanism by which circularly polarised light is emitted — the polarisation of the emitted photon is determined by the valley index of the electron-hole pair that can recombine because intervalley transition is forbidden. The resulting circularly polarised electroluminescence spectra demonstrates valley polarisation, and the intensity of the σ^+ and σ^- polarised light can be controlled by the magnetic field (Fig. 5.6(b)). This is because permalloy is a soft ferromagnetic material which flips its bulk magnetic polarisation at small applied magnetic fields. This results in a majority of spin-up (spin-down) holes being injected through this ferromagnetic contact when a positive (negative) external magnetic field is applied as depicted schematically in Fig. 5.6(e). This paper marks the first demonstration of spin injection resulting in valley polarisation in a heterostructure based on 2D semiconductors.

A valley light-emitting diode (vLED) may also be fabricated to demonstrate valley contrast. In 2016, Yu *et al.* demonstrated electrically tuneable vLED in monolayer WS_2 [19] whereby the valley current is generated electrically and detected optically. In this report, monolayer WS_2 was obtained via CVD instead of mechanical exfoliation, which allows for large-area grown samples (Fig. 5.7(a)). The CVD-grown WS_2 is then integrated into a p^+-Si/i-WS_2/n-ITO (indium tin oxide) heterojunction LED device (Fig. 5.7(b)) to form a PIN diode, where an intrinsic semiconductor is sandwiched between a p-type and n-type semiconductor.

Under forward bias, electrons are injected from the n-ITO which recombine with holes from p^+-Si in the WS_2 monolayer as depicted

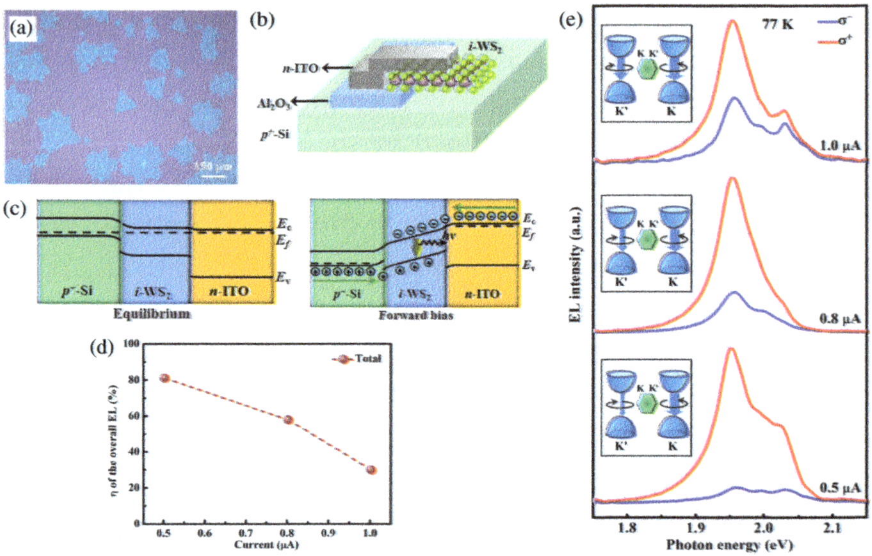

Fig. 5.7. Electrically tunable vLED based on CVD-grown WS_2. (a) Optical image of as-grown WS_2 on SiO_2/Si substrate. (b) Schematic of the p^+-Si/i-WS2/n-ITO heterojunction LED device. (c) Band diagram of the p^+-Si/i-WS_2/n-ITO heterojunction in equilibrium condition (left) and under forward bias (right). (d) Current dependence of valley polarisation η for the overall emission. (e) Current-dependent circularly polarised electroluminescence spectra normalised to the right-handed circular (σ^+) component at each injection current. Inset: Schematic of valley-contrasting circular dichroism under different current injection. Reproduced with permission from Ref. [19].

in Fig. 5.7(c). The recombination releases chiral electroluminescence, whereby the valley polarisation η (Eq. (5.2)) of up to 81% at 0.5 μA was detected and was found to be current dependent as shown in Fig. 5.7(d) and (e). Hence, the valley asymmetry at K and K′ can be attributed to the built-in potential in WS_2 that shifts the anisotropic carrier distributions at K and K′. From the perspective of a practical device, modulation of valley polarisation by injected current alone could be advantageous over other methods involving laser excitation energy or temperature. This report marks the first demonstration of circularly polarised electroluminescence from CVD-grown monolayer WS_2.

The helicity-dependent optical generation technique occupies much of the earlier works involving valleytronics. However, a major disadvantage of this technique is that bound excitons and/or trions are generated, and these bound carriers tend to have lower valley polarisation lifetimes than free carriers such as electrons or holes as the excitons or trions tend to recombine. Hence, the generation of valley current via electrical techniques that produces free carriers with longer lifetime is preferred.

Furthermore, while some methods involve the electrical generation of valley current, the detection of valley polarisation involves optical methods, which makes it challenging to realise complex valleytronics devices and circuits based on integrated optoelectronic components. There is therefore a need for fully electrical methods, involving both electrical generation of valley currents, and an electrical method of detecting these currents.

5.6 All-Electrical Generation and Detection of Valley Current

From the perspective of engineering practical valleytronic devices, electrical methods of control and read-out are highly desirable for direct interfacing with many existing electronic technologies. Spin-valley currents are charge neutral and topologically protected and hence, in principle, they are non-dissipative, allowing for ultralow power devices. It is therefore unsurprising that significant efforts went into the study and development of all-electrical generation and detection of valley-coupled spin currents in 2D materials.

5.6.1 Principle of Operation

The VHE exploits the finite Berry curvatures which has opposite signs in the K and K′ valley to generate transverse spin currents in the presence of an in-plane electric field. Electrons carrying up-spins and down-spins diverge in-plane with opposite velocities. It is possible to recombine such up- and down-spin currents in a clever device geometry known as the non-local measurement of the Hall bar and measure it through the inverse valley Hall effect (IVHE).

The Hall bar structure shown in Fig. 5.8 is a well-known device structure used in magnetotransport measurements because it allows for low error measurement of material properties, such as carrier mobility and resistivity, as a function of magnetic field. In the standard Hall measurement (known here as the local measurement), a current is sourced through the source-drain electrodes in the longitudinal direction, and the longitudinal voltage (V_{xx}) and transverse Hall voltage (V_{xy}) can be measured (Fig. 5.8(a)) along the edge of the conductive channel from the little fingers known as voltage probes. The main advantage of the Hall bar measurement is that voltage drop across contact resistances can be eliminated by using high-input impedance voltmeters which draw no current — providing an accurate measurement of the channel resistance. The R_{xx} and R_{xy} can also be measured on the same point and across multiple points of the sample by fabricating more voltage probes, providing some information on sample homogeneity.

In an inversion-asymmetric material, a local charge current driven by an applied in-plane electric field generates a transverse valley

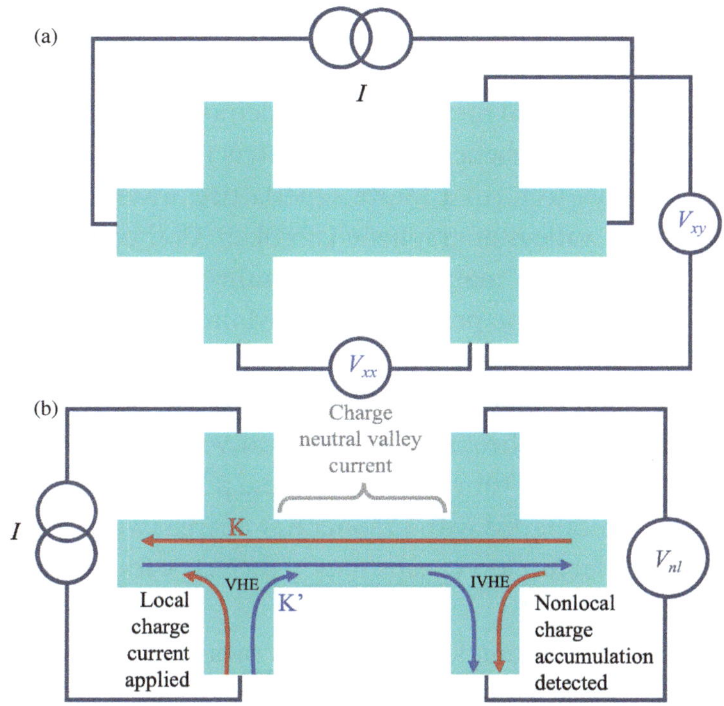

Fig. 5.8. Measurements in a Hall bar. (a) Local measurements to obtain longitudinal voltage (V_{xx}) and Hall voltage (V_{xy}). (b) Non-local measurement resulting in an all-electrical generation and detection of valley current in a Hall bar geometry.

current via the VHE shown schematically in Fig. 5.8(b). Provided the charge transport is ballistic, a long range charge-neutral valley current flows throughout the length of the Hall bar. The valley current develops a *non-local charge accumulation* at the other end of the sample and is detectable through the IVHE. This can be measured as a non-local voltage difference V_{NL} or non-local resistance R_{NL}. This way, an all-electrical generation and detection of valley current can both be done in one device.

5.6.2 Examples of Non-Local IVHE Measurements

In 2014, an all-electrical method was first demonstrated by Gorbachev *et al.* in a hBN/graphene/hBN heterostructure [3]. By aligning graphene to hBN to form a Moiré superlattice, inversion symmetry of the K and K′ valleys of graphene is broken. Consequently, a small bandgap opens at the main charge-neutrality point (CNP) and secondary CNP due to the presence of the Moiré superlattice emerges. These points are also known as Berry curvature 'hot spots' because the non-zero Berry curvature in the material can be probed electrically at these points through the non-local voltage when the Fermi level of the material is tuned to match these points. The top hBN was intentionally misaligned with graphene at a large twist angle ~30° so that it acts as a featureless encapsulation dielectric for graphene [20].

Figure 5.9 shows the non-local transport measurement of resistivity against gate voltage to study valley transport in graphene/hBN superlattices. Gorbachev *et al.* found that two non-local resistance (R_{NL}) peaks at different gate voltages appeared when their device was measured at $T = 20$ K. Changing the gate voltage varies carrier density, and hence the Fermi level. The non-local response was observed when the Fermi level tuned into one of the Berry curvature 'hot spots', where the peak at $V_g \approx 0$ V was due to the 'hot spot' formed at the main CNP, and the peak at $V_g \approx -17$ V was attributed to the secondary CNP. Gorbachev *et al.* concluded that the R_{NL} peaks originated from VHE and IVHE after ruling out the possibilities of R_{NL} peaks occurring due to stray charge currents, charge-neutral flow of spin and/or energy, and gapless edge transport mechanisms.

Soon after this measurement, in 2015, Sui *et al.* and Shimazaki *et al.* conducted similiar non-local measurements in bilayer graphene (BLG)

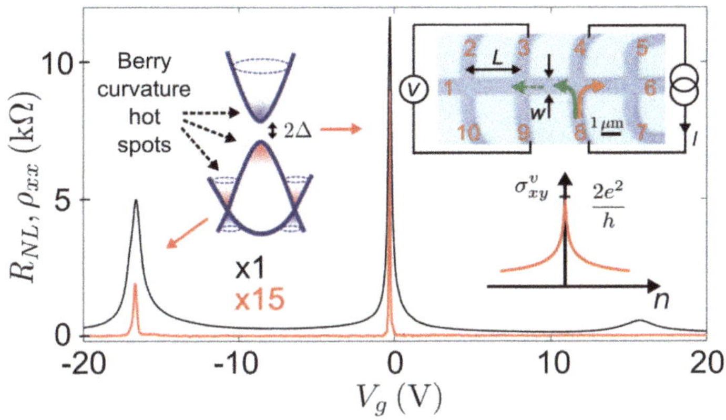

Fig. 5.9. Non-local resistance (red curve) and longitudinal resistivity (black curve) measured in graphene/hBN superlattices. Shown schematically are valley K and K′ currents and the long-range response mechanism. (Left inset) Schematic band structure of graphene superlattices, with Berry curvature hot spots arising near the gap opening and avoiding band-crossing regions. (Bottom right inset) Valley Hall conductivity modelled for gapped Dirac fermions as a function of carrier density. Adapted with permission from Ref. [3].

to demonstrate the electrical generation and detection of pure valley current in BLG [4], [5]. Fig. 5.10 shows the report by Sui *et al.* [5], where they used a perpendicular electric field to break the inversion symmetry in the BLG and observed a non-local resistance peak at 70 K as a result of the topological transport of the valley pseudospin as shown in Fig. 5.10(c)–(h). Similarly, they found that the ohmic resistance contribution was substantially smaller than R_{NL}. The non-local signal also persisted up to room temperature and over long distances (up to 10 μm).

In the same issue of the report by Sui *et al.*, Shimazaki *et al.* independently designed and fabricated the same device in BLG, and also demonstrated gate-tunable valley transport in BLG using a dual-gated BLG field-effect transistor [4]. Since the AB-stacked BLG is an inversion-symmetric crystal lattice, a perpendicular electric field was applied to create an energy difference between the top and bottom

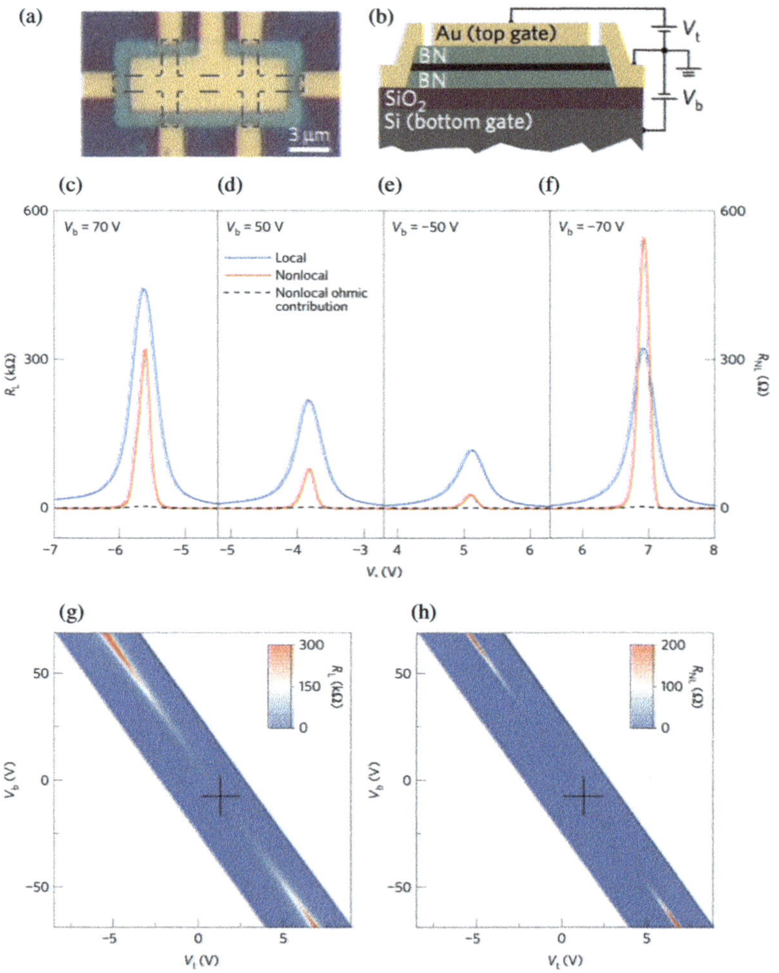

Fig. 5.10. Dual-gated bilayer graphene (BLG) field-effect transistor and its local and non-local characterisation. (a) Optical image of a typical BLG field-effect transistor viewed from the top. The hBN/BLG/hBN stack (blue) was etched to expose the BLG leads at the edge that are subsequently contacted by Au electrodes (orange). (b) Schematic cross-section of the device shown in (a). (c)–(f) Local and non-local resistance as a function of V_t with V_b fixed at varying voltages. (g) Local and (h) non-local resistance measured as a function of both V_t and V_b. The crosses mark the point (V_{t0}, V_{b0}) = (1.3 V, −7.5 V) where no bandgap opened in BLG. Reproduced with permission from Ref. [5].

layers which breaks the inversion symmetry of BLG, while opening a small bandgap at the CNP in graphene's band structure and results in the emergence of non-zero Berry curvature (Fig. 5.11(a)). The valley

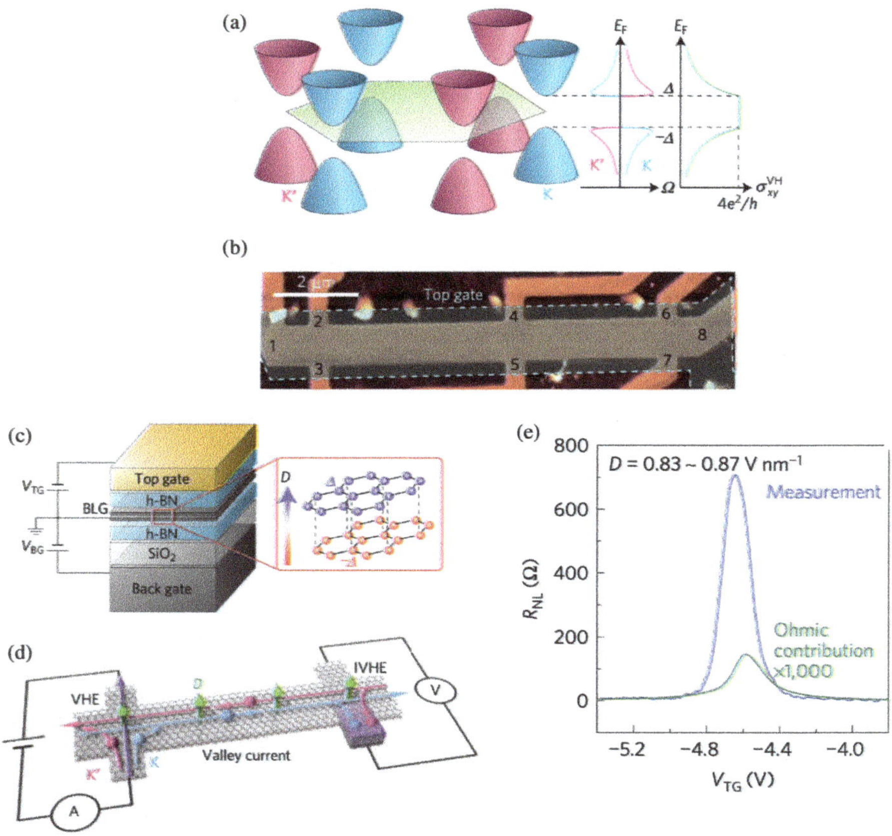

Fig. 5.11. Detection of the non-local resistance due to valley current flow in BLG. (a) Band structure, Berry curvature and valley Hall conductivity of BLG. (b) Atomic force microscopy image of the BLG device without the top hBN. (c) Schematic of the dual-gated BLG device. (d) Schematic of the non-local resistance measurement and the non-local transport mediated by a pure valley current. (e) R_{NL} peak and ohmic contribution peak at the highest displacement field D. Reproduced with permission from Ref. [4].

Hall conductivity, σ_{xy}^{VH}, calculated by integrating the Berry curvature, is constant in the bandgap. Compared to the previous device by Gorbachev *et al.* where monolayer graphene is mechanically aligned with hBN for inversion symmetry breaking, the electrically induced Berry curvature here allows for the tunability of valley current. Furthermore, a dual-gated device (having both top and bottom gates) can be used for independent control of the Berry curvature and carrier density (Figs. 5.11(b) and (c)). In the dual-gated device, the top gate is made by depositing a gold film on top of a hBN gate and encapsulation dielectric, and the back gate is a heavily p-doped silicon substrate insulated from the bilayer graphene by the thermally oxidised SiO_2 film and hBN layers.

Applying an out-of-plane gate bias creates an electric displacement field across the bilayer graphene which breaks inversion symmetry. By measuring the non-local resistance in these devices (Fig. 5.11(d)), they found that the measured R_{NL} was significantly higher than ohmic contribution and the R_{NL} peaks can be attributed to the generation of valley current (Fig. 5.11(e)). Conversely, the non-local signal vanished when the out-of-plane gate bias was not applied, indicating that inversion symmetry was restored. However, external mechanisms to break spatial-inversion symmetry are not always necessary, such as the case of monolayer MoS_2, where inversion symmetry is inherently absent; carriers in the K and K′ valleys therefore have non-zero Berry curvature. It is worth noting that the VHE should not exist in even-layered MoS_2 because inversion symmetry is restored.

In 2019, Wu *et al.* also reported (Fig. 5.12) that intrinsic valley Hall transport had been observed in monolayer and trilayer MoS_2 in a fully electrical device, as indicated by a measurable non-local resistance [21]. They found that the valley Hall signal was present even at room temperature, with a measured a micron-scale valley diffusion length. In contrast, valley Hall signal was absent in a centrosymmetric bilayer MoS_2 device.

All-electrical valley current generation and detection had also been realised in a WSe_2/graphene hybrid device by Hung *et al.* [22]. This work demonstrated that spin currents can be generated in monolayer WSe_2, and it is done by applying an in-plane electric field across

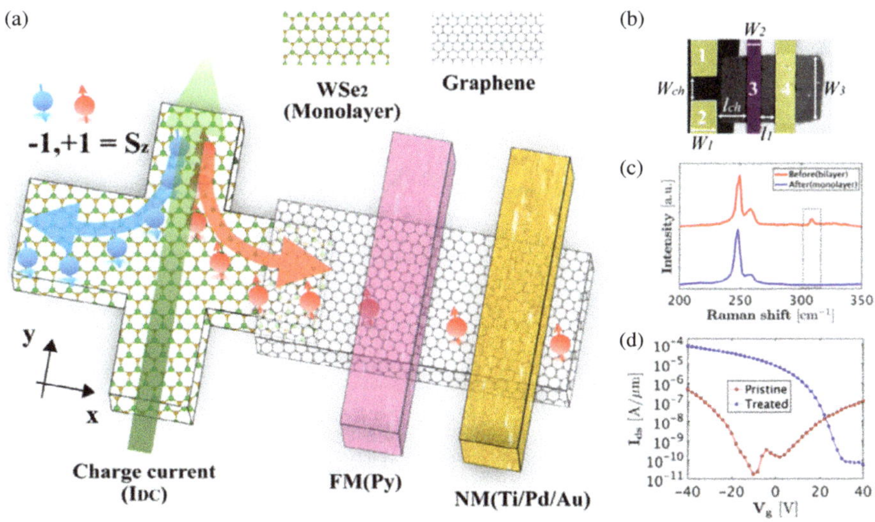

Fig. 5.12. Geometry of the WSe_2 device and characterisations. (a) Schematic of spin-locked valley Hall effect in monolayer WSe_2 and the non-local spin valve detection by the ferromagnet metal (FM) and non-ferromagnet metal (NM). (b) Coloured scanning electron microscopy image with detailed geometry of the devices where yellow electrodes (1, 2, 4) are NM, purple electrode (3) is FM, dark green area is monolayer WSe_2, and grey area is graphene. (c) Raman spectrum measured on WSe_2 after the O_2 plasma exposure shows no peak at ~308 cm^{-1} (dotted box), which indicates the bilayer-to-monolayer conversion. Note that the peak around 308 cm^{-1} is attributed to interlayer interaction and will only show up when it is not a monolayer. (d) Typical transfer characteristic of pristine monolayer WSe_2 (red). Transfer characteristics of monolayer WSe_2 from the bilayer conversion, measured across electrodes 1 and 2 (blue). Reproduced with permission from Ref. [22].

electrodes 1 and 2 in a monolayer WSe_2 structure at room temperature (see Fig. 5.12).

In this work, Hung *et al.* detected valley current by detecting the proxy spin current, which is made possible due to the presence of spin-valley locking in this material. As a result, the VHE generates out-of-plane polarised spins, which can be detected with a spin detector. Spin-polarised holes are injected from the WSe_2 flake to the graphene layer on top, and then diffused towards the ferromagnetic contact.

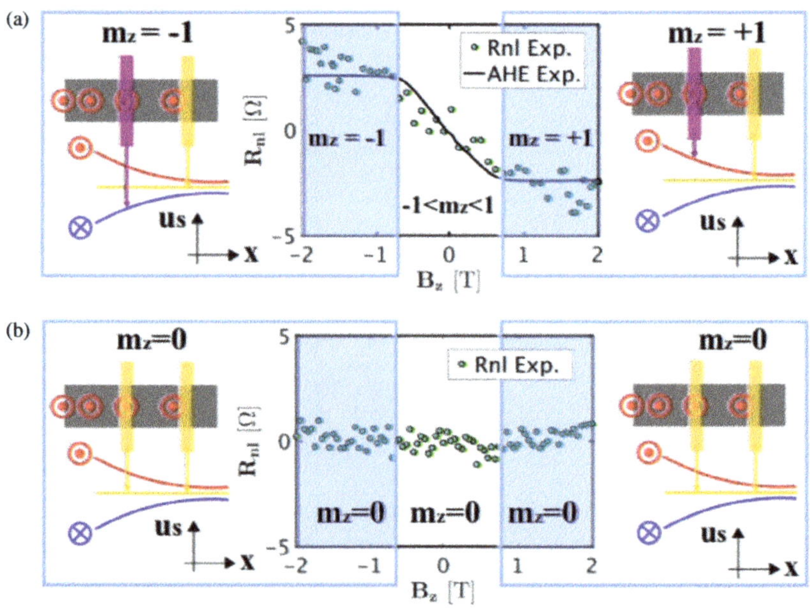

Fig. 5.13. Non-local valve measurements and corresponding probing schemes. (a) Non-local measurement of the device and the corresponding probing schemes at fixed spin current (polarity). Red dots and blue crosses are spin polarisation along +z and –z directions. Red, blue, and yellow lines are spin chemical potential of (m_z = +1), (m_z = –1), and (m_z = 0) respectively. (b) Same measurement with FM electrode (purple) being replaced with NM electrode (yellow). Reproduced with permission from Ref. [22].

Unlike the previously mentioned all-electrical techniques, valley current in this paper is detected using a novel non-local spin valve (Fig. 5.13). This detector consists of a ferromagnetic permalloy film and a non-ferromagnetic metal deposited on graphene, which probes the chemical potential of the diffused spins in the graphene channel. By measuring the non-local voltage across electrodes 3 and 4 as a function of the applied B_z field, one can observe the presence of spin polarisation and hence deduce the presence of valley polarisation.

Similarly, Hung *et al.* also demonstrated non-local resistance which persisted from 4 K up to room temperature, with a measured valley diffusion length at the micron scale in CVD-grown monolayer MoS_2 Hall bar devices [23]. Similar to the previous papers discussed in this section, electrical generation and detection of valley currents in monolayer

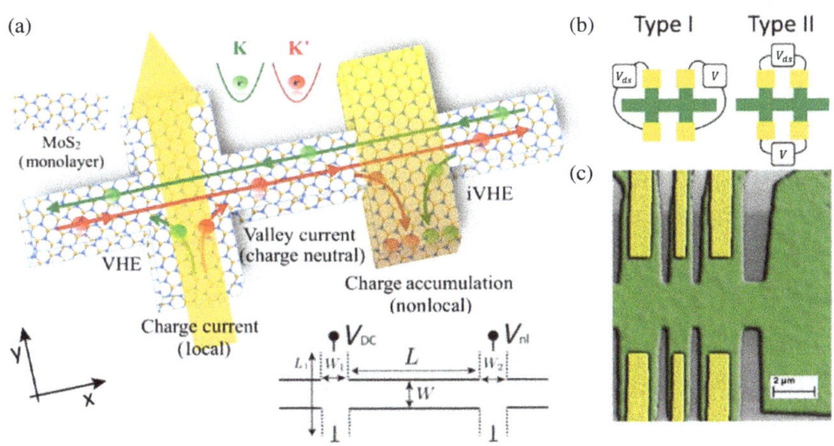

Fig. 5.14. Valley-coupled topological current in MoS_2. (a) Schematic of valley-coupled topological current due to VHE and IVHE in monolayer MoS_2 and the device geometry (bottom), where $W_1 = 1$ mm, $W = W_2 = 2$ mm, $L_1 = 4.5$ mm, and $L = 0.5$ mm. (b) Schematics of two measurement setups: type I (non-local) and type II (local). (c) Coloured scanning electron microscope image of patterned MoS_2 flake (green) and lithographically defined metal electrodes (yellow). Reproduced with permission from Ref. [23].

MoS_2 are demonstrated by combining VHE and IVHE in the non-local Hall bar device geometry (Fig. 5.14). The non-local resistance is detected even at distances more than 4 μm away from the charge current path.

Thus far, all-electrical methods of valley current generation and detection have been demonstrated for graphene/hBN, BLG, MoS_2 and WSe_2 by means of a Hall bar structure. These first demonstrations bode well for the engineering of valleytronic devices that could be fully operated electrically.

5.7 Summary

In this chapter, we discussed the VHE, followed by presenting known methods for valley current generation and detection, including optical, electrical and hybrid opto-electrical techniques. These techniques can be classified into the four categories shown in Fig. 5.15.

Compared to the optical methods, electrical methods of valley control are more challenging, as the injection of pure spin currents from a ferromagnet into semiconductors is a fundamentally prohibited process due to impedance mismatch between the ferromagnet and the semiconductor [24]. While there have been advances in the field of spintronics, a high-efficiency direct spin injection mechanism into semiconductors from a ferromagnet is still elusive. Hence, electrical generation of spin-valley currents is mostly done using the inverse VHE in reported literature. This can be advantageous since spin-valley currents are charge neutral and topologically protected, and hence in principle non-dissipative, allowing for ultralow power devices. However, a sizable charge current, which leads to Joule heating, is needed to drive the transverse spin-valley

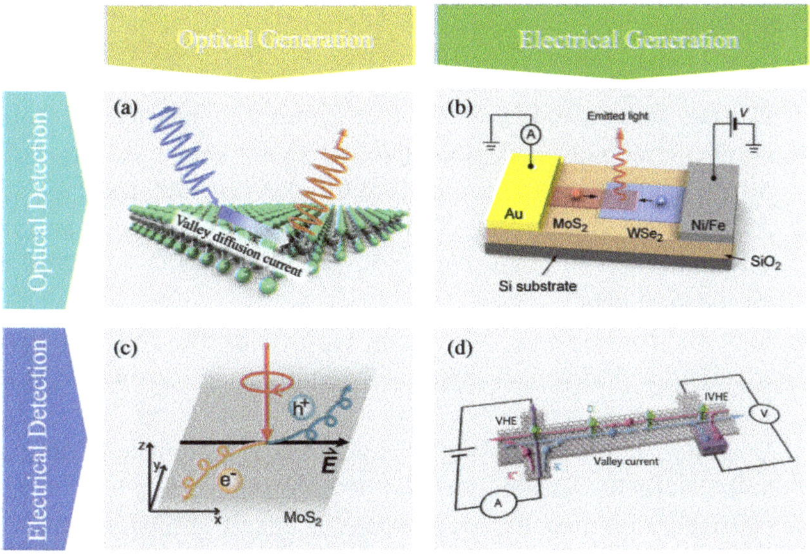

Fig. 5.15. Overview of valley current generation and detection techniques. (a) Optical pumping with circularly polarised light and detection of valley diffusion current. (b) Carrier injection and detection of circularly polarised electroluminescence. (c) Optical pumping with circularly polarised light and Hall voltage measurement. (d) All-electrical devices with non-local resistance measurement in Hall bar structure. (b), (c) and (d) are reproduced with permission from Refs. [18], [16] and [4] respectively.

currents in the material, negating the advantages of a non-dissipative valley current. Also, for a practical device, it can be challenging to integrate the laser source into an integrated circuit chip on a printed circuit board or integrated optoelectronic device. Thus, an all-electrical method is likely more practical.

In terms of fabrication, these valley devices are heavily reliant on the availability of robust device engineering techniques, including interface engineering, band engineering, doping techniques, contacts engineering and gate engineering, which remain in active research today. We shall further elaborate on these in Chapter 7.

References

[1] D. Xiao, W. Yao, and Q. Niu, "Valley-contrasting physics in graphene: Magnetic moment and topological transport," *Phys. Rev. Lett.*, vol. 99, no. 23, p. 236809, 2007, doi: 10.1103/PhysRevLett.99.236809.

[2] A. Rycerz, J. Tworzydło, and C. W. J. Beenakker, "Valley filter and valley valve in graphene," *Nat. Phys.*, vol. 3, no. 3, pp. 172–175, 2007, doi: 10.1038/nphys547.

[3] R. V. Gorbachev *et al.*, "Detecting topological currents in graphene superlattices," *Science*, vol. 346, no. 6208, pp. 448–451, 2014, doi: 10.1126/science.1254966.

[4] Y. Shimazaki, M. Yamamoto, I. V. Borzenets, K. Watanabe, T. Taniguchi, and S. Tarucha, "Generation and detection of pure valley current by electrically induced Berry curvature in bilayer graphene," *Nat. Phys.*, vol. 11, no. 12, pp. 1032–1036, 2015, doi: 10.1038/nphys3551.

[5] M. Sui *et al.*, "Gate-tunable topological valley transport in bilayer graphene," *Nat. Phys.*, vol. 11, no. 12, pp. 1027–1031, 2015, doi: 10.1038/nphys3485.

[6] J. Yin *et al.*, "Tunable and giant valley-selective Hall effect in gapped bilayer graphene," *Science*, vol. 375, no. 6587, pp. 1398–1402, 2022, doi: 10.1126/science.abl4266.

[7] D. Xiao, G.-B. Liu, W. Feng, X. Xu, and W. Yao, "Coupled spin and valley physics in monolayers of MoS_2 and other group-VI dichalcogenides," *Phys. Rev. Lett.*, vol. 108, no. 19, p. 196802, 2012, doi: 10.1103/PhysRevLett.108.196802.

[8] H. Zeng, J. Dai, W. Yao, D. Xiao, and X. Cui, "Valley polarisation in MoS_2 by optical pumping," *Nat. Nanotechnol.*, vol. 7, no. 8, pp. 490–493, 2012, doi: 10.1038/nnano.2012.95.

[9] T. Cao *et al.*, "Valley-selective circular dichroism of monolayer molybdenum disulphide," *Nat. Commun.*, vol. 3, no. May, p. 887, 2012, doi: 10.1038/ncomms1882.

[10] K. F. Mak, K. He, J. Shan, and T. F. Heinz, "Control of valley polarisation in monolayer MoS_2 by optical helicity," *Nat. Nanotechnol.*, vol. 7, no. 8, pp. 494–498, 2012, doi: 10.1038/nnano.2012.96.

[11] W. Yao, D. Xiao, and Q. Niu, "Valley-dependent optoelectronics from inversion symmetry breaking," *Phys. Rev. B*, vol. 77, no. 23, pp. 1–7, 2008, doi: 10.1103/PhysRevB.77.235406.

[12] C. Jin *et al.*, "Imaging of pure spin-valley diffusion current in WS_2 -WSe_2 heterostructures," *Science*, vol. 360, no. 6391, pp. 893–896, 2018, doi: 10.1126/science.aao3503.

[13] P. Rivera *et al.*, "Valley-polarised exciton dynamics in a 2D semiconductor heterostructure," *Science*, vol. 351, no. 6274, pp. 688–691, 2016, doi: 10.1126/science.aac7820.

[14] F. Cadiz, S. Gerl, T. Taniguchi, and K. Watanabe, "Imaging the effect of high photoexcited densities on valley polarisation and coherence in MoS_2 monolayers," *NPJ 2D Mater. Appl.*, vol. 6, no. 1, p. 27, 2022, doi: 10.1038/s41699-022-00303-x.

[15] L. Ren *et al.*, "Optical detection of long electron spin transport lengths in a monolayer semiconductor," *Phys. Rev. Lett.*, vol. 129, no. 2, p. 027402, 2022, doi: 10.1103/PhysRevLett.129.027402.

[16] K. F. Mak, K. L. McGill, J. Park, and P. L. McEuen, "The valley Hall effect in MoS_2 transistors," *Science*, vol. 344, no. 6191, pp. 1489–1492, 2014, doi: 10.1126/science.1250140.

[17] J. Lee, K. F. Mak, and J. Shan, "Electrical control of the valley Hall effect in bilayer MoS_2 transistors," *Nat. Nanotechnol.*, vol. 11, no. 5, pp. 421–425, 2016, doi: 10.1038/nnano.2015.337.

[18] O. L. Sanchez, D. Ovchinnikov, S. Misra, A. Allain, and A. Kis, "Valley polarisation by spin injection in a light-emitting van der Waals heterojunction," *Nano Lett.*, vol. 16, no. 9, pp. 5792–5797, 2016, doi: 10.1021/acs.nanolett.6b02527.

[19] W. Yang *et al.*, "Electrically tunable valley-light emitting diode (vLED) based on CVD-grown monolayer WS_2," *Nano Lett.*, vol. 16, no. 3, pp. 1560–1567, 2016, doi: 10.1021/acs.nanolett.5b04066.

[20] C. R. Dean *et al.*, "Boron nitride substrates for high-quality graphene electronics," *Nat. Nanotechnol.*, vol. 5, no. 10, pp. 722–726, 2010, doi: 10.1038/nnano.2010.172.

[21] Z. Wu *et al.*, "Intrinsic valley Hall transport in atomically thin MoS_2," *Nat. Commun.*, vol. 10, no. 1, p. 611, 2019, doi: 10.1038/s41467-019-08629-9.

[22] T. Y. T. Hung, A. Rustagi, S. Zhang, P. Upadhyaya, and Z. Chen, "Experimental observation of coupled valley and spin Hall effect in p-doped WSe_2 devices," *InfoMat*, vol. 2, no. 5, pp. 968–974, 2020, doi: 10.1002/inf2.12095.
[23] T. Y. T. Hung, K. Y. Camsari, S. Zhang, P. Upadhyaya, and Z. Chen, "Direct observation of valley-coupled topological current in MoS_2," *Sci. Adv.*, vol. 5, no. 4, p. eaau6478, 2019, doi: 10.1126/sciadv.aau6478.
[24] E. I. Rashba, "Theory of electrical spin injection: Tunnel contacts as a solution of the conductivity mismatch problem," *Phys. Rev. B*, vol. 62, no. 24, pp. R16267–R16270, 2000, doi: 10.1103/PhysRevB.62.R16267.

https://doi.org/10.1142/9789811229107_0006

Chapter 6

Valley-Polarised Single Photons

Tong Wang

6.1 Sub-Poissonian Quantum Light Sources

In this chapter, we turn our attention to a rather different flavour of optoelectronic devices — single-photon sources — based on the valleytronic materials of transition metal dichalcogenides (TMDs). Unless one has specifically worked on these non-classical light sources, the science of the relevant photonic concepts could be less familiar to both new and experienced valleytronic researchers alike. Therefore, we begin the chapter by introducing the reader to the basic science of the field, and subsequently explaining the advantages, challenges, and potential of TMD-based single-photon sources. In particular, we will also highlight the importance of the generation of linearly polarised single photons, and how the valleytronic properties of TMDs make this possible.

6.1.1 What is a Single-photon Source?

The very beginning of the explanation of a single-photon light source is surprisingly straightforward. As the word "single-photon" suggests, it is a light source that strictly emits only one photon at a time. However, any further examination of this definition will easily make one bemused. Why "strictly"? How do we define "at a time"? Is there a requirement on how large the time period between two emitted photons must be for the light source to be considered single-photon? In order to answer these questions, we have to take a detour and examine the nature of light sources first.

We know that light, or "bunches" of photons, emitted from a coherent laser and a thermal light bulb can be very different in their properties. The behaviour of these photons can be described by the nature of their emission and the probability distribution of their detection.

For a coherent laser with a constant emission intensity, the photon statistics can be shown to follow the **Poisson distribution.** We do not want to impose the readers with heavy mathematics in derivations but will instead directly mention this result for readers to familiarise with the relevant concepts more quickly. With a perfectly Poissonian distribution, if we define n as the number of photons, and the standard deviation Δn and mean $\bar{n}$ follow $\Delta n = \bar{n}$, it is characteristic of coherent light sources.

In reality, most of the light sources that we come across are not coherent. The photons from our monitor and smartphone screens, LED lights in the living room, and streetlamps along the road all have intensities that fluctuate with time. In these cases, a "bunch" of photons are emitted together at a time, while another "bunch" of more or fewer photons are generated subsequently. Due to these fluctuations, there is nothing that controls the coherence of these emitted photons. Without having to resort to mathematical proofs, it is already intuitive that in such cases, the standard deviation of photons is greater than that of coherent photons during the same period of time. Thus $\Delta n > \bar{n}$, and the photon statistics is described by a **super-Poissonian distribution**. We have mentioned the word "bunch" several times now as it describes the situation of photon emission well, and physicists indeed termed photons that follow such statistics as **bunched** light.

Following this method of categorisation, we have a third, less intuitive and more mind-boggling, flavour of light whose photon statistics follow the relation $\Delta n < \bar{n}$. Such distributions are called **sub-Poissonian** and describe the behaviour of single-photon sources. To contrast with bunched light, single-photon sources are said to exhibit photon **antibunching**. These light sources are much rarer to find and mostly live in the optical test benches of high-end laboratories. A graphical depiction of the three distributions can be seen in Fig. 6.1. To properly understand the concept of antibunching, we need to study the experimental

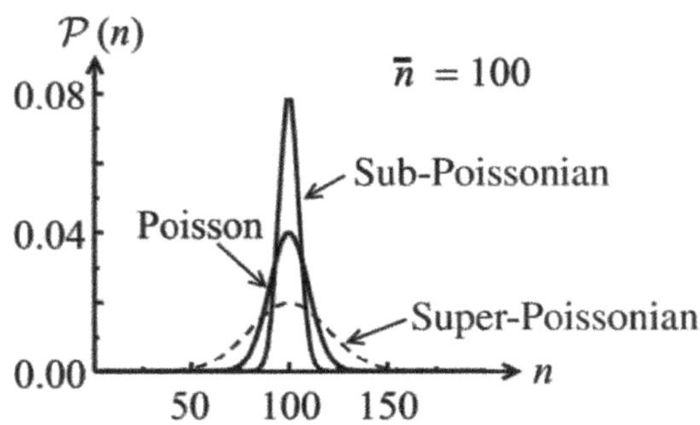

Fig. 6.1. A simple graphic demonstrating the photon statistics of sub-Poissonian, super-Poissonian, and Poissonian distributions. If the standard deviation of the Poissonian distribution is 100, then those of the sub- and super-Poissonian are less and greater than 100 respectively. Graphic obtained from https://physicsopenlab.org under CC-BY licence.

method that investigates the photon statistics of a light source — the Hanbury Brown and Twiss experiment [1] — and examine the non-classical consequence of single photons on the detection results. But before we delve deeper into more photophysics, let us first review the motivations for single-photon research.

6.1.2 Why do We Need this "Exotic" Flavour of Light Source?

In the information age that we live in, secure communication is one of the key demands of society. For instance, no matter how secure a classical information distributor is, there is always a non-zero probability that more sophisticated eavesdropping mechanisms can be developed to hack the transmission without alerting the source or the receiver. As a result, we see numerous cases of wiretapping, login information leaks, and other related news being reported from time to time. With quantum information sciences, this problem can be resolved in a fundamental way.

As a simplified example, the bits of information can be encoded to the optical polarisation states of photons emitted by a single-photon source. No matter how complex the wiretapping method is, a measurement on these single photons cannot be avoided. Based on the rules of quantum mechanics, such an interaction will collapse the quantum state of the transmitted photon and the integrity of information, thereby alerting the receiver that the security of the transmission has been compromised. If we use a bunched or coherent photon source that emits multiple photons at a time, each bit of information will be encoded in several photons. In this case, the wiretapper could in principle make use of the additional photons for theft, with minimal risk of being found. The actual implementation of these quantum information protocols would of course be much more complex than what is described here, but it is clear that single-photon sources are the indispensable hardware and are at the heart of fundamentally secure quantum communication.

As of now, ground-breaking demonstrations of the said optical quantum communication have already been made over a distance exceeding 50 km in daylight conditions [2]. Moreover, researchers have also developed these quantum light sources to operate in satellites, such as the *Mo Zi Hao* satellite from China that is already in operation and generating "unhackable" signals of information in space. These progresses paved the way for the building of future inter-satellite and satellite-to-earth communication networks using quantum information sciences.

However, the importance and usefulness of single-photon sources exceed quantum communication. Another one of its many applications is in quantum computing. Contrary to superconducting qubits or electrostatically confined quantum dots, single-photon sources operate on an all-optical platform. Therefore, issues such as rapid quantum decoherence and the requirement of <1 K operation temperatures do not plague single-photon emitters, giving these non-classical sources a significant head start towards the realistic construction and operation of quantum computing platforms. While different materials suffer from various sources of challenges, particularly in maintaining a reliable mesoscopic environment, there have been several innovative solutions to combat these comparatively less fundamentally challenging issues. The 2019 demonstration of "the quantum advantage" [3], or that a quantum computing operation works faster than a classical solution, has put the superconducting Sycamore chips in the limelight of not only the quantum computing research community, but also the general public. However, it is less well known that the first demonstration of "quantum supremacy" for a single-photon optical quantum computing platform was reported in 2017, where the complexity of boson sampling machines rivalled that of early classical computers for the first time [4]. The bottleneck of copper interconnects is fast approaching and calling for more innovative solutions in quantum computing. The all-optical approach offered by single-photon sources could be one of the most promising and realistic answers.

6.1.3 The Quests for Single Photons

Unlike conventional super-Poissonian light sources, the development of these sub-Poissonian photon emitters has been very slow until recently. Several key difficulties existed during the past five decades, particularly the devising of a method to probe these non-classical light sources, the identification of reliable two-level photon emitters, and the combat against their low emission intensity. The proposal of the right experimental setup came forth in 1956, by Robert Hanbury Brown and Richard Twiss [1]. The experimental detections of photon antibunching were first realised using trapped single atoms [5] a number of years later. It was not till around 2000 when many more systems joined the single-photon research community.

While single atoms and trapped ions continue to be good sources of single photons, they are not naturally occurring and need sophisticated setups to isolate. Coupled with the difficulty in integration into realistic optoelectronic platforms, they face significant fundamental challenges in moving further up the technology readiness level. The surge of single-photon development in the past two or three decades has been mostly made possible due to the success in realising sophisticated material growth and advanced optical detection systems. For instance, a semiconductor quantum dot [6], [7] grown by epitaxy could be thousands of times larger than a single atom, and yet the band offset against its matrix material could result in significant quantum confinement in all three spatial dimensions. Therefore, an atom-like two-level quantum system is achieved by deterministic growth, without the aforementioned difficulty in isolation. Such 3D confined quantum systems are also already integrated in semiconductor platforms. With the right level of doping and electric contacts, a *p-i-n* structure can be built and driven electrically as a single-photon emitting device. The deterministic placement of these devices across a wafer enables large-scale fabrication and the movement towards higher technology readiness levels. Hence, the non-classical light-matter interaction in single-photon sources are no longer limited to a laboratory setting.

Needless to say, there are several challenges in each of these steps forward, but they are being tackled by more advanced fabrication methods, novel nanostructure proposals, and innovative single-photon emitting materials. The 3D confined structures are not only found in epitaxially grown semiconductor quantum dots, but also in various material defects such as those in TMDs [8], diamond nitrogen vacancy centres [9], silicon carbide [10], carbon nanotubes [11], zinc oxide [12], perovskites [13], or even unprocessed wafers of gallium nitride [14].

Each of these systems has their own advantages and challenges. Take epitaxial quantum dots for example: the metal-organic vapour phase epitaxy routines that are needed for growth also introduce complications in the quantum dots' performance as single-photon sources. In particular, most quantum dots are fabricated using the

Stranski-Krastanov island formation mechanism [15], or variants related to it. Such a method still relies heavily on the dots' surrounding material. If the wetting layer is not spectrally separated from the quantum dot's emission, such as the case in ternary nitride alloys [16], the optically active regions emit photons with similar energies that contaminate the sub-Poissonian signature of the two-level system. Furthermore, the process of epitaxy and island formation inevitably create defects and charge-trapping sites in the vicinity of the quantum dots. The disordered instantaneous Coulomb interaction between trapped carriers and oscillator of the exciton undermines the indistinguishability of the emitted single photons.

Solutions to combat the complications arising from the mesoscopic environments have been proposed and made possible with more mature semiconductor platforms [17], such as arsenide-based nanostructures. Pillar structures with distributed Bragg reflectors defining a cavity that overlaps with one single active emitter have shown ultraclean optical emission with near-zero effects from the surrounding matrix materials [18]. When the resonantly excited emission of the active material is tuned thermally [19] or electrically [18] to the energy of the defined cavity, electrodynamic enhancement can be achieved. These developments allow total count rates of hour-long autocorrelation histograms, a concept we will explain in detail later, to be achieved in just one second, and are the building blocks for the aforementioned single-photon sources in satellites and complex boson sampling optical quantum computers.

However, even these systems are still fundamentally limited by their operation temperature at cryogenic conditions, and lack of intrinsic optical polarisation properties. For clarification, the "low" temperatures required here are around 4 K, which are much higher and easier to reach than the temperatures in dilution fridges mentioned earlier. Due to the small bandgap, even though the count rate from these systems exceeds tens of millions at 4 K, the emission is basically quenched to nil above 150 K, while losing their signature of single photons to the additional thermally induced energy uncertainty. The lack of intrinsic polarisation from the material can only be partially resolved using nanostructure engineering, with effects similar to a "built-in" polariser that limit the dipole transition output to one direction. These solutions currently involve the use of horizontally lying nanowires [20], elliptical nanopillars [21], and pyramids with asymmetric truncated top [22]. However, none of these can provide fully dipole manipulation and near-unity polarisation degree, and all of them require very sophisticated methods of fabrication.

This is where defect states in other materials prove to be more useful. Most defects in nitrogen vacancy centres, carbon nanotubes, and even nitride wafers have been shown to operate at ambient conditions. Defects from layered hBN set a record of 800 K for high-temperature single-photon generation [23], with stable emission characteristics unaffected by thermal cycles (that is, from 4 K to 800 K and back to 4 K). The advantages of defects lie in the depth of their trap. The large activation energies that are required to surmount before intensity quenching occurs make them extremely suitable for operations in realistic temperatures, such as those at room temperature (~295 K) and in unstable, heated electronic enclosures (>350 K). In the case of monolayer TMDs, the formation of strain-induced defects creates radiative carrier traps with a deeper confinement potential than the free exciton bandgap. Thus single, sharp emission peaks can often be observed at a higher energy. Since strain could affect the entire wafer, the densities of these single-photon emitting defects are also very high.

Two of the most significant advantages of TMD defects compared to other platforms are their intrinsic optical linear polarisation and ability to achieve electrically pumped single-photon generation. As a valleytronic material, the superposition of the emitted photons from two valleys results in linearly polarised light output, which then allows intrinsic polarisation control at the material level without the need for external nanostructure engineering. For most defect systems, the ability to build an electrically driven device has been the bottleneck of further development. In the case of monolayer TMD, the ability to stack hBN with TMD allows tunnelling of carriers and thus the making of electrically pumped single-photon emitters [24], [25]. Understanding the two-level systems in TMD defects, researchers can now use the tip of atomic force microscopes to artificially create defects, and place single-photon sources at any location of the wafer of their choice [26]. Other attempts have also been made to place TMDs on top of deterministically fabricated nanopillars, an approach similar to those used in epitaxially grown semiconductor quantum dots, and have managed to achieve extremely high quantum efficiency [27].

As we see, the quests for single-photon sources have increased in number at an accelerated pace. Many systems using different materials and nanostructures have, step by step, moved the development of single-photon sources towards higher technology readiness levels. This is especially so in the valleytronic material of TMDs, where several advantages make them one of the ideal candidates with great potential.

6.2 Hanbury Brown and Twiss Experimentation

Having looked at the motivations for single-photon sources and overviewed the current developments of various platforms, we now take a closer look at the detailed science and experimental strategies for investigating and characterising a single-photon source. The theory behind the experimental setup will be explained along with practical challenges in actually running a single-photon experiment. After we go through the details, we will realise that the testing of single photons is far more challenging than the studies of photoluminescence (PL) or polarisation-resolved PL alone, and requires the precise maintenance of several critical factors.

6.2.1 Optical Setup and Second-order Autocorrelation

As mentioned previously, the proposal of the single-photon experiment was first made by Hanbury Brown and Twiss (HBT) in 1956 [1]. Their proposal has been so useful that we are still using the same HBT setup, only with newly developed, more precise and sensitive optical components, in today's single-photon laboratories. An actual HBT setup requires many different and carefully chosen optical components for the specific study, but the basic concept can be simplified and illustrated by Fig. 6.2.

As shown, the light beam generated from the source will first pass through a 50:50 beamsplitter, which will be placed at 45° to the beam. The beamsplitter has a transmission and reflection of exactly 50% each. The incoming light is then split into two perpendicular ones with equal intensity, which is half of the original beam. One of the light beams reaches a photomultiplier tube (PMT), which is a highly sensitive light detector and makes use of the photoelectric effect to convert the light signal into electric signal. The signal is then amplified within the PMT before being directed to a time-correlated single-photon counter, which has a "Start" and a "Stop" switch. The electric signal is then registered on the "Start" with a time stamp,

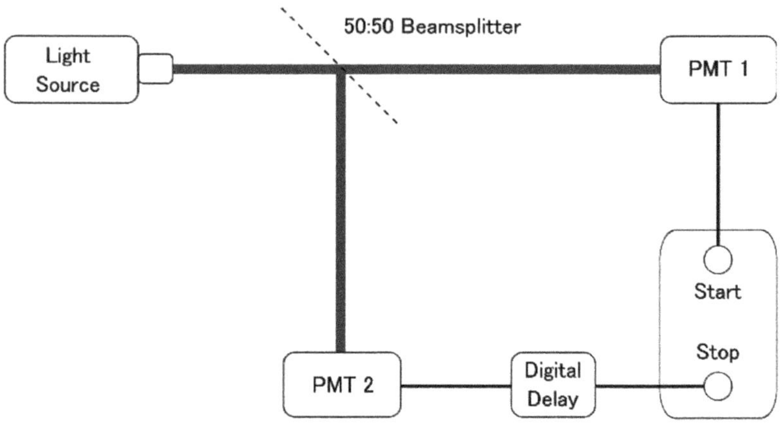

Fig. 6.2. Simplified illustration of a Hanbury Brown and Twiss experimental setup for the study of second-order autocorrelation and examination of photon statistics of a given light source.

within the resolution of the detector. The second beam will also eventually pass through an identical PMT detector, and be registered with a time stamp on the "Stop" switch. The only difference here is the addition of a digital delay on the second beam path. This delay can of course be added physically by altering the optical path length difference between the two arms. In this case, the two path lengths are the same in Fig. 6.2 and only a digital delay is introduced. The delay can be varied to study what is called the **second-order autocorrelation function**, which simply put is a type of relationship between the intensities of two equally split branches of the original light when a time delay is introduced to one of the branches, like we explained in the earlier optical setup. Eq. (6.1) is the exact form of the function that allows one to plot the entire trace of time-dependent histograms accumulated from the single-photon counter [28].

$$g^{(2)}(\tau) = \frac{\langle I(t)I(t+\tau)\rangle}{\langle I(t)\rangle\langle I(t+\tau)\rangle} \tag{6.1}$$

Here, I is the intensity of each branch of light, t is the time, τ is the delay time introduced, and the brackets denote the average. To make the discussion more straightforward for our readers, we only consider one special case of the second-order autocorrelation function — its form at a time delay τ of zero. We define one event as the **simultaneous** registration on both the "Start" and the "Stop" switches. As another simplification without resorting to using mathematical derivations, we define the second-order autocorrelation at time zero, or $g^{(2)}(0)$, as the ratio of the detected events to the maximum possible number of events. In this case, we can determine whether the light source is sub-Poissonian by only looking at the value of $g^{(2)}(0)$ itself.

We first consider a case where the light source is a conventional one, and multiple photons are emitted at the same time. In this scenario, at each given time, half of the photons will reach detector 1 while the other half detector 2. Hence, there will always be events registered at all possible time instances, giving a $g^{(2)}(0)$ of 1.

Now we look at what happens when a single-photon source is used. At each given time, there will only be one photon incident on the beamsplitter. However, as a single indivisible photon, it can only either pass through the beamsplitter and reach detector 1, or be reflected and reach detector 2. The probability is 50% either way. But no matter which scenario it is, we will never see any events registered on the photon counter, because we simply cannot have simultaneous events with one single photon. In this case, the value of $g^{(2)}(0)$ is 0, and is indicative of a perfectly pure single-photon source.

Note that at this instance, the classical wave theory of light completely breaks down. If we consider light as waves, the value of $g^{(2)}(0)$ will still be 1 because the intensity of that "one photon's worth" of light wave would be halved and two beams will still be directed to two detectors. However, in numerous real-world experiments till now, we have observed $g^{(2)}(0) < 1$, thereby confirming the need for the quantum nature of light to explain this non-classical antibunching behaviour.

You might have realised that we said $g^{(2)}(0) < 1$ instead of $g^{(2)}(0) = 0$ when we mentioned real-world experiments. This is because in an actual HBT experiment setting, it is almost impossible to 100% block off all background photon noise that could very easily be picked up by the highly sensitive single-photon detectors. It is equally challenging to make a single-photon source that has absolutely zero factors perturbing its two-level transition. Therefore, the imperfection of existing single-photon sources and limitation of optical setups mean that the detected $g^{(2)}(0)$ value will almost always be greater than 0 to some extent. Nonetheless, as long as $g^{(2)}(0) < 1$, non-classical photon antibunching is confirmed.

In this case, does the observation of antibunching alone confirm the identity of a single-photon emitter? A fully quantum mechanical treatment, which will not be presented here in this book, will demonstrate that the threshold between a one-photon and a two-photon Fock state is at $g^{(2)}(0) = 0.5$. In other words, assuming there are no optical noise signals at all, the emission from the light source itself needs to follow $g^{(2)}(0) \leq 0.5$ to confirm the presence of a single-photon emitter. It becomes tricky to add in the effect of noise signals theoretically, as they could arise from various sources and are most likely not under the same quantum basis in the Hilbert space that we are considering. However, if we move back to our earlier simplified definition of $g^{(2}(0)$, it becomes easier to see that we should still require a condition of $g^{(2)}(0) \leq 0.5$ to claim that less than half of the total possible events (recall: events = not single-photon) are registered. Consequently,

$g^{(2)}(0) \leq 0.5$ would imply that there is a higher probability of single-photon, instead of non-single-photon, emission even in the presence of noise signals. In such a case, of course, the 'true' purity of the single-photon source itself would need to be much smaller than the threshold of 0.5 to compensate for the effect of noise signals.

Here, we put inverted commas over the word 'true', because it begs the question of what constitutes the true purity of a single-photon source. We can argue that the intrinsic performance of the emitter alone should be used as the true benchmark. But without its surrounding material and the imperfect optics, the emitted single photons cannot be detected or used for any potential applications. The performance of the entire optical system that allows the emission and detection of single photons should thus be considered as the true benchmark. We call this performance $g^{(2)}_{raw}(0)$. Many attempts have been made to correct for background noises, which will of course result in a lower value called $g^{(2)}_{corr}(0)$. At this stage, it is crucial to understand that $g^{(2)}_{raw}(0)$ should be the **only** benchmark as it reflects the realistic performance of the single-photon emitting system. On the other hand, $g^{(2)}_{corr}(0)$ should only be used to understand the performance limitations of the system and to gain further insights on the potential areas of improvement. Moreover, most of the post-experiment background correction techniques are without a strong theoretical foundation, prone to large uncertainties, and should be used with extra caution.

To summarise the key points of our autocorrelation discussion, we should know that:

(1) Evaluation of $g^{(2)}(0)$ in an HBT experiment is the necessary measurement to claim whether a system is emitting single photons.

(2) $g^{(2)}_{raw}(0) < 1$ is the necessary condition for the confirmation of non-classical photon antibunching.
(3) $g^{(2)}_{raw}(0) \leq 0.5$ is the necessary condition for the confirmation of a single-photon source.
(4) $g^{(2)}_{corr}(0)$ should not be used as a single-photon performance benchmark.

It is important to stress point (1). One cannot simply claim that an emitter is a single-photon source, just because another on the same sample is. Moreover, one cannot claim to have single-photon emitters just because such emitters have been seen on other similar samples. The words "single photon" can only be used if an accompanying $g^{(2}(0)$ for the exact emitter in question has been performed and the condition of $g^{(2)}_{raw}(0) \leq 0.5$ fulfilled.

For point (2), we used the word "necessary". In realistic single-photon applications such as quantum key distribution and optical quantum computing, $g^{(2)}_{raw}(0) \leq 0.5$ is necessary but not **sufficient**. In most of the usage scenarios, $g^{(2)}_{raw}(0) < 0.1$ or much lower would be required [17]. Therefore, the pursuit of single-photon purity has been the top ongoing challenge for all researchers in this community. However, if we browse through single-photon publications and look at the materials and nanostructures investigated, we will realise that most of the time the results reported are >0.1 and in fact closer to 0.5. Furthermore, we tend to see many more reports of optical properties of these systems other than their single-photon characteristics. This is because HBT experiments are significantly more challenging to perform than other optical PL measurements. For many materials platforms, it becomes exceedingly difficult to even obtain a positive HBT result at all. In the following subsection, we explain what some of these challenges are and how they could potentially be tackled.

6.2.2 Challenges in Performing Real-life HBT Experiments

The most significant challenge in an HBT experiment is spectral isolation and intensity maximisation. In most single-photon emitting platforms, the PL itself usually contains far more than the sharp peaks of single-photon sources. They can be charged excitonic states, broader emission features of underlying materials, radiatively active defects and even the phonon side band spectrally overlapping with the emitter itself. Over the entire spectral detection range, these unwanted signals are usually much stronger than the single-photon signal. In a normal PL study, we could isolate the data of each emitting source by analysing them separately. However, during the HBT experiment, this isolation has to be done *in situ*, prior to the photons reaching the detectors. Some of the commercial spectrographs are capable of selecting individual pixels by limiting the size of the output slit. However, most spectrographs have detection losses and the selected light intensity becomes weak. A workaround is to bypass the spectrometer in the optical collection arm, and build a separate HBT detection optical branch. Before the spectral separation, a pair of tuneable bandpass filters can be used to define the two limits of interest at the bluer and redder ends of the spectrum. While this solution mitigates losses in the spectrometer, even state-of-the-art tuneable bandpass filters do not have perfect transmission at all angles, and do not possess exact boxcar-like transmission profiles. Depending on the optical profile of the emitter, either of these methods could be the better solution and detailed investigations would have to be undertaken to ascertain the best way forward.

An illustration of how either of these two setups can be done is shown in Fig. 6.3, which outlines the simplified detection arm of a PL with HBT capabilities integrated. It is worth noting that only one of the

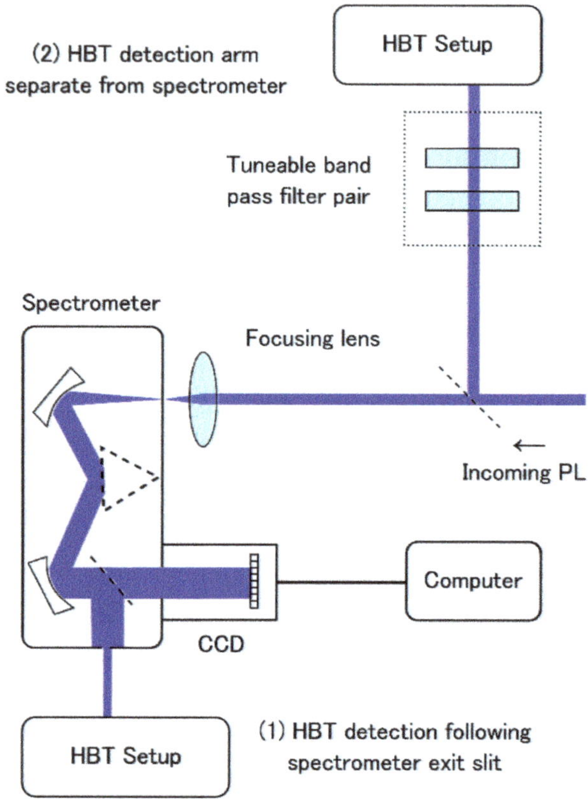

Fig. 6.3. Illustration of a typical detection scheme in a PL system that incorporates a Hanbury Brown and Twiss experimental setup. Two possible ways for this to be done are both illustrated on the same diagram: (1) using the exit slit of the spectrometer, and (2) building a separate branch before the spectrometer.

two is needed — the inclusion of both HBT setups is for illustration purposes and for contrasting their differences. For the setup before the spectrometer, it is also important to note that the pair of tuneable bandpass filters need to be as high quality as possible, for reasons discussed above. Due to the physical thickness of these filters, there might be potential beam path changes when the PL is incident at an angle. The small difference can often cause undesired effects, especially in worsening optical alignment. A way to combat this is to tune the pair of bandpass filters at opposite angles (e.g., ± 30°) in order to compensate for this beam path change.

We should not be confused that the spectral selection *per se* is better with the spectrograph. But the issue that is equally important is with the detected intensity. In a standard PL setup, we could observe the emission profile even if its count rate is <20 cts/s, and do most optical analyses with it. However, such a low count rate would pose significant practical challenges in an HBT experiment. As we recall, we will need to halve the count rate, and have both of the two split beams detected in two detectors. Therefore, the intensity of the accumulated histograms by the event counter is proportional to the square of half the total PL intensity. For 20 cts/s, the maximum would be 10 cts/s at each detector. If the intensity falls to 50%, rate of event registration would fall to 25%. Based on experimental experience, a single-photon emitter with a Gaussian linewidth of 1 meV and a peak count rate of 1000 cts/s would likely require a minimum of 1 h histogram accumulation with typical commercial PMTs to ascertain the presence or absence of a dip at time zero. Hence, maximising the intensity in an HBT experiment is as important as maximising the desired spectral selection.

One obvious solution to increase brightness is at the light source itself. As discussed in Chapter 3, the ideal situation of PL is one in which 1 incoming photon of excitation generates exactly 1 photon in emission. However, in realistic systems, especially semiconductor platforms, the underlying material photophysics can be very complex. Various known and unknown mechanisms happen during light-matter interaction and decrease the quantum efficiency. For instance, the mesoscopic environment of a semiconductor is never pristine — there is always a finite defect density or threading dislocation count even in the best commercial samples and substrates. In these cases, it is highly likely that charge-trapping sites can form due to these defects, or localised strain fields. These sites could possibly act as non-radiative decay centres, meaning that it is more energetically favourable for an exciton to undergo these pathways, without recombining radiatively and emitting a photon that we can detect. These centres usually have their own activation energies, which can be different across different samples and material systems. While it is not possible to eradicate them, they can be minimised by improving the quality of the materials. For instance, if the material has a high uniformity and smooth morphology, it is less likely for these non-radiative decay sites to form in large quantities. With higher-quality materials, it is therefore more likely to have radiatively emitted photons and thus higher brightness for single-photon experimentations. Again, due to the more stringent requirement of brightness, the requirement of the corresponding material quality is also significantly higher than that for simpler PL or circular dichroic PL characterisation.

Apart from the development and growth of more optically active materials, there have also been other revenues to increase the total brightness of emission, particularly in nanostructure engineering. Most of these utilise the physics of cavity quantum electrodynamics (cQED) — the manipulation of quantum systems through their coupling to optical cavities [29]. There are several regimes of coupling through cQED, and the one that is useful in our scenario is called "weak coupling". In this regime, when the optical mode of a cavity is in resonance with the two-level single-photon emitter, the rate of the exciton transition increases. From another perspective, the weak coupling process increases the intrinsic energy uncertainty of the single-photon transition. Due to Heisenberg's uncertainty principle, the time uncertainty of this transition decreases, resulting in faster radiative recombination lifetimes. This is a simplified manner to view the Purcell Effect. In PL studies, the Purcell Effect is visually seen as the narrowing of the exciton transition linewidth in time-integrated PL, as well as the increase of exciton population decay speed in time-resolved PL. Through this process of Purcell enhancement, excitons recombine at a faster rate. At the nanoscopic level, radiative and non-radiative recombinations are two competing processes when an exciton is formed. With faster radiative recombination rate, it is less likely for the excitons to undergo non-radiative recombination. At the macroscopic level, what we observe is an increased PL intensity count rate with the same excitation and collection conditions. In turn, the efficiency of detection and single-photon brightness can be significantly enhanced.

Many single-photon systems, especially those built with the mature arsenide semiconducting platforms, use thermally or electrically tuned cavities to achieve this weak coupling [18], [19]. For instance, a popular cavity that is widely used is a nanopillar structure with top and bottom distributed Bragg reflectors (DBRs). By designing the exact height, width, and number of DBR layers, one single-photon emitter can be embedded in a cavity with an optical mode that is spectrally very close to its exciton transition. The emitter's wavelength is then tuned, e.g., thermally following Varshni's equation, to match the cavity mode exactly, thereby achieving weak coupling and Purcell enhancement. With the nanopillar structure, any excessive material that is also optically active and might adulterate the single-photon signal has also been removed. A well-optimised system can often achieve near-unity brightness, and near-perfect single-photon purity.

For most other systems, especially TMD-based ones, the development of single-photon sources is still at an initial stage. In these cases, the linewidth of the desired exciton transition can often be significantly broader than the theoretical values, and it becomes very hard to achieve weak coupling with meaningful Purcell factors. The issue of an imperfect mesoscopic environment also plagues these systems to a greater extent. However, it is worth noting that even though meaningful cQED enhancement has not been realised yet, the use of cavity-like nanostructures, e.g., micropillars without top-and-bottom DBRs, can also enhance the PL count rate considerably. The reason for this is threefold:

(1) Photon extraction efficiency: Even without the Purcell effect, the vertical nanostructure is still able to guide the light, so that the radially emitted PL is directed upward. This significantly increases the light extraction efficiency. In systems where the efficiency is $\ll 1\%$, the use of nanopillars can often increase the count rate by several times. For this purpose alone (without the need of site control), there are many ways in which the desired pillar structures can be fabricated. The simplest manner is the spin-coat silica nanospheres, or similar particulate masks, onto the sample, followed by etching of the sample directly. Nanopillar structures will hence form at sites where these masks are present. The simplicity of processing and significance in the brightness increase make this method highly attractive for a variety of systems under initial development.
(2) Removal of materials: The material surrounding the single-photon emitter could increase the likelihood of non-radiative decay, in systems where such pathways are abundant. Furthermore, if these surrounding materials are also optically active under the same laser excitation, it would adversely affect the single-photon purity of the system as well. Nanopillars remove most of these surrounding materials, thereby making these issues much less severe.

(3) Site control: Nanopillars can be fabricated in well-defined arrays. This is especially useful in TMDs due to the 2D nature of the materials. Using these nanopillars, the randomness of the occurrence of single-photon emitters can be significantly reduced — the emitters would mostly become site controlled. We will discuss these developments in more detail in the next section.

6.3 Single Quantum Emitters in Valleytronic Materials

6.3.1 Development of TMD-based Single-photon Sources

The momentum of single-photon research in TMD-based valleytronic materials picked up from 2015, when a series of these non-classical light sources were reported in WSe_2 [30]–[33]. In low-temperature PL studies, additional sharp emission features are sometimes observed at the redder end of the spectrum, with higher emission energies than the exciton and trion peaks. When the PL from some of these peaks was directed to an HBT setup, researchers found that they are actually single-photon sources. The frequent appearances of these sharp peaks also indicate that they exist with a very high density of 0.05 μm^{-2}. The origin of these sharp peaks was ascribed to the localised defects.

The increase of single-photon research interest in TMD defects was also partly due to the breakthroughs in these light sources. Around these few years, many other systems have also demonstrated single-photon emission with very interesting characteristics. In particular, it is very common for an imperfect-bound defect state to emit as a two-level system. The confinement potential of these localised defects is similar to those of a semiconductor quantum dot, and produces atom-like delta function-type density of states. This is a very significant advantage, because we do not need complex optical designs to isolate single atoms or ions, nor do we require challenging fabrication methods to grow 0D quantum structures. The emitters are naturally found in the TMD samples. If we want to control the formation of these single-photon emitters, researchers have also shown that we can use the tip of atomic force microscopes to artificially create such imperfect sites in the material, and deterministically fabricate single-photon sources. Electrically driven single-photon emitting devices have been built [24], [25], and large-scale site-controlled nanopillar enhanced single-photon emitters have also been realised [27].

6.3.2 Optical Properties of TMD-based Single-photon Sources

To understand the performance of these TMD single-photon emitters, we firstly need to examine their optical properties. For an ideal single-photon two-level emitter, we expect their optical transition to exhibit a Lorentzian profile, with very narrow linewidth around ~10 μeV (based on current arsenide systems). The non-zero linewidth is caused by the energy uncertainty of the transition of a quantum system, which is directly related to the radiative recombination lifetime. The linewidth gives us two important insights:

(1) Fast-timescale spectral diffusion. This occurs at ~ns timescales where carriers passing by the single-photon source could perturb the two-level transition and increase its energy uncertainty. As a result, the PL peak location instantaneously shifts. Under typical time-integrated PL studies, the normally distributed Lorentzian peak shifts produce a Voigt (Gaussian + Lorentzian) profile with

a broadened linewidth. The degree of broadening compared to the theoretical Lorentzian linewidth is thus an indication of this spectral diffusion.

(2) Photon indistinguishability. This is also related to the previous point on spectral diffusion. For application in optical quantum computing, we need each photon to be indistinguishable in their emission energy, phase, and other optical properties. A broadened linewidth will compromise photon indistinguishability and the potential applications of these single-photon emitters.

Figure 6.4 shows the PL characteristics of typical WSe_2 single-photon sources [31]. At the redder end of the spectra (lower energy) relative to the neutral exciton peak, there are several sharp peaks which correspond to emissions for localised material imperfection sites. The detection of these peaks usually requires a high-resolution spectrometer. The typical ones used in the study of TMD neutral and charged excitons might not have the right combination of length and grating needed to resolve these peaks, which are typically around 100 μeV.

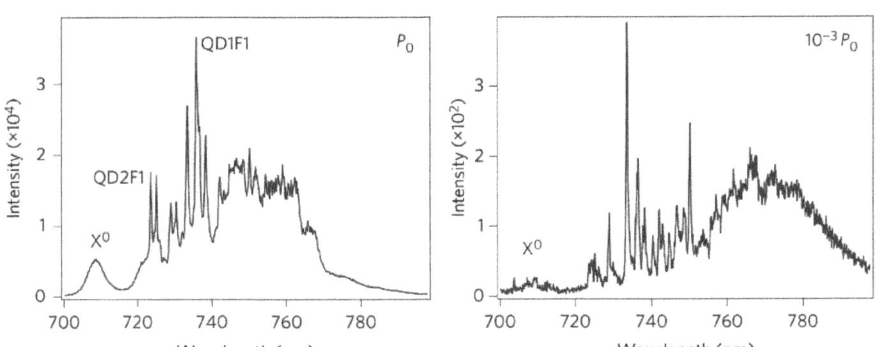

Fig. 6.4. PL spectra of typical WSe_2 single-photon sources based on defect-localised states. There are a number of sharp peaks at the lower energy end of the exciton peak. Reproduced with permission from Ref. [31].

At the same time, most of these sharp peaks are observed at cryogenic temperatures. Although the free exciton transitions of TMDs persist to room temperatures, the same usually cannot be expected of these defect-bound single-photon transitions. The quantum confinement required for these two-level systems is usually surmounted by the increase of thermal energies. Over a certain barrier determined by the specific sample and material, the probability of carriers being confined in these states and emitting radiatively approaches zero. This thermal quenching of TMD single-photon sources is an ongoing challenge that needs to be resolved.

The transition energies of these single-photon sources are also not unique but varied across a small range of wavelengths. Under the same instant of laser excitation, several of these sharp peaks can be detected. Some of these peaks are overlapping, and they all have different emission count rates. Therefore, for single-photon investigations, selection of suitable emitters and the corresponding spectral isolation would be needed. It is also worth noting that most of these sharp peaks are spectrally overlapping with a broad background, which is likely not single-photon emitting (or not in the same basis/from the same source). Therefore, the selection of suitable candidates for HBT experimentations can be quite challenging, and is likely to occur only once in many. Despite these challenges, it is promising to note that the ~100 μeV linewidth is very narrow, and already better than many other materials where the linewidth can be as high as ~1 meV. Nonetheless, the narrow linewidth is still higher than the theoretical Lorentzian linewidth, indicating that there is non-zero spectral diffusion in the system.

At this point, it is important to point out that while fast-timescale spectral diffusion is captured in time-integrated PL linewidth, slow-timescale diffusion still requires a separate time-dependent study. These spectral wandering effects are caused by carriers trapped in the vicinity of the emitter and are especially relevant in defect-localised single-photon sources. The fact that we are seeing multiple sharp peaks in the same spectrum is likely an indication that many other imperfection sites are present. In this case, the trapped carriers, radiative or non-radiative, can persist from milliseconds to seconds, and perturb the single-photon transition for a much longer time period. As shown in Fig. 6.5, the ~100 μeV peak actually fluctuates over a range of ~1 meV during the course of 2 min [31]. For applications that require photon indistinguishability, these spectral wandering effects are also undesirable. Potential ways to reduce these slow-timescale spectral diffusion is to use cavity-enhanced systems, as we discussed earlier.

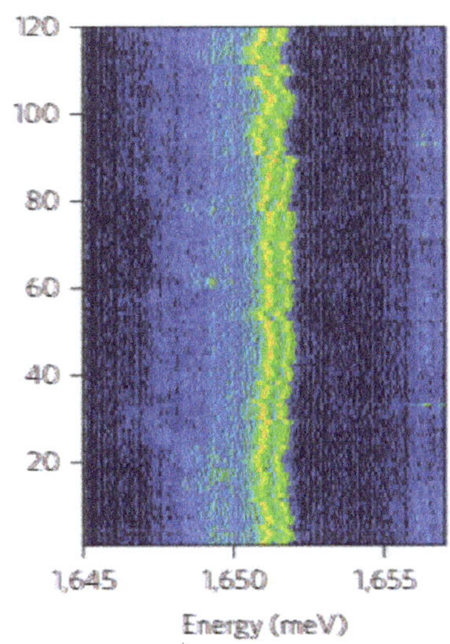

Fig. 6.5. Slow-timescale spectral diffusion of a WSe_2 single-photon source based on defect-localised states. The range of spectral wandering is around 1 meV over a course of 2 min. Reproduced with permission from Ref. [31].

Despite these challenges, it is still an important milestone that proof-of-concept of single-photon generation has been demonstrated in selected emitters in these systems. The initial results from TMD-based single-photon sources showed purities of ~0.2, which is well below the threshold of 0.5. However, it is important to note that while a $g^{(2)}(0)$ of 0.5 is the criteria for the successful demonstration of a single-photon emitter, it is not sufficient for application in quantum information sciences or optical quantum computing, both of which require much higher single-photon purity. For quantum key distribution, a $g^{(2)}(0)$ of <0.1 is usually needed. The requirement for optical quantum computing or quantum memory is much more stringent, and a $g^{(2)}(0)$ of <0.001 would be necessary.

It is good to see the fast pace of development in TMD-based single-photon sources, because emitter with $g^{(2)}(0)$ of <0.1 has already been reported in nanopillar arrays of WSe_2 single-photon sources. As shown in Fig. 6.6, a $g^{(2)}(0)$ measurement of 0.03 has been reported

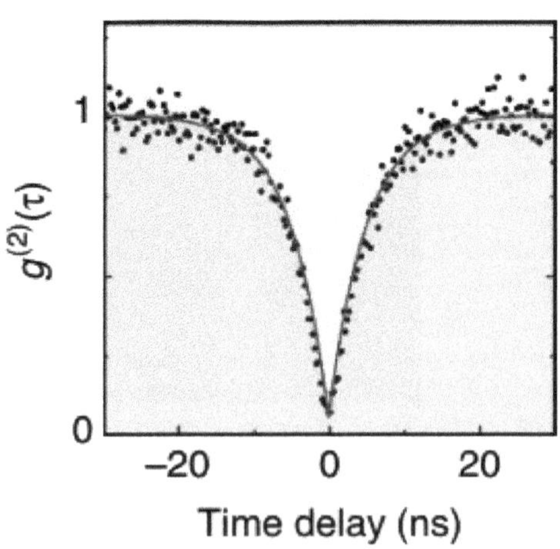

Fig. 6.6. Photon autocorrelation results of a representative WSe_2 defect-localised single-photon source, under continuous wave laser excitation. The $g^{(2)}(0)$ is 0.03, demonstrating excellent single-photon purity. Reproduced with permission from Ref. [27].

and it in principle fulfils the requirement of quantum key distribution [27]. Although 0.03 is still not <0.001, and higher than most arsenide-based systems, it already meets the $g^{(2)}(0)$ of <0.1 requirement needed for quantum information applications [17].

One other key optical property is the radiative recombination lifetime, which is also indicated by the slope of the antibunching dip in autocorrelation data. Separate time-resolved PL measurement demonstrated radiative lifetimes between 1 and 3 ns. These experimentations can be easily done with a slightly modified HBT system. Instead of splitting the PL and detecting the number of coincidences on two detectors, we can direct a small portion of the excitation laser to one of the detectors. In this case, the laser would act as a "source", while the actual PL signal would act as the input. The signals are still directed to the same time-correlated single-photon counting cards. In this case, the count rate will only linearly depend on the PL signal, instead of quadratically, making the measurement requirement on PL intensity significantly less stringent than in an HBT autocorrelation.

It is important to note the difference between radiative lifetime and valley lifetime, both of which have been discussed in Chapter 3. In this case, the radiative lifetime of the TMD-based single-photon sources (~1 ns) are much longer than that of the excitons (<100 ps). This is expected because we are not looking at the free exciton transitions themselves. Instead, we are focusing on the localised states, e.g., a particular radiatively active carrier trap, or an imperfect material site. Carriers are bound in these states by potential barriers and tend to stay there for a longer time than the free excitons.

For valley lifetime, we would like it to be long because we need the valleytronic properties to persist. On the contrary, for single-photon applications, we need the radiative lifetime to be short. This is because most applications in quantum information sciences and optical quantum computing require these light sources to work as "repeaters". Thus, the repetition rate of the single-photon sources directly determines their potential performance and usefulness in these applications. No matter how fast our excitation is, optical or electrical, the radiative recombination lifetime is an upper limit. The two-level system can only accommodate one exciton at a time before it recombines and emits a single photon. If the radiative lifetime is too long, the repetition rate will inevitably become low. In this case, a 1 ns radiative lifetime will in principle translate into a 1 GHz repetition rate. Hence, the current TMD-based single-photon sources have near GHz repetition rates, which can potentially be further increased by cavity coupling and Purcell enhancement.

6.4 Optical Linear Polarisation, and Other Advantages and Disadvantages of TMDs

6.4.1 Achieving Linear Polarisation in Single-photon Sources

In the beginning of the chapter, we discussed the motivations in single-photon research with applications in quantum information sciences and optical quantum computing. If we look more closely at both application scenarios, we will realise that achieving optical polarisation control in single-photon sources is crucial. For truly secure communication, we need to employ fundamentally unhackable quantum key distribution protocols. An example of this is the BB84 protocol, which dictates that the information be encoded into the polarisation state of each single photon. In this case, a pair of orthogonally polarised states should exist, and each of them should correspond to 0 and 1 respectively. For applications in optical quantum computing, such as demonstrations using boson samplings, linearly polarised photons are also required.

The simplest method to achieve linear polarisation is to use a polariser on the PL output of the single-photon system. In this case, the output can be adjusted to a desired linear polarisation with a deterministic and controllable axis. However, one critical issue of this setup is the sacrifice of 50% brightness, for PL that is unpolarised. This sacrifice could be critical in limiting the potential performance of single-photon sources. In order to circumvent this problem, researchers tend to build single-photon platforms that are polarisation controlled. There are two ways in which this has been done — nanostructure engineering, and materials with intrinsic polarisation properties.

The first is through nanostructure engineering. Similar to the concept of nanopillars enhancing photon extraction efficiencies, we can modify the geometry of the nanostructure containing the single-photon emitters, and thus control its dipole's oscillation direction. For example, nanocolumns with elliptical tops have been built, so as to preferentially enhance the dipole transition strength along the long axis of the ellipse. However, there is a finite limit to the degree of ellipticity and possibility of fabrication. Other innovative designs have used a pyramidal structure with a specially truncated top. In this case, the pyramid supports the structure in a more stable manner than nanocolumns, but the truncated top can have a much larger difference in the x-y dimensions. Both of these structures have shown polarisation-controlled single-photon generation in different materials. Another way to make use of already engineered nanostructures is to control the location of growth. Researchers have found innovative ways to grow these quantum emitters on the sidewalls of horizontally lying nanowires. In this case, the large aspect ratio of the nanowire enhances the dipole strength along its length, resulting in near-unity optical linear polarisation degrees.

However, these innovative solutions have one challenge in common — futureproofing. After these initial demonstrations, it becomes much harder to engineer contacts and make realistic devices out of them. This is why an increasing number of researchers are also looking at the use of materials that have intrinsic polarisation properties. The advantage of this is to avoid complex engineering, and thus making it less restrictive to design devices and contact solutions. The intrinsic polarisation properties also tend to have very high polarisation degrees. However, most materials do not possess such properties. TMDs in this case are the lucky ones, and do have intrinsic optical polarisation characteristics thanks to their valleytronic properties.

6.4.2 Valleytronic Origin of Optical Linear Polarisation

In TMDs, optical linear polarisation can be achieved because of the valleytronic properties of the emitters. When we pump the system with unpolarised light, both valleys are excited. As a result, an equal amount of co-circular and counter-circular light would be absorbed by the two valleys. Subsequently, PL with the same circular polarisations would be emitted. The superposition of equal amounts of co- and counter-circular light results in optical linear polarisation, and is the primary reason for the generation of linearly polarised single photons.

This optical linear polarisation is also "protected" from valley depolarisation, and is not affected by the valley coherence lifetimes. In this case, even though valley depolarisation is bound to happen, we are only looking at the superposition of the two valleys. No matter how they change, the valley population would still be identical if we used unpolarised or linearly polarised excitation. Even if we used circularly polarised light for excitation, the depolarisation mechanism will only make the population of the two valleys more even. As a result, the superposition will still give us optical linear polarisation degrees close to unity.

6.5 Summary

In this chapter, we have discussed the basic science of single-photon sources, their potential applications in quantum information sciences and optical quantum computing. The concept of HBT experimentation and the importance of $g^{(2)}(0)$ have been discussed in detail. We have also looked at some of the highlights of the current single-photon research in TMDs and examined their optical properties. Specific considerations in practically carrying out related experiments have also been explained.

As discussed, there are several advantages that TMD-based single-photon emitters have over most other systems. These include the natural occurrence in typical samples, narrow transition linewidth, relatively good single-photon purity achieved, and the linearly polarised emission. However, it is also important to note that some of these initial advantages might turn out to be the intrinsic limiting factors for future development. For instance, while at the current stage it is easy to find or even engineer defect-localised emission sites in imperfect materials, it could be challenging to further improve the emission characteristics due to the high defect densities of the material itself. As discussed earlier, for realistic applications, we will eventually need the material quality to be extremely high. But we have already seen that such defects are already having an adverse effect on the spectral stability by introducing large spectral diffusion. Nonetheless, the realisation of $g^{(2)}(0) < 0.1$, electrically driven single-photon emission, and scalable site-controlled nanopillar fabrication over the span of merely half a decade demonstrate the tremendous potential of TMDs in single-photon research and their applications in quantum information sciences.

References

[1] R. H. Brown and R. Q. Twiss, "Correlation between photons in two coherent beams of light," *Nature*, vol. 177, no. 4497, pp. 27–29, 1956, doi: 10.1038/177027a0.
[2] S.-K. Liao *et al.*, "Long-distance free-space quantum key distribution in daylight towards inter-satellite communication," *Nat. Photonics*, vol. 11, no. 8, pp. 509–513, 2017, doi: 10.1038/nphoton.2017.116.
[3] F. Arute *et al.*, "Quantum supremacy using a programmable superconducting processor," *Nature*, vol. 574, no. 7779, pp. 505–510, 2019, doi: 10.1038/s41586-019-1666-5.
[4] H. Wang *et al.*, "High-efficiency multiphoton boson sampling," *Nat. Photonics*, vol. 11, no. 6, pp. 361–365, 2017, doi: 10.1038/nphoton.2017.63.
[5] A. Ashkin, "Trapping of atoms by resonance radiation pressure," *Phys. Rev. Lett.*, vol. 40, no. 12, pp. 729–732, 1978, doi: 10.1103/PhysRevLett.40.729.
[6] P. Michler *et al.*, "A quantum dot single-photon turnstile device," *Science*, vol. 290, no. 5500, pp. 2282–2285, 2000, doi: 10.1126/science.290.5500.2282.
[7] C. Santori, M. Pelton, G. Solomon, Y. Dale, and Y. Yamamoto, "Triggered single photons from a quantum dot," *Phys. Rev. Lett.*, vol. 86, no. 8, pp. 1502–1505, 2001, doi: 10.1103/PhysRevLett.86.1502.

[8] X. Liu and M. C. Hersam, "2D materials for quantum information science," *Nat. Rev. Mater.*, vol. 4, no. 10, pp. 669–684, 2019, doi: 10.1038/s41578-019-0136-x.

[9] N. Mizuochi *et al.*, "Electrically driven single-photon source at room temperature in diamond," *Nat. Photonics*, vol. 6, pp. 299–303, 2012, doi: 10.1038/nphoton.2012.75.

[10] S. Castelletto *et al.*, "A silicon carbide room-temperature single-photon source," *Nat. Mater.*, vol. 13, no. 2, pp. 151–156, 2014, doi: 10.1038/nmat3806.

[11] X. He *et al.*, "Tunable room-temperature single-photon emission at telecom wavelengths from sp3 defects in carbon nanotubes," *Nat. Photonics*, vol. 11, no. 9, pp. 577–582, 2017, doi: 10.1038/nphoton.2017.119.

[12] A. J. Morfa *et al.*, "Single-photon emission and quantum characterization of zinc oxide defects," *Nano Lett.*, vol. 12, no. 2, pp. 949–954, 2012, doi: 10.1021/nl204010e.

[13] Y.-S. Park, S. Guo, N. S. Makarov, and V. I. Klimov, "Room temperature single-photon emission from individual perovskite quantum dots," *ACS Nano*, vol. 9, no. 10, pp. 10386–10393, 2015, doi: 10.1021/acsnano.5b04584.

[14] A. M. Berhane *et al.*, "Bright room-temperature single-photon emission from defects in gallium nitride," *Adv. Mater.*, vol. 29, no. 12, p. 1605092, 2017, doi: 10.1002/adma.201605092.

[15] A. Baskaran and P. Smereka, "Mechanisms of Stranski-Krastanov growth," *J. Appl. Phys.*, vol. 111, no. 4, p. 044321, 2012, doi: 10.1063/1.3679068.

[16] M. J. Holmes, M. Arita, and Y. Arakawa, "III-nitride quantum dots as single photon emitters," *Semicond. Sci. Technol.*, vol. 34, no. 3, p. 033001, 2019, doi: 10.1088/1361-6641/ab02c8.

[17] I. Aharonovich, D. Englund, and M. Toth, "Solid-state single-photon emitters," *Nat. Photonics*, vol. 10, no. 10, pp. 631–641, 2016, doi: 10.1038/nphoton.2016.186.

[18] N. Somaschi *et al.*, "Near-optimal single-photon sources in the solid state," *Nat. Photonics*, vol. 10, no. 5, pp. 340–345, 2016, doi: 10.1038/nphoton.2016.23.

[19] X. Ding *et al.*, "On-demand single photons with high extraction efficiency and near-unity indistinguishability from a resonantly driven quantum dot in a micropillar," *Phys. Rev. Lett.*, vol. 116, no. 2, p. 020401, 2016, doi: 10.1103/PhysRevLett.116.020401.

[20] T. J. Puchtler *et al.*, "Ultrafast, polarized, single-photon emission from m-plane InGaN quantum dots on GaN nanowires," *Nano Lett.*, vol. 16, no. 12, pp. 7779–7785, 2016, doi: 10.1021/acs.nanolett.6b03980.

[21] C.-H. Teng, B. Demory, P.-C. Ku, L. Zhang, T. A. Hill, and H. Deng, "Elliptical quantum dots as on-demand single photons sources with deterministic polarization states," *Appl. Phys. Lett.*, vol. 107, no. 19, pp. 191105-191105.4, 2015, doi: 10.1063/1.4935463.

[22] A. Lundskog *et al.*, "Direct generation of linearly polarized photon emission with designated orientations from site-controlled InGaN quantum dots," *Light Sci. Appl.*, vol. 3, no. 1, p. e139, 2014, doi: 10.1038/lsa.2014.20.

[23] M. Kianinia *et al.*, "Robust solid-state quantum system operating at 800 K," *ACS Photonics*, vol. 4, no. 4, pp. 768–773, 2017, doi: 10.1021/acsphotonics.7b00086.

[24] C. Palacios-Berraquero *et al.*, "Atomically thin quantum light-emitting diodes," *Nat. Commun.*, vol. 7, no. 1, p. 12978, 2016, doi: 10.1038/ncomms12978.

[25] G. Clark *et al.*, "Single defect light-emitting diode in a van der Waals heterostructure," *Nano Lett.*, vol. 16, no. 6, pp. 3944–3948, 2016, doi: 10.1021/acs.nanolett.6b01580.

[26] M. R. Rosenberger *et al.*, "Quantum calligraphy: writing single-photon emitters in a two-dimensional materials platform," *ACS Nano*, vol. 13, no. 1, pp. 904–912, 2019, doi: 10.1021/acsnano.8b08730.

[27] A. Branny, S. Kumar, R. Proux, and B. D. Gerardot, "Deterministic strain-induced arrays of quantum emitters in a two-dimensional semiconductor," *Nat. Commun.*, vol. 8, no. 1, p. 15053, 2017, doi: 10.1038/ncomms15053.

[28] B. Lounis and M. Orrit, "Single-photon sources," *Rep. Prog. Phys.*, vol. 68, no. 5, pp. 1129–1179, 2005, doi: 10.1088/0034-4885/68/5/R04.

[29] S. Haroche and D. Kleppner, "Cavity quantum electrodynamics," *Phys. Today*, vol. 42, no. 1, pp. 24–30, 1989, doi: 10.1063/1.881201.

[30] C. Chakraborty, L. Kinnischtzke, K. M. Goodfellow, R. Beams, and A. N. Vamivakas, "Voltage-controlled quantum light from an atomically thin semiconductor," *Nat. Nanotechnol.*, vol. 10, no. 6, pp. 507–511, 2015, doi: 10.1038/nnano.2015.79.

[31] A. Srivastava, M. Sidler, A. V. Allain, D. S. Lembke, A. Kis, and A. Imamoğlu, "Optically active quantum dots in monolayer WSe_2," *Nat. Nanotechnol.*, vol. 10, no. 6, pp. 491–496, 2015, doi: 10.1038/nnano.2015.60.

[32] Y.-M. He *et al.*, "Single quantum emitters in monolayer semiconductors," *Nat. Nanotechnol.*, vol. 10, no. 6, pp. 497–502, 2015, doi: 10.1038/nnano.2015.75.

[33] M. Koperski *et al.*, "Single photon emitters in exfoliated WSe_2 structures," *Nat. Nanotechnol.*, vol. 10, no. 6, pp. 503–506, 2015, doi: 10.1038/nnano.2015.67.

https://doi.org/10.1142/9789811229107_0007

Chapter 7

Valley Device Engineering

Qian Ling Kee, Calvin Pei Yu Wong, and Kuan Eng Johnson Goh

7.1 Introduction

In the context of what we have discussed in this book, the basic building blocks of valleytronic devices (or valley devices in short) are 2D materials, often transferred to or grown on specific substrates using the variety of techniques discussed in Chapter 4. 2D materials are central to valleytronics due to their broken inversion symmetry (either intrinsic or induced) which allow them to acquire a non-zero Berry curvature. As discussed in Chapter 2, and exemplified in Chapter 5, the non-zero Berry curvature behaves like an effective out-of-plane magnetic field. In particular, the Berry curvature for K and K′ valleys are opposite due to the time-reversal symmetry. Hence, in the presence of an in-plane electric field, electrons acquire a transverse velocity proportional to the Berry curvature, causing electrons from the K and K′ valleys to move in opposite directions and thus generating valley currents.

Van der Waals heterostructures may be formed by stacking different 2D materials on top of each other. Geim and Grigorieva initially suggested that this is analogous to building with Lego blocks, where there are unique rules of construction for distinct sets of materials [1] but it is now understood that van der Waals materials can stack in many degrees of freedom. Examples of parameters that can be controlled include the rotation angle between 2D layers, strain to the material, doping of the material, and geometry of the material. Hence, the number of ways that one can form heterostructures based on 2D materials is almost infinite. To realise the construction possibilities that these 'Lego blocks' offer, there is a need to control construction parameters guided by valley device engineering, which is the focus of this chapter.

We first discuss the key engineering processes needed to realise practical valleytronics, including 2D semiconductor engineering of interfaces, heterostructures, doping, contacts, and control gates. These engineering techniques allow us to create concept valley devices including valley valves, beam splitters, and transistors, which will be discussed near the end of the chapter. In addition, devices involving valleytronics also enable other behaviours such as valley-polarised exciton currents, valley magnetoelectricity, and valley Zeeman spin-orbit coupling-induced spin precession which can exist at room temperature. Such valley-enabled characteristics further expand the possibilities for scientific discoveries and technological innovations.

7.2 Interfaces and Grain Boundaries

One distinct characteristic of 2D materials is their exceptionally high surface area to volume ratio due to their atomic thinness. As a result, the interfaces in a 2D stack, surface functionalisation and encapsulation

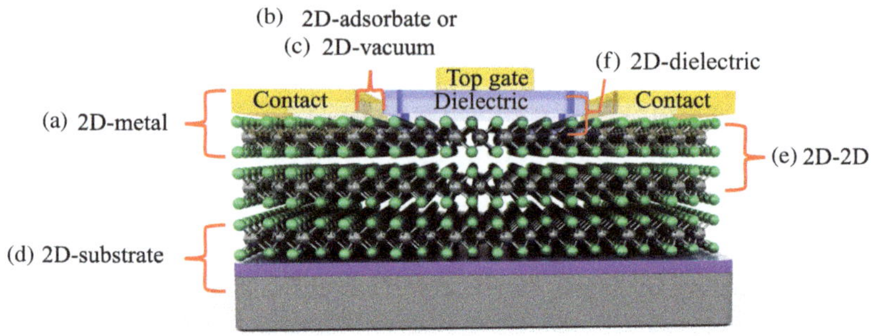

Fig. 7.1. Schematic of the types of interfaces in a typical 2D material-based device including (a) 2D-metal, (b) 2D-adsorbate, (c) 2D-vacuum, (d) 2D-substrate, (e) 2D-2D and (f) 2D-dielectric interfaces.

layers play a significant role in determining the properties of the heterostructure [2]. Fig. 7.1 shows the different interfaces within a 2D device which could influence the behaviour of the device. This includes the (a) 2D-metal, (b) 2D-adsorbate, (c) 2D-vacuum, (d) 2D-substrate, (e) 2D-2D and (f) 2D-dielectric interfaces. These interfaces can become contaminated during the fabrication processes, which include stacking and growth of these materials. The contaminated interfaces can form bubbles and wrinkles, and thereby impair device performance.

Grain boundaries are also an important consideration in 2D material devices; they often originate from the poor stitching of different nuclei during the less-than-ideal growth conditions or because of unintended contamination during growth. In 2D materials, grain boundaries are 1D line defects separating two grains at an angle. Fig. 7.2 shows two examples of how grain boundaries can look like with grains at a relative angle of 21.8° and 32.3° in graphene. Like interfaces, grain boundaries can change the properties of atomically thin materials significantly. This section explores the cause and effect of interface contamination and grain boundaries, and how their engineering can improve the performance of devices.

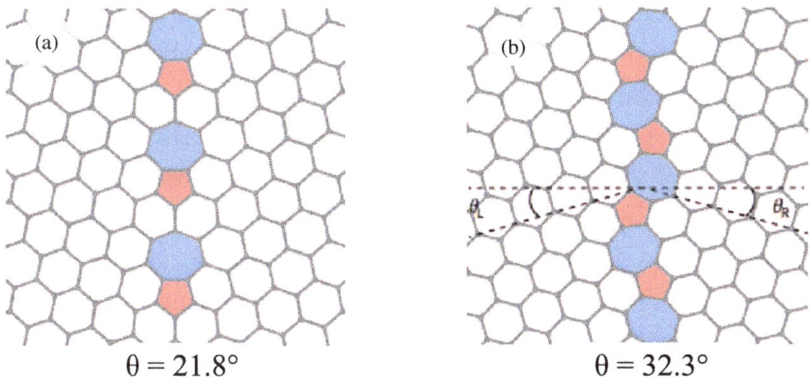

Fig. 7.2. Examples of grain boundaries in a hexagonal 2D material at a relative angle of (a) 21.8° and (b) 32.3°. Reproduced with permission from Ref. [3].

7.2.1 Interface Engineering

Heterostructures formed by mechanically stacked 2D materials require atomically clean out-of-plane interfaces. This is challenging because during stacking, contaminants such as hydrocarbons, air, or water can be trapped between the layers, forming bubbles or blisters [4]. Lithographic polymer-resists and solvents used during device processing are also sources of contamination. They can aggregate to form spatially localised pockets with sizes ranging from a few nanometres to a few micrometres in diameter. This is due to the presence of a self-cleansing effect which occurs at the interfaces of van der Waals materials, where contaminants aggregate into randomly distributed bubbles, leaving the bubble-free regions of the heterostructure with atomically clean interfaces [5].

The presence of contaminant bubbles in the 2D material heterostructure can locally increase unintended doping, degrade carrier mobility, and result in poor device reproducibility and reliability.

This is detrimental to the performance of the devices, which includes valleytronic devices, that are built using those heterostructures. Hence, a clean fabrication technique which can avoid contaminating the critical device interfaces is required. This is usually done by transfer stacking of 2D materials in a controlled environment such as a glovebox [6], or by growing and processing materials in ultrahigh vacuum conditions [7]. Furthermore, several post-fabrication cleaning techniques have also been developed to remove these contaminant bubbles/blisters [4]. We will summarise some of these cleaning techniques developed such as the hot pick-up technique, thermal annealing and tip-based cleaning and discuss their effectiveness.

The hot pick-up technique developed by Pizzocchero *et al.* has been shown to obtain blister-free areas of more than 10 μm [8]. In this technique, bubbles were eliminated between graphene and hBN flakes when the heterostructure is assembled at 110°C. A polymer stack consisting of polypropylene carbonate (PPC), mounted on a block of polydimethylsiloxane (PDMS) for mechanical support, is used to pick up and drop down 2D flakes. PPC is chosen due to its thermoplasticity — the material becomes very soft at high temperatures and flows like a viscous liquid with low adhesivity, but hardens at room temperature and has high adhesivity. Thus the PPC/PDMS stamp can be used for determistic and reliable pick up and release of 2D material flakes. Fig. 7.3 shows schematically how this process is carried out. Using the PPC/PDMS stamp, exfoliated hBN flakes on a SiO_2 substrate are first picked up at 40°C, below the glass transition temperature of PPC, where the adhesion between the 2D flake and PPC is known to be strong [9]. Next, the hBN is dropped onto graphene at 110°C to form a bubble-free interface depicted in Fig. 7.3(b) and (c). The hBN/graphene is then cooled down to 40°C before the stack is picked up and the process is repeated as shown in Fig. 7.3(d)–(f) to assemble the 2D heterostructure shown in Fig. 7.3(g). Pizzocchero *et al.* proposed

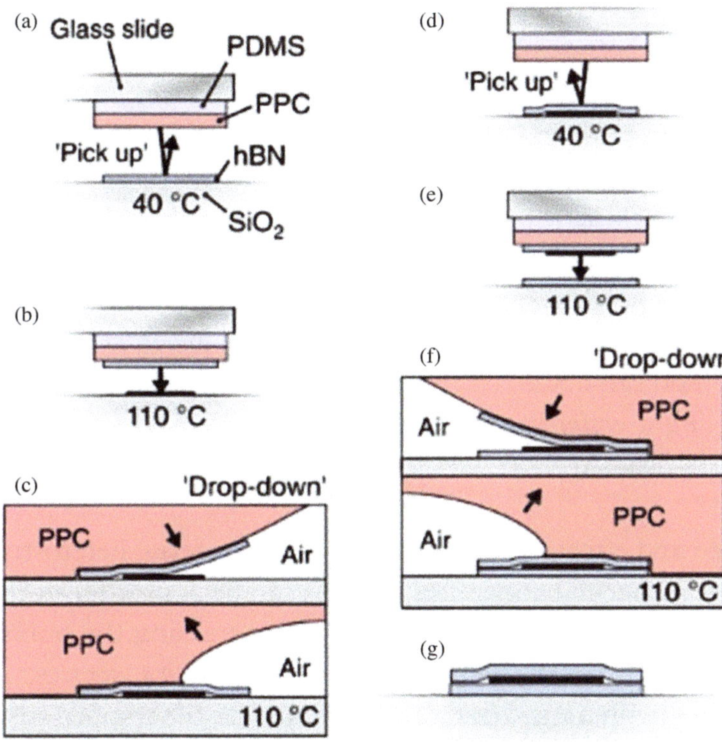

Fig. 7.3. Schematic process flow for assembly of hBN/graphene/hBN heterostructure by the hot pick-up and drop-down technique: (a) Pick up hBN using PPC/PDMS stack, (b, c) drop down hBN deterministically on graphene, (d) pick up hBN/graphene heterostructure, and (e, f) drop down on hBN to complete encapsulation. (g) Assembled 2D heterostructure. Reproduced with permission from Ref. [8].

that the working principle of this technique is similar to thermal annealing — adsorbates present on the surface have higher mobility at elevated temperatures, allowing them to diffuse out of the sample from the edges of hBN during encapsulation. They showed that when the heterostructure was assembled at 40°C, large bubbles of contamination could be seen trapped between the 2D materials but were eliminated for heterostuctures assembled at 110°C. It should be noted that an oxygen plasma treatment on PDMS helps to increase the adhesion between PDMS and PPC, preventing their delamination at high temperature, increasing the fabrication yield.

A limitation to this technique is that clean interfaces are only achieved when the stacking is performed slowly at a speed of $<1\ \mu m/s$ by controlling the z-height of the stage. This process is also difficult to scale since it is dependent on the operator and the greater the number of interfaces within a heterostructure stack, the lower the yield. While this method can obtain clean interfaces over areas larger than 20 μm, assembling a wafer-sized heterostructure is still an extremely challenging endeavour.

A simple and effective way to clean the 2D material heterostructure interface is thermal annealing and this can be done after stacking to significantly reduce bubble density [10], increasing the bubble-free area available for the fabrication of high-quality devices. At the annealing temperature (typically 200–500°C), the trapped adsorbate material diffuse along the interface and small bubbles are desorbed and merge with their neighbouring bubbles to form larger bubbles via the Ostwald ripening process. Some bubbles diffuse to the edge of the material and are expelled, thereby increasing the clean area. Fig. 7.4 shows the atomic force micrograph of how annealing a graphene/hBN/MoS_2/hBN heterostructure reduced surface roughness and bubble density, shown by the reduction of white spots and reduced variations in the height profile, from which one can conclude that the interface is now cleaner than before annealing. However, because annealing relies on the random diffusion of trapped adsorbates coalescing into larger interface bubbles, it cannot deterministically and reliably remove trapped contaminants from specific interface regions. Decomposition of contaminants during annealing may also produce radicals that could damage the 2D material.

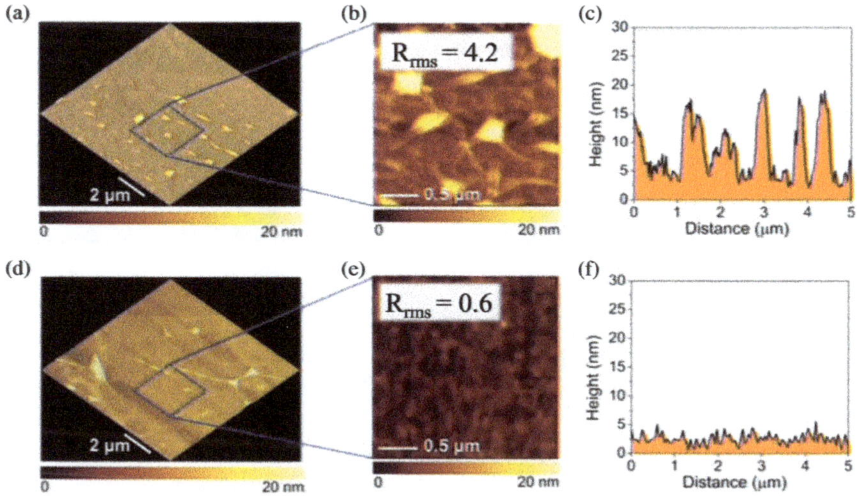

Fig. 7.4. Effect of annealing on the interface of the van der Waals heterostructure. (a, d) AFM images, (b, e) surface roughness as well as the height profile of graphene/hBN/MoS_2/hBN stacking (c) before and (f) after annealing. White dot spots represent the air trap bubbles. Reproduced with permission from Ref. [11].

To address these drawbacks, the "nano-squeegee" tip-based cleaning method which involves using an atomic force microscope (AFM) tip to controllably squeeze contaminants out from between the 2D layers was developed. Rosenberger *et al.* showed that this method is highly reproducible and precise in removing bubbles trapped in chemical vapour deposition (CVD)-grown WSe_2 on hBN [12]. The nano-squeegee technique relies on the force applied by the tip of an AFM in contact mode on the 2D material to push out contaminants trapped between the 2D material heterostructure as the tip raster scans across the surface represented schematically in Fig. 7.5(a) and (b). These contaminant bubbles accumulate into large pockets and are forced out at the edge of the material, as shown in the AFM images in Fig. 7.5(c) and (d). This tip-based cleaning method is highly effective in cleaning graphene/hBN interfaces and have shown to increase the mobility of bilayer graphene (BLG) on hBN by a factor of 60–250% [13]. This method

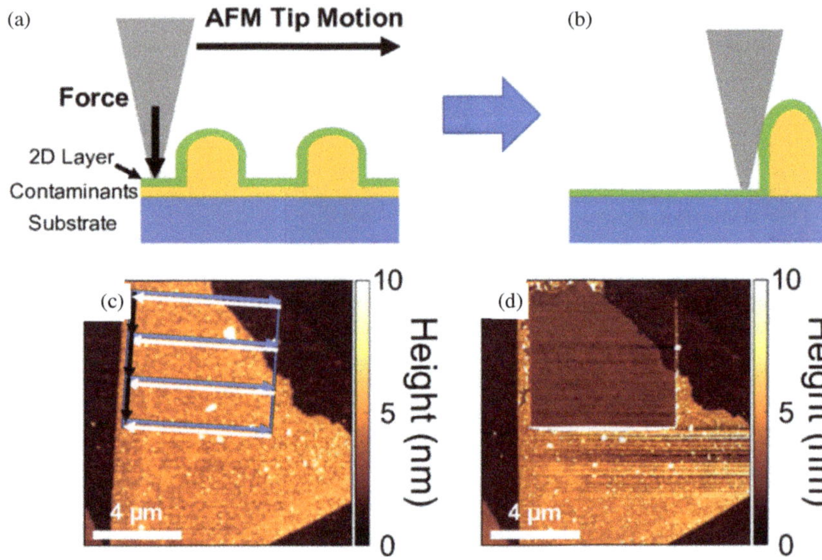

Fig. 7.5. (a, b) Illustration of flattening of material by AFM. (c) AFM image of WSe_2 on hBN before flattening and (d) after flattening. Reproduced with permission from Ref. [12].

is generalisable to other 2D materials and have proven to also be effective for few-layer MoS_2 devices, which showed improvement in their magneto-transport properties after tip-based cleaning [10]. However, this nano-squeegee method can be challenging to an inexperienced practitioner because high lateral forces exerted on the sample in contact mode due to a wrong cantilever choice or force setpoint can result in sample damage. The tip can also be easily damaged by contaminants which attach themselves to the tip during scanning, reducing scanning resolution and may require frequent replacements. Careful recipe development and tip calibration before the cleaning process is therefore required for individual AFMs.

Contamination between 2D materials and adsorbates present in the ambient environment are not limited only to physical adsorbates, and interactions between 2D materials and their substrates. Some

2D materials such as black phosphorus or CVD-grown MoS_2 are unstable in air and can easily be oxidised over time. Fortunately, hBN is a suitable 2D insulator that can function as the protection encapsulation layer and does not perturb the electronic properties of the 2D material [14]. hBN is an air-stable, defect-free, dangling bond-free, and atomically flat insulating material, and has shown to be a superior atomically flat substrate and encapsulation material for 2D materials such as graphene and MoS_2 [15]. These properties also allow hBN to be put in between 2D materials and SiO_2 substrates to screen the charge impurities on the SiO_2 surface and to reduce the substrate roughness. Indeed, the use of hBN has been shown to improve carrier mobility and channel quality significantly through reducing the effect of charged scatter centres from SiO_2. For example, in the works of Shimazaki *et al.* [16] and Sui *et al.* [17], a BLG Hall bar device encapsulated between two hBN layers was fabricated and excellent graphene quality was preserved as evidenced by high carrier mobilities, and the use of hBN layers also allowed the fabrication of top gates where the hBN was used as a gate dielectric.

Besides ensuring the cleanliness of interfaces in a 2D heterostructure, avoiding or reducing grain boundaries within the 2D material is also needed for good device performance. In the next section, the effect of grain boundaries will be discussed.

7.2.2 Grain Boundaries

Besides the use of mechanical exfoliation and transfer stacking to form 2D devices and heterostructures, CVD may also be used for large-area 2D material growth. However, this process typically results in the formation of grain boundaries when the different nuclei stitch together imperfectly during growth, which may affect the device performance. As CVD is a complex process involving adsorption and decomposition of precursor species on the surface, the mechanisms of surface diffusion, nucleation, and growth are coupled in a complex manner, which can often lead to poor reproducibility over the whole substrate area. Consequently, the grains and grain boundary patterns are complex and can differ from sample to sample. For example, as graphene films grow on polycrystalline Cu, initially isolated graphene domains may finally connect to form a continuous polycrystalline film, consisting of small domains and grain boundaries. The presence of grain boundaries can bring about changes to the material through electrical, chemical, thermal, and mechanical effects.

Reports have shown that the presence of grain boundaries worsen the conductivity and carrier mobility of graphene as compared to those of monocrystalline graphene domains [18]. Among polycrystalline graphene samples, those with larger domain sizes are generally found to have a higher mobility as the grain boundaries which act as scatter centres are spaced further apart. Song *et al.* found that while point defects and impurities adsorbed on the surface of graphene grown by CVD had little effect on mobility, the presence of grain boundaries impacted the mobility drastically [19]. Using scanning tunnelling microscopy, Clark *et al.* discovered that there is a sudden change in electrochemical potential and a decrease of electrical conductivity at the grain boundary of graphene [20], which correlates with experimental data showing that an increase in the density of grain boundaries decreases carrier mobility.

Similarly, grain boundaries can also impact the transport properties of TMDs. For example, while transport across mirror twin grain boundaries in MoS_2 remains rather unaffected, transport across tilt grain boundaries shows a decrease in conductivity [21], indicating that not all grain boundaries are equal. Huang *et al.* discovered using scanning tunnelling microscopy a misorientation-angle dependent decrease in bandgap at the grain boundary in monolayer MoS_2 [22]. Fig. 7.6(a) shows the grain boundaries formed in the CVD-grown MoS_2, and Fig. 7.6(b)–(d) show the images at the grain boundary taken at different tip biases (V_{Tip}). The misorientation between grains I and II is 18° as determined in Fig. 7.6(c). Fig. 7.6(e)–(f) show a pronounced change in bandgap with the distance from the grain boundary where the bandgap is 1.55 V at the boundary, and 2.4 V at 3.5 nm away from the boundary. Hence, grain boundary engineering can perhaps be used to tweak the bandgap of TMDs.

The presence of grain boundaries in the material may also bring about chemical effects, such as an increased chemical reactivity at the

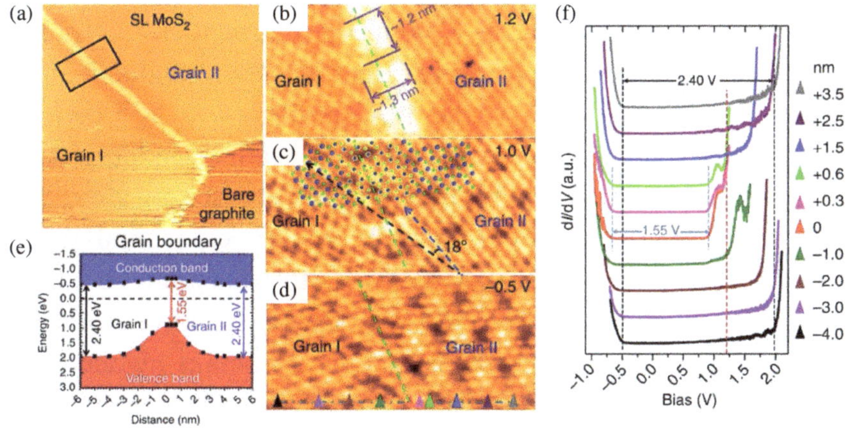

Fig. 7.6. (a) Large-scale STM image showing a single-layer (SL) MoS_2 island composed of two crystal grains, where the grain boundary (GB) appears as bright protrusions ($75 \times 75\,nm^2$; $V_{Tip} = 1.5\,V$). (b)–(d) are bias-dependent images recorded at the boundary region highlighted by the $8 \times 4\,nm^2$ black rectangle in panel (a) with respective tip biases given by (b) $V_{Tip} = 1.2\,V$; (c) $V_{Tip} = 1.0\,V$; (d) $V_{Tip} = -0.5\,V$. (e) Schematic diagram shows the bandgap change at the grain boundary. (f) dI/dV spectra recorded at the positions denoted by the triangles in panel (d) with the following set point conditions: $V_{Tip} = 1.0\,V$, $I_{Tip} = 100\,pA$. Reproduced with permission from Ref. [22].

boundary. For instance, graphene etching was found to occur easily at defective sites [23]. Grain boundaries can also enhance catalytic activity compared to its basal plane, which is the plane parallel to the lateral axis. For example, the 2H-2H and 2H-1T domain grain boundaries in the basal plane of MoS_2 are stable and highly active catalytic sites for hydrogen evolution reaction [24].

Grain boundaries can also change the thermal properties of 2D materials. Thermal transport is typically hindered across graphene grain boundaries as they scatter phonons [25]. Interestingly, the elastic stiffness and mechanical strength of polycrystalline graphene are found to be comparable to those of pristine graphene despite the presence of grain boundaries [26] due to the presence of well-stitched covalently bonded pentagon-heptagon ring defects along the straight grain boundary in graphene. Among the multiple effects that the presence of grain boundaries brings, their effects on conductivity and carrier mobility are arguably the most significant when it comes to device engineering. To reduce the effects of grain boundaries on the transport properties of materials, it is possible to simply avoid them by making devices far away from grain boundaries, or small devices within a single grain but it is often difficult to avoid grain boundaries as they are usually invisible to the basic characterisation techniques. Hence controlling the growth parameters to obtain higher-quality crystals with lesser grain boundaries is preferred. While interface and grain boundary engineering may be done to improve device performance, it is still a challenging feat to implement these techniques to realise a practical valleytronic device that is reliable, scalable, and has long-term stability.

It is perhaps interesting to also note that pristine grains stitched together by grain boundaries have broken inversion symmetry, which makes them a possible candidate for valley devices. Indeed, 2D materials with grain boundaries or line defects have been the subject of several theoretical works in valleytronics. One example is the 558-type line defect in graphene (see Fig. 7.7(a)) which have been proposed to act as a valley filter [27]. This means that the line defect can filter out electrons belonging to a certain valley, while allowing electrons from another valley to be transmitted. At incident angle α, the line defect permits (forbids) the quasiparticle with valley index $\tau = +1$ (–1)

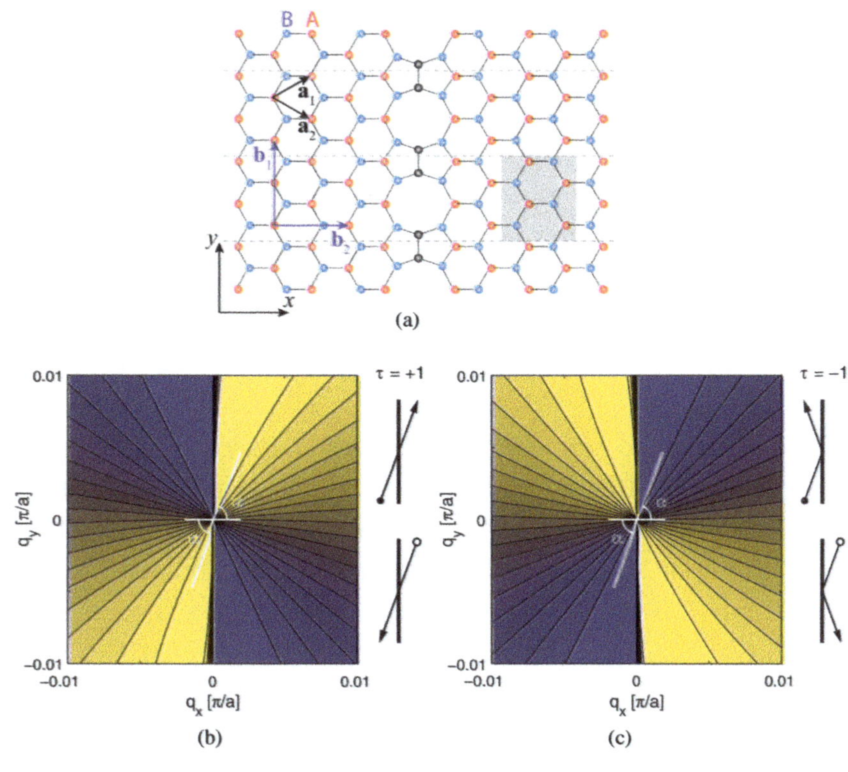

Fig. 7.7. (a) Schematic drawing of a 558-type line defect in graphene. (b) The transmission probability (indicated by the brightness) for a quasiparticle with finite energy approaching the line defect at an angle of incidence α from the K valley and (c) K′ valley. Reproduced with permission from Ref. [27].

to pass through (Fig. 7.7(b) and (c)). More of these concept devices will be discussed in Section 7.7.

7.3 Band Engineering in van der Waals Structures

Band engineering is a key consideration in the design of nearly all modern semiconductor devices. It is highly desirable to not only tune semiconductor properties using band engineering, but also to miniaturise the devices towards nearly atomically thin dimensions using van der Waals materials. These materials, especially TMDs, are particularly useful for valleytronics due to their intrinsic absence of inversion symmetry, which allows for a non-zero Berry curvature and thereby allowing the observation of the valley Hall effect. An external electric field may also be applied to BLG to break its inversion symmetry. Given the importance of 2D materials in valley devices, band engineering in van der Waals structure is particularly important.

In heterostructure engineering, materials may be selected and arranged based on their conductivity and their properties at heterojunctions. TMDs with chemical composition MX_2 (M = transition metals, and X = S, Se, and Te) exhibit various structural phases, including 2H, 1T, and 1T'. Fig. 7.8 shows various 2D materials, arranged according to their bandgap. Specifically, the semiconducting 2H TMDs (e.g., MoS_2) have bandgaps in the 1–2 eV range. In contrast, TMDs with 1T or 1T' phases are generally metallic or semi-metallic. Large-bandgap materials such as hBN play a critical role in heterostructures as well. Other layered materials include topological insulators (e.g., Bi_2Se_3 and Sb_2Se_3) which are used for their topologically protected and spin momentum-locked electronic transport.

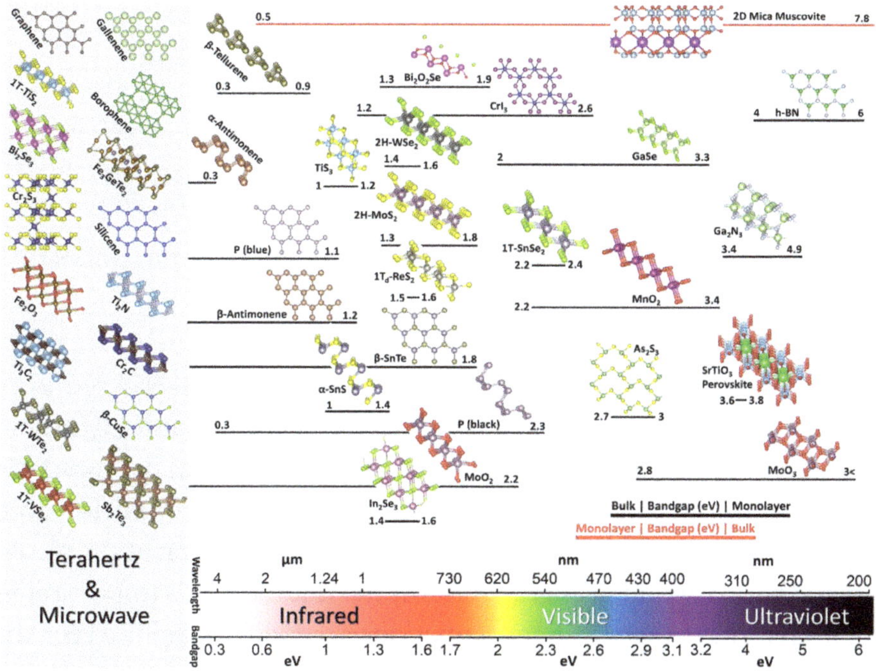

Fig. 7.8. Selected family of 2D materials, arranged according to their bandgap. Materials in the grey box have zero or near-zero bandgap. Reproduced with permission from Ref. [29].

Heterostructures may also include the stacking of semiconducting 2D materials, such as different TMDs. The behaviour of a semiconductor junction depends on the alignment of energy bands at the interface. Depending on the bandgap of the 2D materials stacked into a heterostructure, straddling gap (Type I), staggered gap (Type II) or broken gap (Type III) heterojunctions may be formed (see Fig. 7.9). In Type I, the conduction band of the second semiconductor is lower than that of the first, and its valence band is higher than that of the first. This means the bandgap of the second semiconductor is smaller than the first. Type I heterojunctions allow the flow of current in only one direction and are usually used as a light-emitting diode. In Type II,

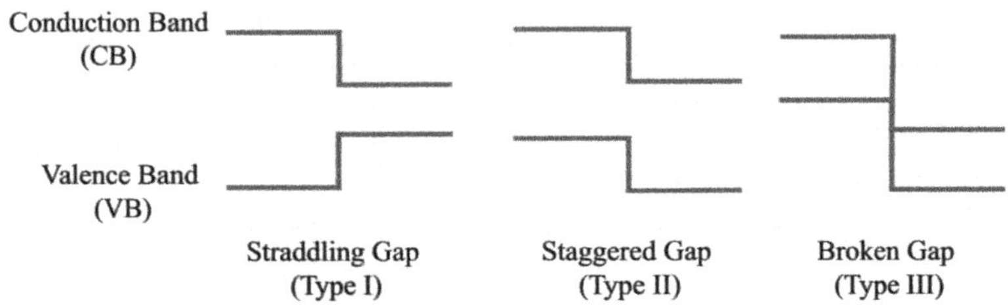

Fig. 7.9. Type I, Type II and Type III heterojunctions. VB and CB represent the valence and conduction bands, respectively.

both the conduction band and valence band of the second semiconductor are lower than that of the first. This type of band alignment can easily allow for electron and hole carrier transfer at the interface. In Type III, the conduction band of the second semiconductor overlaps with the valence band of the first semiconductor. As a result, there are no forbidden energy states at the interface. It is important to note that these types of heterojunctions are highly simplified for the purpose of broad classification; they do not take into account band bending, interface states and other interface-related physics.

Van der Waals structures may be classified into heterostructures and homostructures. As the name suggests, a heterostructure consists of different types of 2D materials that are joined or stacked together, laterally or vertically. In a lateral heterostructure, the different materials are located side by side, such as the 2H-MoS_2 and 1T-MoS_2 interface in Fig. 7.10(a). 2D materials may also be stacked in a vertical configuration as in Fig. 7.10(b). In Chapter 5, we have discussed an example of graphene on hBN structure, which is an interesting valleytronic device structure due to the presence of a broken inversion symmetry in this device [28].

Van der Waals homostructures may also be formed, such as the case of bilayer MoS_2, WS_2 or graphene. The layers can also be assembled

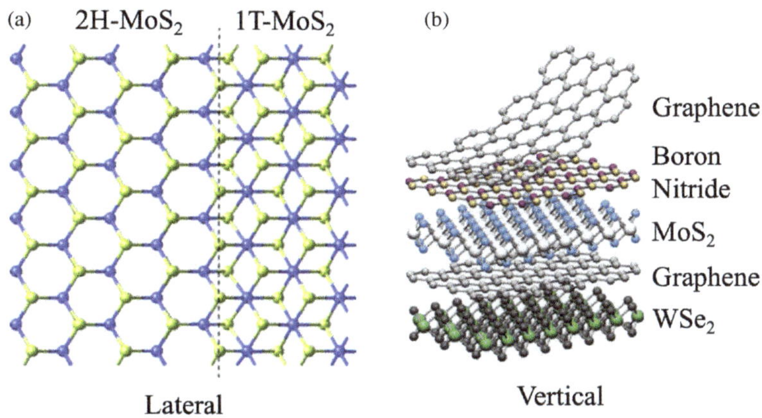

Fig. 7.10. Schematic of (a) lateral heterostructure and (b) vertical heterostructure. Reproduced with permission from Refs. [29] and [1], respectively.

with a twist angle with respect to each other which endows the materials with many interesting properties. In this section, we discuss methods for band alignment engineering, bandgap engineering, along with some examples of how band engineering may be done in valley devices.

7.3.1 Band Alignment Engineering

As mentioned earlier, the behaviour of a junction depends crucially on the alignment of energy bands at a semiconductor heterojunction, and there are three types of band alignments (Fig. 7.9) — straddling gap (Type I), staggered gap (Type II) or broken gap (Type III) alignment. Fig. 7.11 provides an overview of band alignment between various 2D metals and semiconductors. Band alignment engineering is especially needed at the metal/semiconductor interface, since a mismatch between the band edges can reduce carrier injection efficiency at the interface

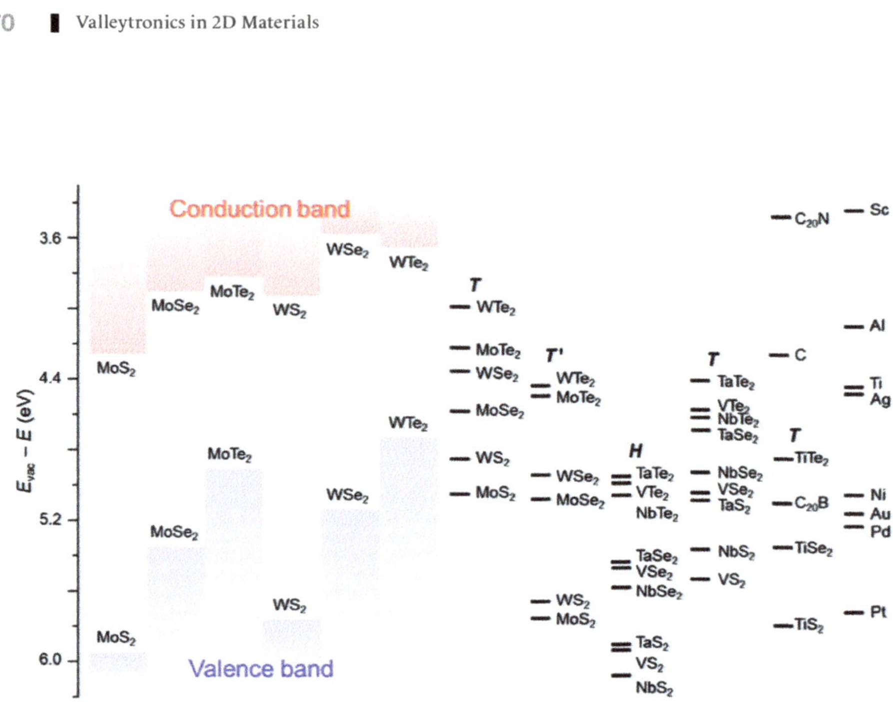

Fig. 7.11. Band alignment between semiconducting (band edges, left block) and metallic (Fermi levels, right block) 2D TMDs. Left bars show the conduction and valence bands of semiconductors. Right columns show the work function of metals. For comparison, the work functions of some commonly used 3D metals are shown at the far right. Reproduced with permission from Ref. [30].

due to the presence of a Schottky barrier. Using a simplistic first-order Schottky-Mott approximation model, the Schottky barrier can be reduced by aligning the metal work function with the band edge of the semiconductor. The work function is defined as the energy needed to remove an electron from the Fermi level of the metal to the vacuum level, and examples of high work function metals are Au (5.1 eV) and Pt (approximately 5.8 eV), while examples of low work function metals are Al (4.1 eV) and Ti (4.3 eV). These metals are chosen such that the Fermi level is close to the valence (conduction) band of the semiconductor for easy hole (electron) injection. This simplistic model does not consider metal reactivity but provides a basic guide for contacts engineering. Band alignment engineering is thus crucial to contacts engineering, which will be discussed in more detail in Section 7.5.

7.3.2 Bandgap Engineering

Within heterostructures, one can also tweak the bandgap of different materials. 2D materials are found to exhibit highly tuneable bandgaps, achieved by controlling number of layers, strain engineering, chemical doping, alloying, intercalation, substrate engineering as well as the use of an external electric field [31]. Materials may also be stacked with a twist angle relative to each other to form Moiré superlattices, which can have novel superlattice properties. Furthermore, 2D materials, unlike traditional quantum well heterostructures, can be stacked in an arbitrary way since they are unconstrained by lattice-matching requirements. This means that within the heterostructures, the bandgap can be controlled by tuning a number of degrees of freedom such as stacking layers, twist angle, interlayer coupling and growth conditions.

As 2H TMDs are typically semiconducting, and 1T (or 1T') TMDs are usually metallic or semi-metallic, phase engineering can be used to change the bandgap of TMDs. For example, 2H-MoS_2 have been shown to convert to its 1T' phase by treating the sample with *n*-butyl lithium [32]. The metallic 1T'-MoS_2 can then be used as an electrode for contacting 2H-MoS_2 to obtain a lower contact resistance compared to conventional metal electrodes. However, this method of phase engineering relies on an aggressive chemical reaction to the TMD layer which is difficult to control. Thus, the main drawback of this method is its low reproducibility and limited material compatibility.

The number of layers and stacking order also play a huge role in tuning the bandgap of 2D materials. A prominent example is that of single-layer graphene (SLG) and BLG. SLG is a semi-metal with zero bandgap, which limits its applications in conventional electronic

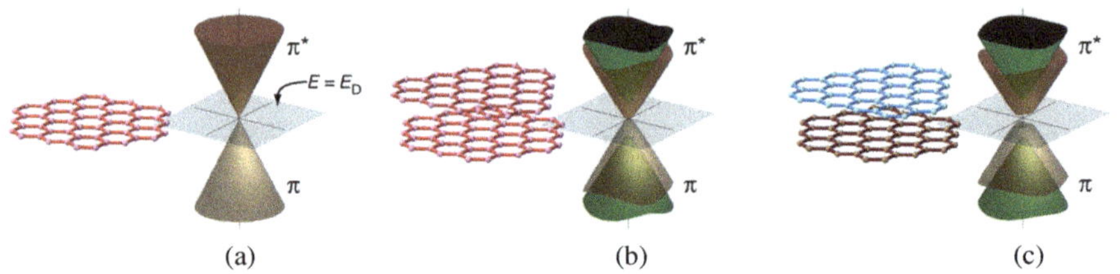

Fig. 7.12. Electronic structure of (a) monolayer, (b) AA-stacked bilayer, and (c) AB-stacked (Bernal-stacked) bilayer graphene. Reproduced with permission from Ref. [35].

device applications based on semiconductors (see Fig. 7.12(a)). BLG, on the other hand, can retain most of the advantages of SLG such as high mobility while allowing a small bandgap opening when a sufficient electric field is applied to break inversion symmetry between graphene layers. This was utilised by Zhang *et al.* in demonstrating a gate-controlled, continuous tuning of bandgap up to 250 meV [33]. Stacking two graphene layers in a Bernal stacking (AB) fashion can break its symmetry as suggested by McCann and Fal'ko [34]. Since the unit cell of a bilayer contains four atoms, its band structure has two additional bands, π and π^* states in each valley split by interlayer coupling, and two bands with lower energy [35] (Fig. 7.12(b)). If the graphene layers are AB-stacked as shown in Fig. 7.12(c), then a small energy gap forms between low-energy bands at the Dirac point.

For graphene in particular, a bandgap can be opened by stacking it on hBN because a Moiré superlattice is formed between the lattice constants of graphene and hBN. A bandgap as large as 40 meV emerges for 0°-aligned graphene/hBN, and the gap varies with the twisting angle between the two layers [36]. Heterostructures with a twist angle can also form a Moiré superlattice, where two copies of a periodic

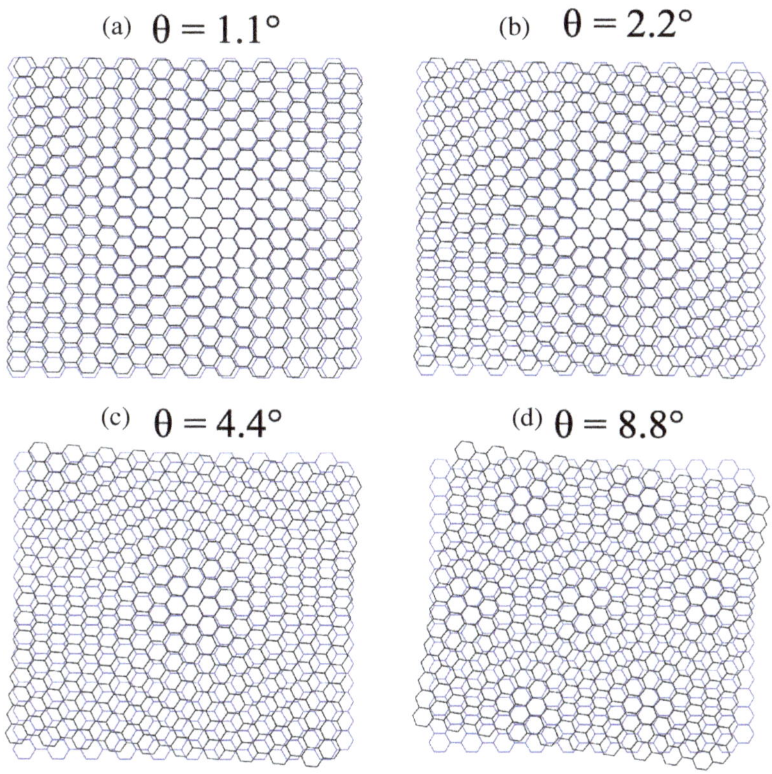

Fig. 7.13. Moiré patterns of graphene/hBN at twist angle of (a) 1.1° (b) 2.2° (c) 4.4° (d) 8.8°.

pattern are overlaid with a relative twist. Fig. 7.13(a)–(d) shows various twist angles in the graphene/hBN Moiré pattern, including twist angles of 8.8°, 4.4°, 2.2°, and 1.1° which are known as magic angles. A notable feature is that the superlattice constant of the Moiré unit cell decreases as the twist angle increases. At these magic angles, the unique feature of a nearly flat band is observed in the electronic band structure and can result in novel properties such as superconductivity in twisted bilayer graphene [37].

In TMDs, the bandgap increases as the number of layers decreases from the bulk material to the monolayer due to quantum confinement. Theoretical calculations have shown that the bandgap of 2H-MoS_2 increases from 1.44 eV to 2.22 eV when the layer number decreases from bulk layers to the monolayer [38]. The bandgap is even more tuneable for TMDs with strong interlayer coupling such as $PdSe_2$, $PtSe_2$, PdS_2, and PtS_2 and their indirect bandgap is preserved when thinned from their bulk to monolayer form. The bandgap of $PdSe_2$, for instance, can vary from 0.5 eV in bulk form to 1.37 eV in monolayer form [39]. Monolayer $PtSe_2$ has a bandgap of 1.2 eV, which drastically drops to only 0.3 eV in bilayer form [40]. Thus, layer number is a good degree of freedom to control the bandgap of the material.

7.3.3 Examples in Valleytronic Devices

As 2D materials such as graphene and MoS_2 have played a key role in the development of valleytronics, the concepts of interface engineering, band engineering and contacts engineering have been applied to these devices. In the early works by Gorbachev *et al.* [28], band engineering using hBN/graphene superlattice was used to break inversion symmetry in graphene, allowing the detection of the valley current through a non-local resistance peak. In the works of Sui *et al.* [17] and Shimazaki *et al.* [16] valley currents are detected in BLG, when the inversion symmetry is broken by electrical gate control. These concepts also apply to the other all-electrical methods of valley current generation and detection mentioned in Chapter 5.

Another example of a device used for valley generation and detection is the WSe_2–graphene–Bi_2Se_3 heterostructure [41]. The device comprises a thick semiconducting WSe_2 for optical generation, conductive monolayer graphene (due to its high spin diffusion length) for the transport channel, and Bi_2Se_3 topological insulator for non-local detection of spin-polarised electrons as shown in Fig. 7.14. By choosing these materials and arranging them in this manner, Cha *et al.* demonstrated the generation, transport, and detection of valley-locked spin photocurrent [41].

Band alignment engineering has also been done to realise valley current generation and detection in a WSe_2/MoS_2 heterostructure [42]

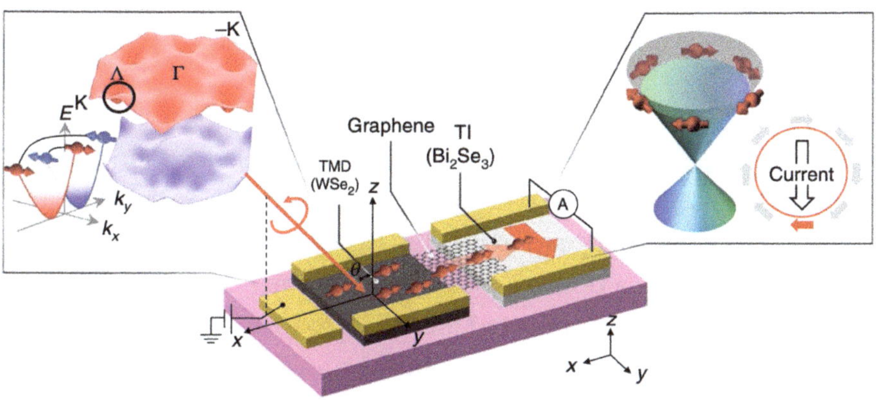

Fig. 7.14. Schematic representation of the non-local circular photogalvanic spectroscopy measurement in WSe_2–graphene–Bi_2Se_3 heterostructures. Circularly polarised optical excitation with an oblique angle of incidence θ generates photocarriers with valley-coupled spins in the WSe_2 transistor, and electrons with 'in-plane' spin textures at the $\Lambda(\Lambda')$ valley diffuse into the topological insulator surface across the adjacent sub-micrometre-long graphene. Spin-momentum locking of the topological insulator surface detects only the spin degree of freedom of the photocarriers. Reproduced with permission from Ref. [41].

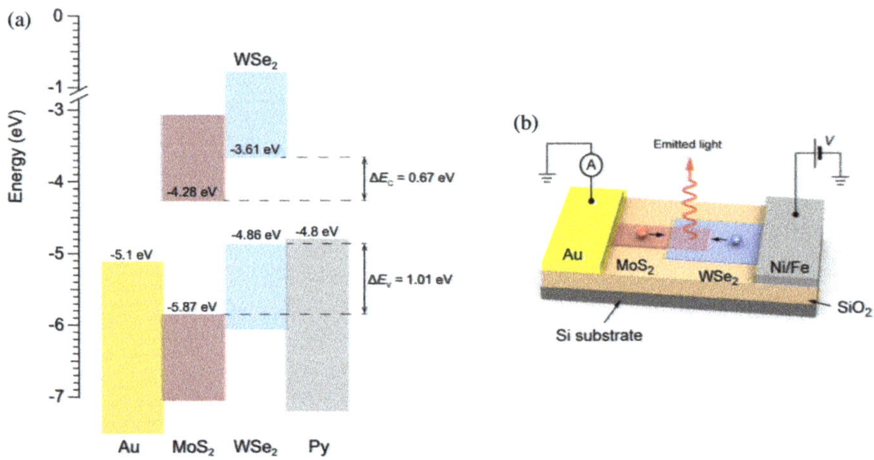

Fig. 7.15. (a) Band profile of the type II WSe_2/MoS_2 heterojunction LED in the lateral direction. (b) Schematic drawing of WSe_2/MoS_2 valley LED device that emits circularly polarised light. Reproduced with permission from Ref. [42].

shown in Fig. 7.15(b), as mentioned in Section 5.5. Specifically, the materials chosen, namely WSe_2 and MoS_2, form a type II heterojunction (Fig. 7.15(a)) that can realise a valley-polarised light-emitting diode. By applying a positive bias voltage to the permalloy electrode, spin-polarised holes are injected into WSe_2 to form negatively charged excitons or trions in WSe_2. Since optical transitions are allowed only for electron-hole pairs in the same valley, when the trion recombines, circularly polarised light is emitted along with an extra electron at the type II heterojunction. The intensity of right circularly polarised light as a proportion of left circularly polarised light thereby gives the degree of valley polarisation in WSe_2, which indicates the presence of spin-polarised charge injection from the permalloy contact. Therefore, obtaining a type II heterostructure through band engineering has enabled the detection of spin-valley polarised LED through the detection of circularly polarised light.

7.4 Doping Techniques for TMDs

To expand the potential of 2D TMDs, it is essential to precisely control the electrical properties of the material. This can be done using doping, which is well known in conventional 3D semiconductors such as silicon where the substitution of silicon atoms with boron or phosphorus tunes the dominant carrier type to holes (p-type) or electrons (n-type) respectively. This can be done using dopant diffusion or ion implantation, which are well-known industrial processes. Doping of 2D semiconductors is more challenging due to the atomically thin nature of the material, thus new and different doping mechanisms are required. For example, traditional semiconductor fabrication methods such as ion implantation would damage the crystal structure of 2D materials [43], or have little to no effect on these material as the energetic ions go straight through them. Hence, new doping methods are needed. This section presents two such methods — substitutional doping and charge transfer doping — which are well-suited for 2D materials.

Like traditional semiconductors such as Si or Ge, TMDs can be substitutionally doped, but the dopants such as Re, Nb and N atoms need to be introduced into the crystal lattice during the bulk crystal growth or 2D material growth. The p-type doping of MoS_2 by Nb was realised by CVD, where MoS_2 was doped using a Mo/Nb/Mo sandwich structure precursor [44] as shown in Fig. 7.16(a). Using a similar method, the n-type doping of MoS_2 by Re was also demonstrated, causing three orders of magnitude increase in electron concentration after doping [45]. Bulk Nb-doped MoS_2 have also been grown using chemical vapour transport in sealed quartz ampoules, which can then be exfoliated to demonstrate p-type degenerate doped 2D MoS_2 [46]. At present, substitutional doping of 2D material is still a challenging task

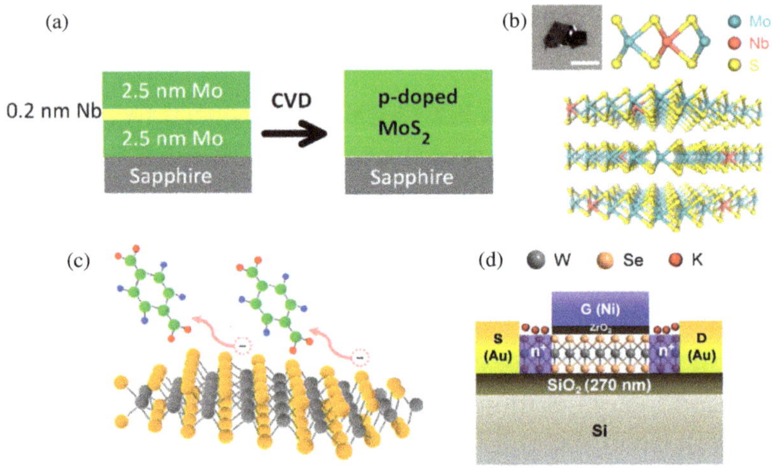

Fig. 7.16. Selected examples of doping techniques for TMDs. (a) Process of Nb doping in CVD-grown MoS_2. (b) Cross-section illustration of van der Waals coupled MoS_2:Nb layers where Nb dopants replace the Mo host atoms in a substitutional manner. A digital photograph inset shows a synthesised MoS_2:Nb crystal with a scale bar of 3 mm. (c) Adsorption of p-type dopant F_4TCNQ on monolayer MoS_2. (d) Device schematic showing n-doping of few-layer WSe_2 by surface charge transfer using potassium. Panels (a, b), (c) and (d) are reproduced with permission from Refs. [44], [48] and [49], respectively.

due to the need to control the dopant concentration consistently and precisely; moreover, poor reproducibility still plagues the processes and there is currently no universal standard to identify the doping concentration for 2D materials.

As 2D materials have a large surface to volume ratio, when 2D materials are exposed to the environment, there is also unintentional doping from ambient molecules such as H_2O and O_2 which physisorb and transfer charges to the 2D material. Tongay *et al.* showed that in n-type MoS_2, exposing the material to H_2O and O_2 increases the photoluminescence (PL) intensity, while in p-type WSe_2, PL intensity decreases [47] from which one can conclude that ambient adsorbate can dope 2D materials. However, these unintentional dopant molecules are typically not considered as doping by the community because these dopants are not controllable. However, deliberate introduction of molecules can be considered as charge transfer dopants. For example, the exposure of monolayer MoS_2 to a well-known electron acceptor molecule, 2,3,5,6-Tetrafluoro-7,7,8,8-tetracyanoquinodimethane (F_4TCNQ), blue-shifted and enhanced the photoluminescence of MoS_2 which suggests significant p-type doping [48] (Fig. 7.16(c)). Similarly, n-type doping is observed when MoS_2 is exposed to the electron donor molecule, nicotinamide adenine dinucleotide [48]. Hence, it can be concluded that charge transfer doping occurs between these electron donor or acceptor molecules and MoS_2, which can change the carrier concentration of monolayer MoS_2. Alkali metal ions such as K^+ physiosorbed onto 2D materials such as WSe_2 have also been shown to degenerately n-dope the material through charge transfer [49] (Fig. 7.16(d)). Other methods to induce charge transfer doping include solvent-based intercalation of atoms, molecules or ions into layered materials. For example, in contrast to naturally grown n-type SnS_2 with S vacancies, Cu-intercalated bilayer SnS_2 was observed to have a hole field-effect mobility of ~40 $cm^2V^{-1}s^{-1}$ [50]. However, unlike substitutional doping of the material, charge transfer doping from physisorbed molecules or ions tend to be less stable and it is not clear if the 2D material remains doped when the physisorbed molecules or ions are degraded due to air exposure or oxidation.

The presence of defects may also introduce more charge carriers in a material. For example, the conductivity type in monolayer WS_2 and MoS_2 may be tuned via the concentration of sulfur vacancies [51]. For low vacancy densities, the sulfur vacancies serve as an electron acceptor but when the sulfur vacancy density increases, they become electron donors and the conductivity type of the WS_2 and MoS_2 can be changed from p-type to n-type.

In the context of valleytronics, doping has also been shown to induce or enhance valley polarisation. Vanadium-doped $CrSe_2$ and MoS_2 were shown to exhibit a high degree of valley polarisation associated with strong spin-valley coupling [52]. In V-doped MoS_2, valley polarisation is found to increase from <12% to 42% after substitutional doping with a low concentration of vanadium [53].

At present, there is still no dominant doping strategy for 2D materials due to the limitations discussed. Unlike established recipes used in traditional semiconductors, there is still no standard recipe to dope 2D semiconductors to tune their transport properties. The challenge lies in the poor control over doping consistency and stability of the dopants, and the difficulty in controlling environmental impacts on the material. Unintentional doping can occur due to exposure to the ambient environment or contact with various chemicals throughout the fabrication of the device. Given the importance of doping in developing next-generation 2D devices, and the present challenges, the doping of 2D semiconductors remains an active area of research.

7.5 Contacts Engineering

The performance of valley devices based on 2D-layered crystals, including graphene, hBN and TMDs, are significantly affected by the electrical contacts that connect these materials with external circuitry. The major challenge in contacting 2D semiconductors is the presence of high contact resistance and Schottky barrier at the 2D semiconductor/metal interface, which drastically impede the source-drain current and induce non-linearity in the current-voltage response of the device. In the field-effect transistor, which is one of the most common electrical devices, low contact resistance is needed to achieve a high 'on' current and high frequency operation. In the 2D regime, the properties of the interface, including the chemical interaction between the metal and the semiconductor, govern the contact quality. The absence of dangling bonds in a 2D material makes it challenging to form strong interface bonds with a metal, and the atomically thin nature of the 2D material is easily damaged by deposition of the metal, which can result in increased contact resistance. Fig. 7.17 shows the contact resistance of 2D semiconductors found in literature so far, plotted against sheet resistivity. At a constant sheet resistivity, contact

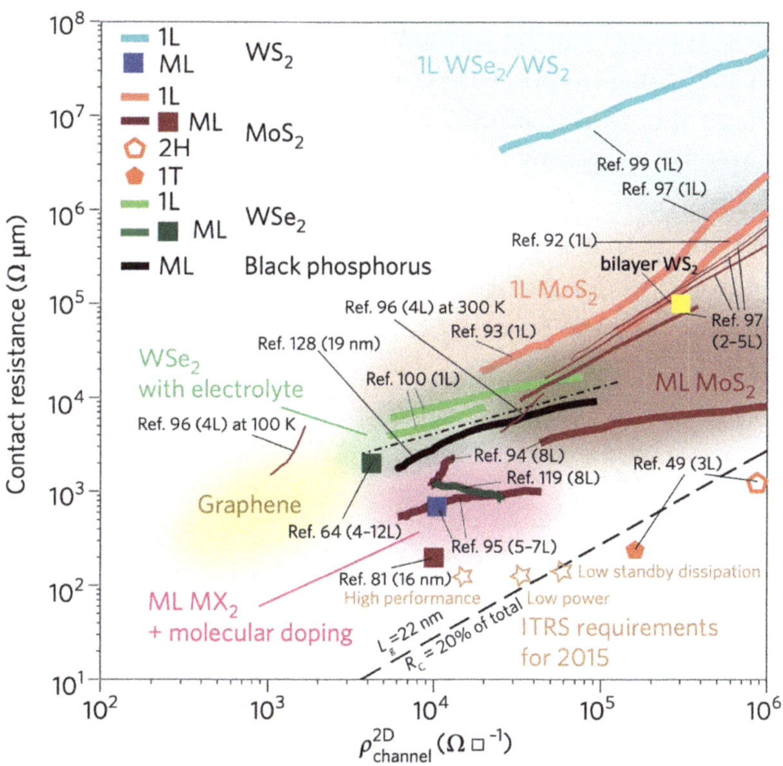

Fig. 7.17. Contact resistance for different values of channel sheet resistivity in several different studies on TMDs, black phosphorus and graphene. Note that the references in this figure refer to those in [54]. Adapted with permission from Ref. [54]. Additional data point using In/Au alloy (yellow square) is obtained from Ref. [59].

resistance decreases with increasing MoS_2 thickness and the lowest contact resistances have been obtained using phase engineering on few-layer and monolayer MoS_2 (see pentagonal markers on Fig. 7.17). By converting MoS_2 from the semiconducting 2H phase to the metallic 1T phase, one can form a seamless contact with contact resistance as low as 200 Ω μm, which pushes the performance of TMD-based transistors closer to the quantum limit.

Fig. 7.18 shows two basic interface geometries between 2D materials and bulk metal — the top contact and the edge contact. Experimentally, pure top contacts can be formed by simply avoiding edge contacts. However, forming a pure edge contact using lithographic techniques is difficult due to the atomically thin nature of single- or few-layer 2D materials. Hence in most experiments, a combination of these two geometries is used. An important thing to consider in contacts engineering is that a Schottky barrier, which is a potential barrier for electron injection, is formed at the metal-semiconductor junction. In the ideal case, the Schottky barrier can be tailored or reduced by

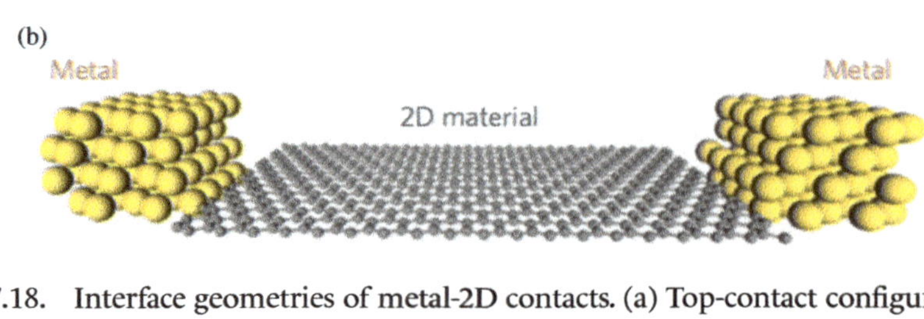

Fig. 7.18. Interface geometries of metal-2D contacts. (a) Top-contact configuration. (b) Edge-contact configuration. Reproduced with permission from Ref. [54].

an appropriate matching/alignment of the metal work function to the semiconductor's conduction or valence band depending on the carrier type. In practice, it is frequently encountered that this barrier does not change, even though metals of different work functions are put in contact with the same semiconductor. It is as if there is a "bunch of ghost states" in the bandgap of the semiconductor that pins the Fermi energy of the metal. This phenomenon, where the Fermi level of the metal is pinned to a position within the bandgap of the semiconductor regardless of the selection of metal, is called *Fermi level pinning* and is attributed to the complex interfacial states at the contacts which could be introduced by impurities, defects or orbital hybridisation, although no consensus has yet been reached regarding the origin of Fermi level pinning [54]. This means that for charges to be injected from the metal into the semiconductor, they often need to overcome such an unintended Schottky barrier. There are two mechanisms through which charges can be injected into a semiconductor: thermionic emission and field emission (tunnelling), as shown in Fig. 7.19. These two mechanisms may also occur together

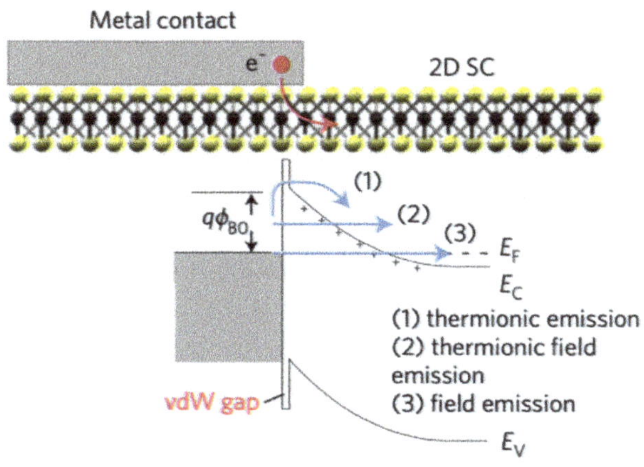

Fig. 7.19. Energy diagram displaying the charge-injection mechanisms between a metal contact and 2D semiconductor with van der Waals gap. Reproduced with permission from Ref. [54].

forming thermionic field emission. In thermionic emission, carriers gain enough thermal energy to go over the Schottky barrier at thermodynamic equilibrium. In field emission, carriers can tunnel through the Schottky barrier into the conduction channel when the barrier is sufficiently narrowed, e.g., under high electrostatic doping. Thermionic field emission may also occur when electrons do not gain enough energy to go over the barrier but they tunnel through the barrier near the top where the tunnel barrier width is narrow. At low temperature, most electrons tunnel near the metal Fermi level through a wide energy barrier. Furthermore, if there is a van der Waals gap between the metal and 2D semiconductor, an additional tunnelling barrier appears in the energy diagram. Knowledge of the dominant mechanism of charge injection is useful as they can be used to engineer device structures to tackle these problems. Strategies to decrease contact resistance and improve carrier injection efficiency include conventional metal contact strategies like doping only the TMD regions to be contacted [54], and 2D metallic contacts strategies such as using graphene as van der Waals contacts [55], phase engineering the TMD contact regions to induce a metallic state [56], using van der Waals metal-semiconductor junctions [57] or a metal alloy contact strategy [58], [59].

7.5.1 Conventional Metal Contacts

Electrically interfacing atomically thin TMDs with metal electrodes is inherently problematic given the presence of undesired metal-semiconductor interface barriers, which suppresses carrier-injection efficiency and introduces large contact resistances. To overcome these problems, device engineering methods which include choice of metal, metal transfer approaches and contact geometry may be employed.

Firstly, the choice of metal is important since there is a need to consider fundamental physical properties of different metals, including their melting point and electrical properties. Al, Ti, Cr, Ni, Cu, Pd, Ag, In, Pt, and Au are usually suitable as contact metals as they can be deposited using standard electron-beam evaporation or sputtering methods [60]. However, the contact metal for TMDs needs to have either appropriately low work function to achieve small n-type Schottky barriers or suitably high work function to achieve small p-type Schottky barriers. The Schottky barrier calculated by DFT [60] for the different bulk metals are listed in Fig. 7.20. For ideal cases, Fig. 7.20 shows that Mo would form the lowest barrier contact for n-type MoS_2 and Pd would form the lowest barrier contact for p-type WSe_2. In reality, a low barrier Mo/MoS_2 contact is hardly observed because Mo is a refractory metal with a high melting point of >2600°C and evaporation of Mo is difficult.

In 2019, Wang *et al.* showed that a low contact resistance of (800 ± 200) Ω μm was achieved for few-layered MoS_2 when an In/Au alloy electrode is used for a bottom-gate field-effect transistor (FET) device

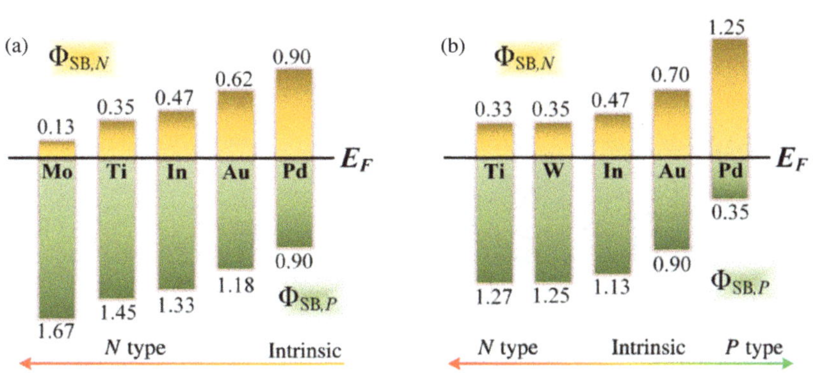

Fig. 7.20. Φ_{SB} (in eV) of various top contact metals with (a) MoS_2 and (b) WSe_2. $\Phi_{SB;N}$ ($\Phi_{SB;P}$) denotes n-type (p-type) Schottky barriers for electrons (holes). Reproduced with permission from Ref. [60].

[58] shown in Fig. 7.21(a). This value is among the lowest observed for bulk metal leads evaporated onto MoS_2, as can be seen in Fig. 7.21(c). In addition, the carrier mobility reaches about 167 cm^2/Vs for a monolayer MoS_2-based back-gated FET. The electrode consisted of a thermally evaporated 10 nm In capped with 100 nm Au. After evaporation, the contact was annealed at 200°C which caused the In and Au layers to alloy together to form a solid solution. They explained that because In has a low melting point of 165°C, it deposited gently on MoS_2 without causing any thermally induced damage (Fig. 7.21(b)) and formed a van der Waals-type metal film. Thereafter, the In layer formed a protective buffer layer for subsequent Au evaporation. Annealing the In-Au films created a low work function metal alloy, facilitating band alignment. Hence, by creating van der Waals-type bonding between 3D metal and 2D semiconductor, the Fermi level pinning effect at

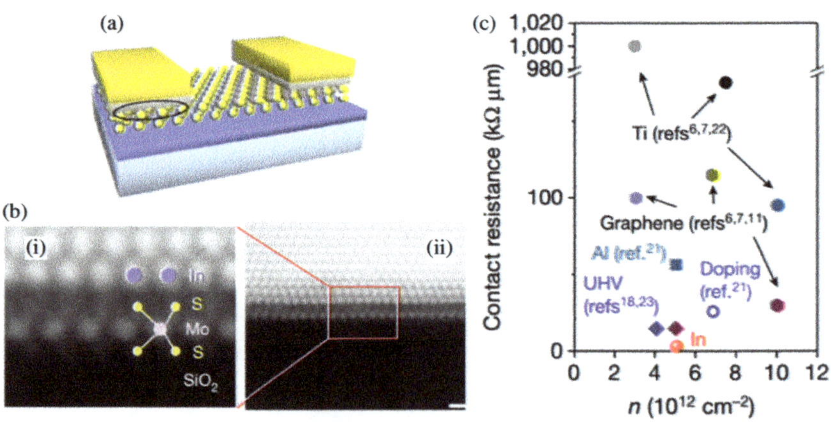

Fig. 7.21. (a) Device structure of the bottom-gate FET. (b) Atomic-resolution images of In/Au on monolayer MoS_2. (ii) Low pass-filtered annular dark-field scanning transmission electron microscope image showing MoS_2 and In/Au atoms as indicated by the enlarged image (i). Scale bar 5 Å. (c) Comparison of contact resistance from different types of electrode materials. Note that the references in this figure refer to those in [59]. Reproduced with permission from Ref. [58].

the metal-2D semiconductor interface can be significantly reduced. Subsequent work using In-Au alloy contacts to TMD materials showed that the ultralow Schottky barrier heights and contact resistances are sustained down to cryogenic temperatures (4 K), and carrier mobility *vs* carrier density trends [58] and quantum effects [59] can now be clearly identified. The advantage of such indium alloy contacts has also been successfully employed in CVD-grown WS_2 single- and bilayer devices where 3/40 nm of In/Au was deposited at 0.5/1 Å s^{-1} [61].

Nearly barrier-free electrical contacts may also be achieved with few-layer TMDs by a local bonding distortion strategy [62]. Cai *et al.* showed that a low-power oxygen plasma step can trigger local bonding distortion in the TMD material, which distorts the TM-C_6 polyhedral bonding geometry from trigonal-prismatic coordination to octahedral and thereby making the material semi-metallic. By exposing only the contact region of the TMD material to the low-power oxygen plasma, the local bonding distortion region is localised to the contact region and acts as a semi-metallic bridge between the metal lead and the TMD channel, facilitating ohmic transport behaviours from room to cryogenic temperatures.

High-performance p-type devices on MoS_2 and WSe_2 can also be fabricated by careful electron beam evaporation of high work function metals such as Pd and Pt [63] where the substrate temperature is kept at room temperature as much as possible through multiple short evaporation times. Transport measurements show that the Fermi level is unpinned with a low contact resistance of 3.3 kΩ μm, high mobility values of ~190 $cm^2\ V^{-1}\ s^{-1}$ at room temperature, and saturation currents in excess of >10^{-5} A-μm^{-1} and on/off ratio of 10^7.

Another approach to improve the performance of 2D TMD FETs is to transfer the metal electrode to the surface of the TMD to form a van der Waals contact without chemical disorder and Fermi level pinning [57]. Figure 7.22 shows the fabrication process of the transferred electrode method. The metal was first deposited on a sacrificial substrate, peeled off with PDMS, then aligned to the TMD and transferred onto the TMD material. Liu *et al.* showed that the transfer of an Au electrode did not damage the underlying MoS_2, whereas the evaporation of Au led to defects, diffusion, bonds and strains shown in Fig. 7.22(b)–(g). Furthermore, the transferred metal can be peeled

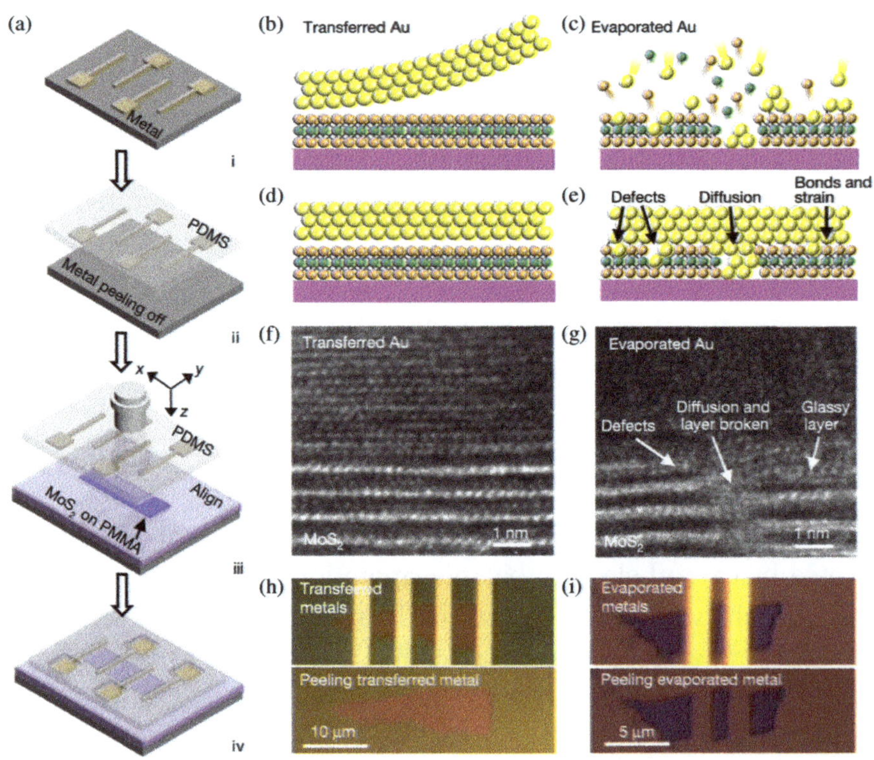

Fig. 7.22. Transferred metal electrode contacted MoS_2 FET. (a) Schematic illustrations of van der Waals integration of metal-semiconductor junctions. (b–e) Cross-sectional schematics of transferred Au and evaporated Au on MoS_2. (f, g) Cross-sectional TEM images of transferred Au and evaporated Au on MoS_2. (h, i) Optical images of MoS_2 flakes with peeling of transferred and evaporated metal. Reproduced with permission from Ref. [57].

off without damaging MoS_2, suggesting that the metal-semiconductor interface does not have direct chemical bonding (see Fig. 7.22(h) and (i)). Using the transferred electrode method, the Schottky barrier height approaches the Schottky-Mott limit, where the Schottky barrier height is dictated only by the work function of the metal and the semiconductor electron affinity. They also showed that by transferring silver or platinum films with a work function that matches the conduction band or valence band edges of molybdenum sulfide MoS_2, transistors with a two-terminal electron mobility at room temperature of 260 $cm^2\ V\ s^{-1}$ and a hole mobility of 175 $cm^2\ V\ s^{-1}$ respectively can be obtained.

Another key consideration in contacts engineering is contact geometry. In particular, the carrier injection efficiency is known to be higher for 3D edge contacts compared to 3D top contacts. Wang *et al.* achieved edge contact to graphene by capping hBN on graphene and etching the hBN to expose the graphene edges before depositing metal leads (1 nm Cr/15 nm Pd/60 nm Au) [64]. The contact resistance obtained was as low as 100 $\Omega\ \mu m$ and the room temperature mobility reached the theoretical phonon scattering limit, which is the fundamental upper limit for the intrinsic mobility of carriers [65]. This method has also been demonstrated for hBN-capped MoS_2 with graphene-metal edge contacts depicted in Fig. 7.23, resulting in a high Hall mobility of 34 000 $cm^2\ V^{-1}\ s^{-1}$ in six-layer MoS_2 at a low temperature of 2 K, and 120 $cm^2\ V^{-1}\ s^{-1}$ at room temperature, demonstrating high contact quality [66].

7.5.2 Metallic 2D Material Contacts

Besides the use of conventional metal contacts, integrating 2D metallic materials (e.g., 1T'-WTe_2, 1T-MoS_2, graphene) on semiconductors to form a van der Waals contact is also a viable way of avoiding the Fermi

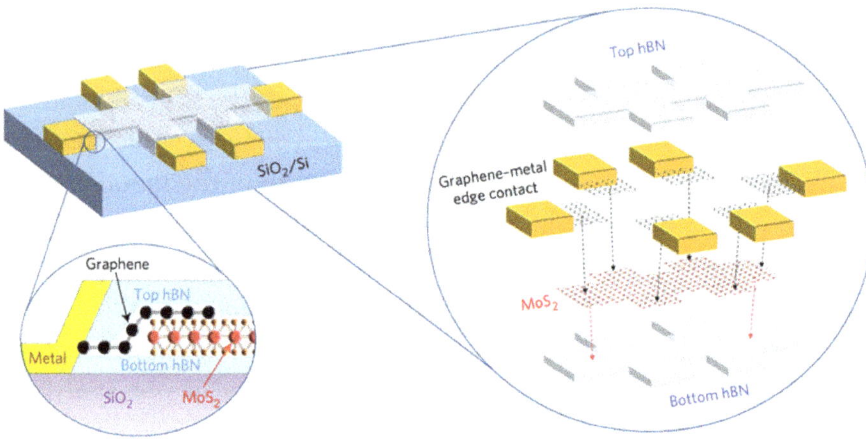

Fig. 7.23. Schematic of the hBN-encapsulated MoS_2 multi-terminal device. The use of graphene to interface MoS_2 with the metal contact resulted in high change high to higher carrier mobilities in the device because edge contacts to graphene have been shown to approach the quantum limit and van der Waals graphene contacts to MoS_2 are defect free and ohmic. Adapted with permission from Ref. [66].

level pinning effect and to decrease the contact resistance [67]. Similar to the In/Au and graphene contacts discussed in the previous section, a 2D metallic top contact also forms van der Waals interface with the 2D semiconductor which facilitates the carrier injection across the interface.

This can be achieved by CVD of the 2D metal (e.g., 1T'-WTe_2, 1T-$NbTe_2$, VS_2) onto the 2D semiconductor typically at temperatures of 700–1150°C. Lee *et al.* demonstrated a heteroepitaxial growth of metallic monolayer 1T'-WTe_2 on semiconducting 2H-WSe_2, where they measured a low Schottky barrier of 100 meV formed across the epitaxial van der Waals gap [68]. They first grew 2H-WTe_2 using vapour-phase reactions of high-purity WO_3 and Se precursors on SiO_2/p$^+$-Si substrates at a growth temperature of 850°C. Subsequently, to reduce intermixing of the grown 2H-WTe_2 with the metallic 1T'-WTe_2 layer, the growth temperature was lowered to 700°C and by using KI as a growth agent, metallic 1T'-WTe_2 was grown on top of the semiconducting 2H-WTe_2 at the edges of the crystal, facilitating subsequent metal contacts fabrication.

CVD may also be used to form a combination of top and edge contacts between 2D semiconductors and metallic 2D materials to form lateral heterostructures, such as a graphene-MoS_2 heterostructure [69]. In this method, Zhao *et al.* first transferred monolayer graphene onto a silica substrate and etched it with oxygen plasma. The sample was then moved into a quartz tube for CVD growth of MoS_2, where MoS_2 began to nucleate around the edges and within the channel. The grown MoS_2 was found to overlap graphene by 100 nm (Fig. 7.24), facilitating charge transport along the graphene/MoS_2 interface. Zhao *et al.* showed that their FET device grown using this method achieved a high transconductance (10 μS), on-off ratio (~10^6) and mobility (~17 $cm^2\ V^{-1}\ s^{-1}$).

Phase engineering may also be done to turn the 2D semiconductor into the metallic phase. The trigonal prismatic (2H) phase of MoS_2 is a semiconductor while the octahedral (1T) phase of MoS_2 is a metastable phase with a metallic nature. This means that by partially converting an exposed area to metallic 1T phase while keeping the channel region covered at 2H phase, an atomically seamless ohmic contact can be engineered between 2H-MoS_2 and 1T-MoS_2. To transform 2D semiconductors into its metallic state, one can perform electron donation-induced phase transitions using (i) *n*-butyl lithium treatment, or (ii) generating chalcogen vacancies by laser irradiation or Ar^+ plasma treatment. Kappera *et al.* used *n*-butyl lithium treatment to locally induce 1T-MoS_2 on 2H-MoS_2 to achieve a low contact resistance of 200–300 $\Omega\ \mu m$ at zero gate bias [32]. Their FETs were fabricated and measured in air, and electrically they showed mobility values of ~50 $cm^2\ V^{-1}\ s^{-1}$, subthreshold swing values below 100 mV per decade, on/off ratios of >10^7, drive currents approaching ~100 $\mu A\ \mu m^{-1}$, and excellent current saturation levels.

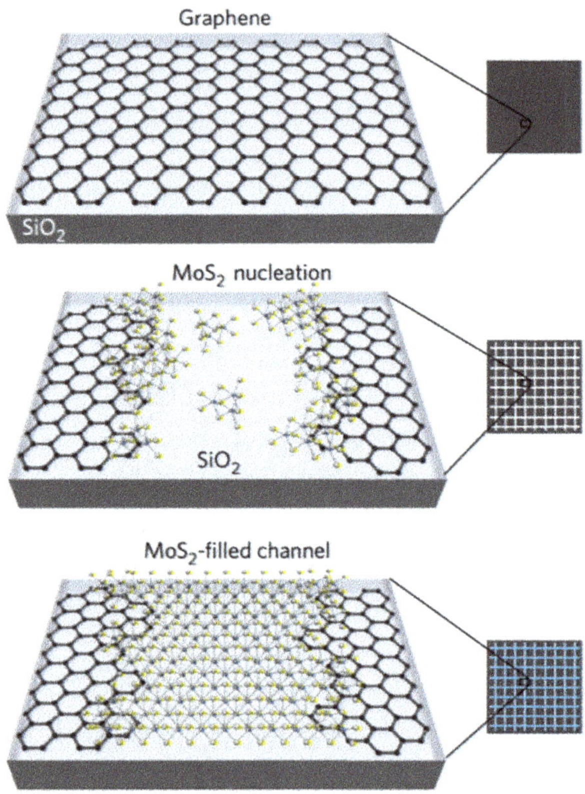

Fig. 7.24. Schematic of the MoS_2-graphene heterostructure growth process showing MoS_2 nucleating around the edges of graphene, eventually filling the channels. Reproduced with permission from Ref. [69].

A metallic state may also be induced via intercalation of van der Waals materials, which involves insertion of molecules or atoms in between layers of a layered structure. n-type SnS_2 can be converted to a p-type semiconductor by Cu intercalation, or into a conductive metal by Co intercalation [50], where Cu or Co atoms are placed in between the van der Waals gap as shown in Fig. 7.25(a) and (b). This leads to changes in the colour of the flake seen, for example, in the optical images in Fig. 7.25(c)–(f) and a corresponding change in the electronic properties of the material reported in Ref. [50].

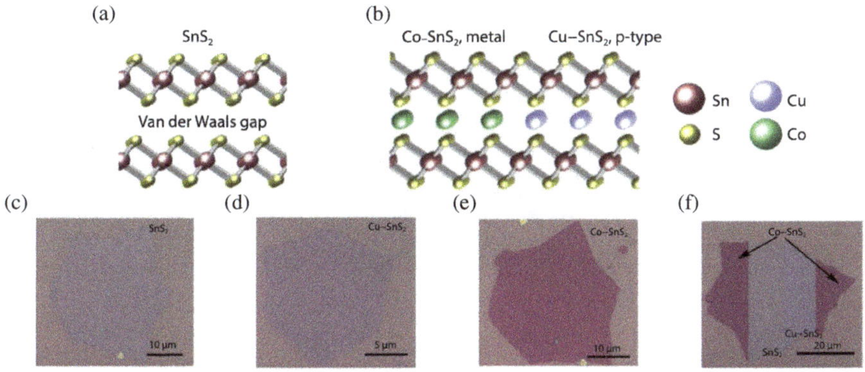

Fig. 7.25. Intercalation-induced phase transition of SnS_2. Atomic structures of (a) pristine, (b) Co-intercalated, and (c) Cu-intercalated SnS_2. (c–f) Optical images of pure SnS_2, Cu, Co and Cu–Co intercalated SnS_2. Reproduced with permission from [50].

Perhaps one of the most interesting methods involves using graphene as a tunable contact to achieve ohmic contact to MoS_2 at cryogenic temperatures. Liu *et al.* achieved a transparent contact to MoS_2 with zero contact barrier [70], using the device shown schematically in Fig. 7.26(a). Specifically, it was found that these devices exhibit linear I–V output, indicating ohmic behaviour down to a temperature of 1.9 K (Fig. 7.26(b) and (c)). A slight non-linearity is observed at low gate voltages (<40 V) but linearity becomes more prominent at high gate voltages (80 V for monolayer and 60 V for multilayer). This is possible as the gate voltage can be used to shift the Fermi level in graphene, to achieve a perfect band alignment with MoS_2.

7.6 Gate Engineering

In most of the fabricated 2D semiconductor devices, back-gating or bottom-gating have been the predominant strategy for tuning the carrier density in the transport channel [71]. Apart from devices fabricated on graphene back gates, most devices tend to use the degenerately

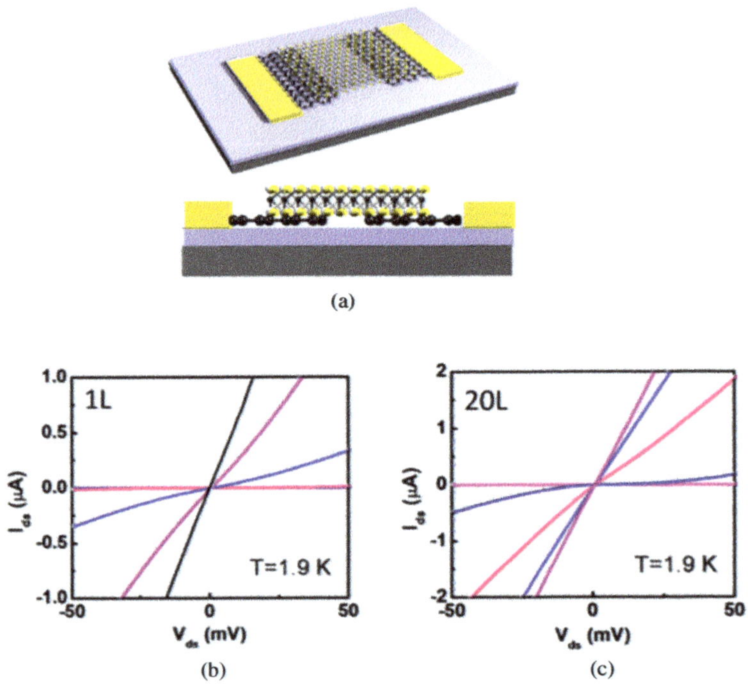

Fig. 7.26. (a) Schematic of MoS_2 device structure with bottom-graphene electrodes. (b) Output characteristics of a monolayer MoS_2 device at various gate voltages at 1.9 K. Gate voltages range from –60 V (pink) to 80 V (maroon) with a 20 V step. Adapted with permission from Ref. [70].

doped silicon substrate as the back gate with the terminating silicon oxide (typically about 90 nm or 280 nm) providing the required optical contrast to identify the 2D material flakes during the fabrication process. While back-gating proved convenient for device fabrication, the several tens of volts typically required to provide any significant tuning of carrier density implies an inefficient gating action.

Gate engineering for valleytronic devices is still a relatively immature subject matter despite some device demonstrations with back-gate action. The most critical aspect of the gate engineering appears to be the challenge of finding a method to interface the 2D material channel with a compatible dielectric that minimises interface defects, maximises gate action, and is robust to electrical breakdown in the operating regime.

To implement efficient valleytronic devices, more efficient gating strategies are required. The following options are available or are under further developments:

1) Bottom electrodes separated by a 2D dielectric (e.g., hBN stack) from the 2D channel material above. The bottom electrodes are usually prefabricated onto a silicon oxide surface of a silicon wafer. The electrode could be made of metal or semi-metallic graphene.
2) Top electrodes (also known as top-gating) separated by a thin dielectric layer (e.g., hBN stack, thin high-K dielectric such as Al_2O_3 or HfO_2) from the 2D channel material below.

Fig. 7.27 shows the schematic of a dual-gated device as previously discussed in Section 5.6. Here, both the top-gate voltage (V_{TG}) and back-gate voltage (V_{BG}) can be tuned independently. The bottom gate is separated by SiO_2 and hBN from the BLG, and the top gate is separated by a hBN stack from the BLG. With two gates, the carrier density and perpendicular electric field (displacement field) can be independently controlled [16].

Ionic-liquid gating has also been demonstrated to work with 2D materials allowing tuning of carrier density to levels much higher than the typical solid-state gate dielectrics [72]. However, this technique does not allow dynamic gate control and carrier density tuning and is difficult to operate at cryogenic temperatures as the ionic liquid freezes. While this method of gating has shown some novel gate-induced effects [73], it is difficult to implement in a practical device and therefore unlikely to useful for eventual device engineering. Hence, we do not elaborate on this method further.

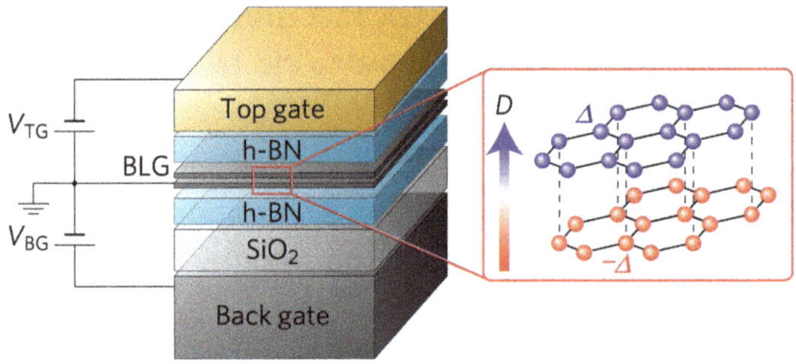

Fig. 7.27. Schematic of the dual-gated BLG device. The top gate is a gold film, and the back gate is a p-doped silicon substrate. Two h-BN layers and a SiO_2 film are used as gate insulators. Reproduced with permission from Ref. [16].

7.6.1 Bottom-gate Examples

Bottom gating or back gating of 2D materials through the substate oxide to tune carrier densities have been shown to be effective since the discovery of graphene in 2004 [74]. It is convenient and easily achievable using typical sourcemetres and have been shown to work for a wide variety of 2D materials [71] such as CVD-grown monolayer WS_2 and MoS_2 FETs [75]. In this example, monolayer WS_2 and MoS_2 were grown by CVD on sapphire substrate and then wet-transferred onto Si/SiO_2 (280 nm oxide) wafer with prepatterned Ti/Au electrodes at 4 μm channel lengths. The FETs were operated at a gate voltage range of –100 V to 100 V, which provided carrier modulation from approximately 1×10^{10} cm^{-2} to 7.2×10^{12} cm^{-2} using 280 nm SiO_2 as the gate oxide. Figure 7.28 shows a FET based on multilayer WS_2 fabricated on 300 nm of SiO_2 [76]. For this device, a 35 nm Au metal contact is deposited on the WS_2 flake and Fig. 7.28(b) shows an AFM image of a device made from a 7.5 nm-thick WS_2 crystal.

While the back gate is convenient, it has some drawbacks. The thick 280 nm SiO_2 layer needed for optical identification of the 2D material

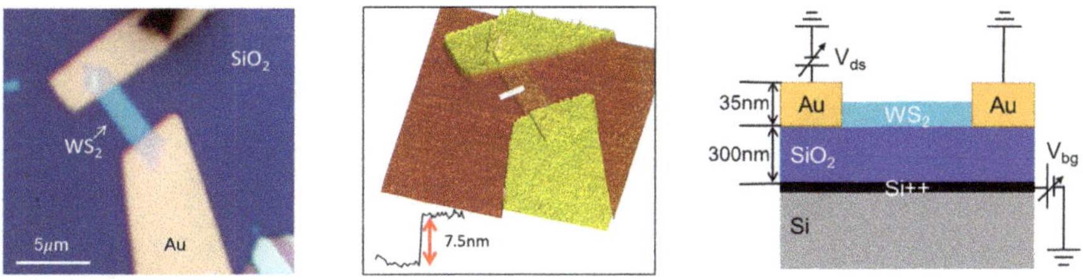

(a) (b) (c)

Fig. 7.28. Device fabrication of WS_2 FET. (a) Optical image of a multilayer WS_2 device with gold as the contact metal and Si/SiO_2 as the substrate. (b) Atomic force microscope image of the device shown in (a). The inset shows the line trace of 7.5 nm-thick WS_2 crystal obtained at the location of white line. (c) Schematic drawing of the measurement setup for electrical transport characterisation. Reproduced with permission from Ref. [76].

flakes renders the gating action inefficient. This limits the subthreshold swing of typical TMD devices to around >80–100 mV/decade [77]. For more efficient gating and novel gating geometries, the top gate is used instead.

7.6.2 Top-gate Examples

Top-gated or dual-gated (both top and back gates) devices have been discussed several times in this book in the context of valleytronics. One of the earlier papers showing this are papers from Shimazaki *et al.* [16] and Sui *et al.* [17] where the top and back gates are swept independently. The advantage of having independent top- and back-gate control is that the carrier density n (green axis in Fig. 7.29) of the 2D material can be tuned independently of the displacement field D (red axis in Fig. 7.29). When dual electrical gating fields are applied to the BLG, the top and bottom electrical displacement fields D_t and D_b produce two main effects: (i) the difference of the two displacement fields, $\delta D = D_b - D_t$, cause carrier density modulation and (ii) the

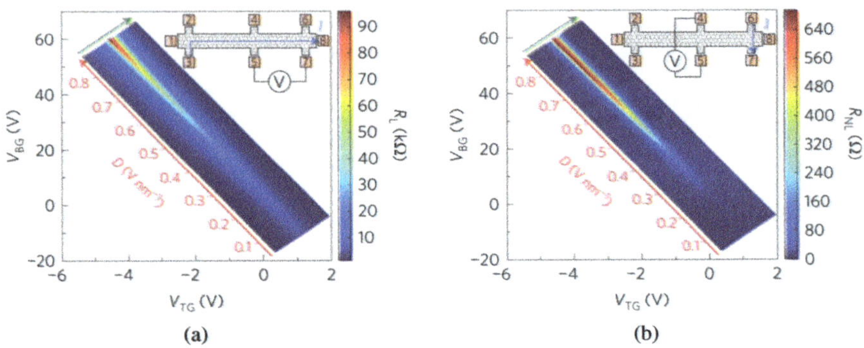

Fig. 7.29. Gate voltage dependence of (a) R_L and (b) R_{NL}. The displacement fields from the back and top gates are defined as $D_{BG} = \varepsilon_{BG}(V_{BG} - V_{BG,0})/d_{BG}$ and $D_{TG} = -\varepsilon_{TG}(V_{TG} - V_{TG,0})/d_{TG}$, respectively, where $\varepsilon_{BG}(\varepsilon_{TG})$ and $d_{BG}(d_{TG})$ are the relative dielectric constant and thickness of the back (top) gate, respectively. $V_{BG,0}(V_{TG,0})$ is the offset of charge neutrality point from the back (top) gate voltage under the top-gated region due to environmental doping. Inversion symmetry breaking is increased as D increases. Reproduced with permission from Ref. [16].

average of the two, $D = (D_b + D_t)/2$, breaks the inversion symmetry of the bilayer and generates a non-zero bandgap. These two degrees of freedom are coupled together when a common back gate is used. Fig. 7.29 shows that the ability to control displacement field D independent of carrier density n ensures that the inversion symmetry can be broken in BLG, enabling the detection of valley currents through the non-local resistance R_{NL} measurement (Fig. 7.29(b)).

An example of a device where the top gates play an important role is the array of top gates demonstrating gate-defined quantum confinement [61] (Fig. 7.30). In this device, WS_2 flakes were grown by CVD and transferred onto a 290 nm SiO_2/Si substrate that also serves as a back gate for tuning the overall carrier densities in the devices. After depositing the source/drain electrodes, HfO_2 was deposited using atomic layer deposition (ALD) to serve as the gate dielectric and encapsulation layer. An array of top gates was subsequently deposited

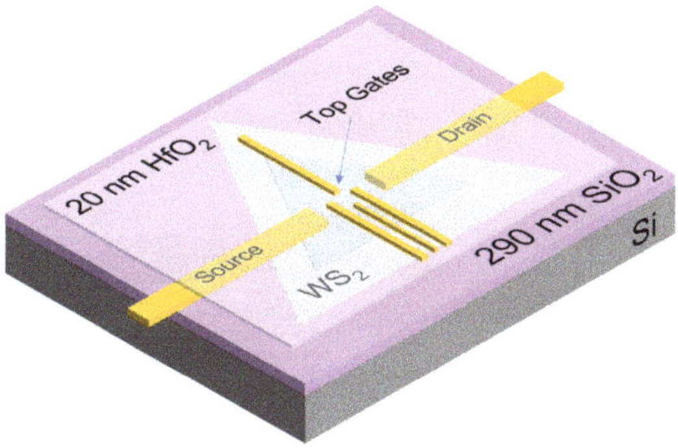

Fig. 7.30. Schematic of the gated quantum dot device on bilayer CVD WS_2. Reproduced with permission from Ref. [61].

on the HfO_2 gate oxide, which allowed electrostatic definition of the quantum dot region by depleting carriers under the top gates. Coulomb blockade has been shown in such a device in BLG [78] and bilayer WS_2 [61].

Similarly, top-gate defined quantum confinement in InSe-based 2D heterostructures was also achieved by Hamer *et al.* (Fig. 7.31) [79]. Here, the dielectric separating the top gate from InSe is a thick hBN layer (20–40 nm) instead of a high-κ oxide. The hBN layer also doubles as an encapsulation layer preserving the pristine qualities of InSe. The top gates are used to electrostatically define the quantum dot region and, also to tune the carrier density at the graphene/InSe interface to ensure band alignment between the graphene and InSe, whereas the global back gate is used to control the total charge density.

Most of the high-quality devices benefit from the synthesis/preparation of high-quality 2D hBN as the gate dielectric. However, the use of hBN is still largely limited to mechanical exfoliation from a high-quality bulk crystal which is not amenable to device scale-up. Controlled growth of large-area high-κ dielectrics (e.g., Al_2O_3, HfO_2)

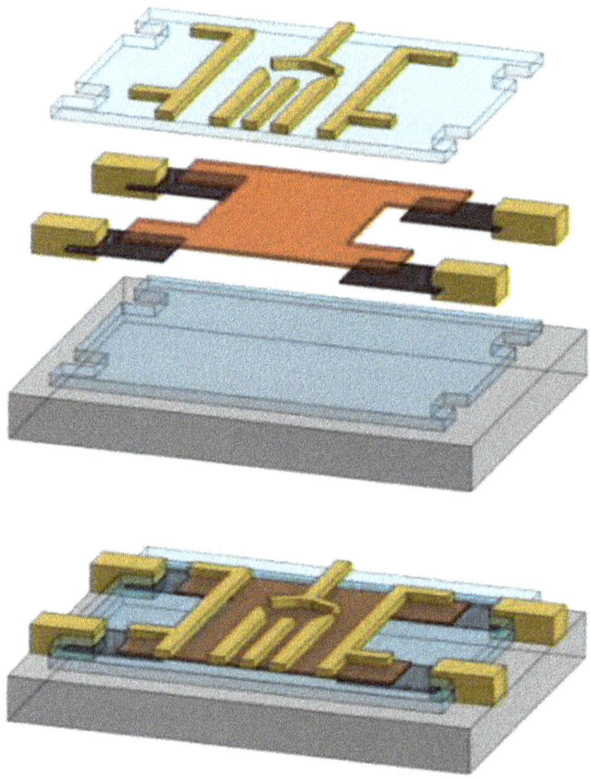

Fig. 7.31. hBN/InSe/hBN heterostructure with graphene contacts and multiple top gates. Schematic showing individual layers: 2D InSe (red), graphene (dark grey), hBN (blue), Cr/Au contacts, top gates (yellow), and Si/SiO_2 substrate (light grey). Reproduced with permission from Ref. [79].

via ALD has been demonstrated, but the ALD process may damage the 2D material channel or create undesirable interface roughness that impact device performance [61]. In addition, the long term stability and reliability of these ALD-grown dielectrics remain to be validated and presents opportunities for further research.

The above gating strategies remain far from being optimised, and new gate architectures are sought for more efficient gating strategies. Large-area direct growth of high-quality compatible dielectrics for 2D devices (including valleytronics) is still sorely lacking and urgently needed for the reproducible engineering of robust devices.

7.7 Concept Valleytronic Devices

Thus far, the experimental techniques for fabricating high-quality devices based on van der Waals heterostructures have been discussed. These techniques include interface and grain boundary engineering, band engineering, doping techniques, and contacts engineering. Broadly speaking, these methods are useful for any device that involves 2D materials, but they are particularly useful for valleytronic devices, which are often based on 2D heterostructures. With the maturing of these techniques, valleytronic devices can now be made a reality in the lab, paving the way for future research and development in the field of valleytronics. This section seeks to present a variety of valleytronic devices, including valley valves, beam splitters, and valley transistors demonstrated to date. In addition, devices involving valleytronics also enable other behaviours such as valley-polarised exciton currents, valley magnetoelectricity, and valley Zeeman spin-orbit coupling-induced spin precession.

7.7.1 Valley Valve and Electron Beam Splitter

In 2018, Li *et al.* demonstrated a valley valve and electron beam splitter [80]. This was enabled by the presence of ballistic valley Hall kink states in BLG, which allowed the group to demonstrate gate-controlled current transmission in a four-kink router device. In the quad-split-gated valley router device shown in Fig. 7.32(b) and (c), the four top-gate voltages can be set with different polarities, which can change the direction of the source current flowing in between

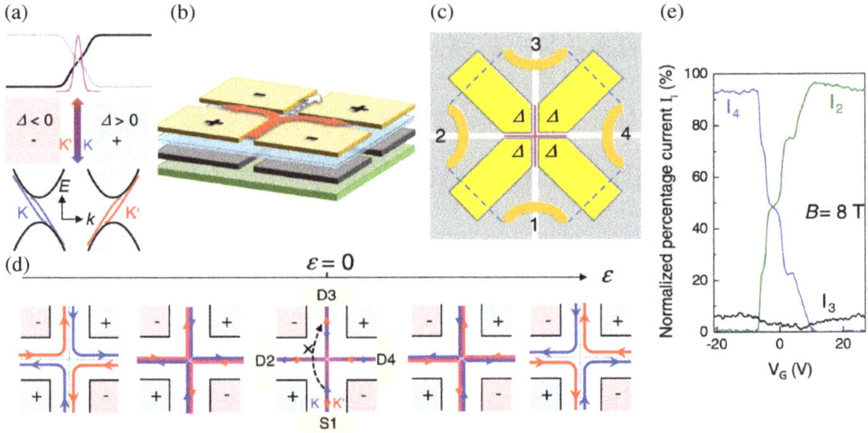

Fig. 7.32. Valley valve and electron beam splitter. (a) Potential and wavefunction (magenta) of the valley-momentum locked kink states in the kink channel. (b, c) Schematics of quad-split-gated valley router device with graphite bottom gates (grey), top gates (yellow), bilayer graphene (blue in (b) and dashed lines in (c)), and Cr/Au side contacts (gold arcs). In (b), the red and white arrows represent valley valve and beam splitting. The magenta cross in (c) represents the four kink channels. (d) Evolution of the K and K′ kink state wavefunction centre as a function of Fermi level energy ε in a moderate magnetic field. (e) Measurements of the normalised percentage current I_i received at terminals 2 to 4 with terminal 1 as the current source as a function of gate voltage V_G. Reproduced with permission from Ref. [80].

the gates either straight through (from 1 to 3), left turn (from 1 to 2) or right turn from (1 to 4) by relying on the valley-locked 1D channels (called Kink states; Fig. 7.32(a)) located in between the split gates. The valley-locked 1D channel can be turned on by applying an opposite polarity (+ –) on the split gate and turned off when the gates have equal polarity (+ +) or (– –). By applying the even polarities $\begin{pmatrix} - & + \\ - & + \end{pmatrix}$, the valley valve (from 1 to 3) can be turned on and when odd polarities are applied $\begin{pmatrix} - & + \\ + & - \end{pmatrix}$, the valley valve is turned off as there are no valley-matched states in channel 3 to accommodate the valley-locked carriers. Instead, the device becomes an electron beam splitter on channels 2 and 4 instead. Furthermore, by changing the degenerately-doped Si global back gate voltage (V_G), the current collected at terminals 2 and 4 can be varied as a function of the carrier density, which demonstrates a tunable electron beam splitter. Fig. 7.32(e) shows the normalised percentage current received at terminal 2(4) is high(low) at a positive gate voltage, and low(high) at a negative gate voltage because the K and K′ kink state wavefunction changes with the global back-gate voltage (V_G) and flips direction when the charge neutrality point is crossed, as shown in Fig. 7.32(d). The valley-momentum locking property of the kink states enabled a valley valve with an ON/OFF ratio of 8 at zero magnetic field and up to 100 at high magnetic fields above 5 T.

7.7.2 Room-Temperature Valleytronic Transistors

Valleytronic transistors which can generate, propagate, detect and manipulate valley information at room temperature have been demonstrated by Li *et al.* [81] showing that valley-based information carriers persist up to room temperature. Fig. 7.33(a) shows a schematic of their device where a chiral nanocrescent plasmonic antennae was used to inject valley-selective charge carriers into monolayer MoS_2. Using this construct, they demonstrated efficient valley polarisation of free carriers in monolayer MoS_2 through a two-stage photoelectric process. In the first stage, the asymmetry of the surface plasmon nanocrescents (NPs) aids to separate photons of opposite helicity spatially to the source and drain. In the second stage, the helical photons selectively excite carriers only in one of the K valleys in the MoS_2 channel via helicity-conserved hot carrier injection. Thus, carriers from opposite valley are generated at the source and drain Au/MoS_2 junctions. From two probe FET measurements, they found that the valley-polarised free carriers have a long lifetime of 100–300 ns. Using Hall probes at the channel edges, they also measured valley polarisation

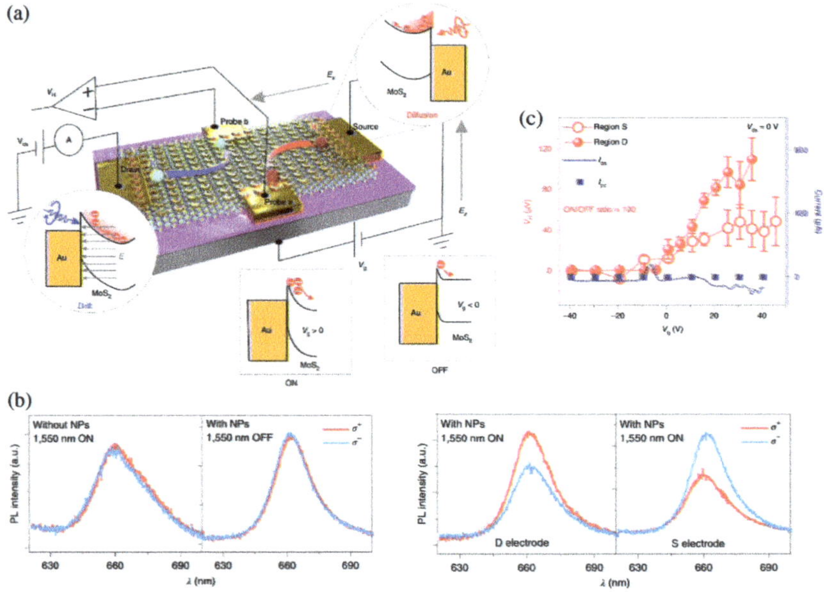

Fig. 7.33. Valleytronic transistor. (a) Schematic and operation mechanism of the device, consisting of an MoS_2 transistor, two lateral Hall probes and a chiral plasmonic antennae array of gold nanoparticles (NPs) in the shape of golden nanocrescents. (b) Circular-polarised PL achieved with infrared-excited nanoantennae with NPs on MoS_2 (right) in contrast with unpolarised PL in pristine MoS_2 and MoS_2 with NPs given an infrared OFF-excitation (left). (c) Hall voltage as a function of gate voltage V_g at V_{ds} = 0 V. NPs, nanoparticles (referring to the gold nanoantennae). Reproduced with permission from Ref. [81].

which persist without a charge current, and they found that the valley current can drift over 18 μm. Circular polarisation-resolved PL measured with an infrared-excited nanoantennae array reported a polarisation ratio of 85%, compared to negligible polarisation in pristine MoS_2 (Fig. 7.33(b)). Furthermore, electrostatic gating can vary the valley Hall voltage, which allows the device to be used as a pure valley transistor. Specifically, Li *et al.* achieved a sufficiently large ON/OFF ratio of up to ~10^3 with a considerable valley Hall output of 110 μV (Fig. 7.33(c)). This pioneering demonstration of the modulation of valley information enabled by the transistor with no net charge current makes the prospective of low-power valleytronic logic devices compelling.

7.7.3 Valley-polarised Exciton Current Generator

In Chapter 5, we discussed several strategies to generate valley current, including optical and electrical methods. Unuchek *et al.* presents a new interesting concept – valley current generation using interlayer excitons. Fig. 7.34 shows a TMD heterostructure device that allows the encoding of data in the valley degree of freedom using valley-polarised exciton currents in the form of a valley-polarised excitonic transistor [82]. While TMDs are ideal hosts for valleytronic devices due to their valley selectivity, stacking them together to form heterostructures brings about a further advantage — they can now host interlayer excitons, which are bound electron-hole pairs spatially separated in opposite layers spaced by a van der Waals gap. These interlayer excitons have shown much higher valley lifetimes in the hundreds of nanoseconds compared to excitons in a single-layer TMD material. As the interlayer excitons have an out-of-plane electric dipole moment, they allow electrical control that persists in these devices up to room temperature. To create a wider spaced (~0.9 nm) WSe_2/hBN/$MoSe_2$ device (Fig. 7.34(a) and (b)) they added an additional monolayer hBN layer in between the WSe_2 and $MoSe_2$ heterostructure. The trilayer heterostructure then allowed valley-polarised excitons to be generated and transported over mesoscopic distances. Fig. 7.34(c) shows the PL intensity of two devices — WSe_2/hbN/$MoSe_2$ heterotrilayer and the WSe_2/$MoSe_2$ heterobilayer — after excitation with right circularly polarised light. The trilayer heterostructure shows a single low-energy interlayer transition at 1.39 eV with preserved circular polarisation of incoming light, and the bilayer heterostructure shows a characteristic interlayer doublet with opposite helicities for the two peaks, typical of

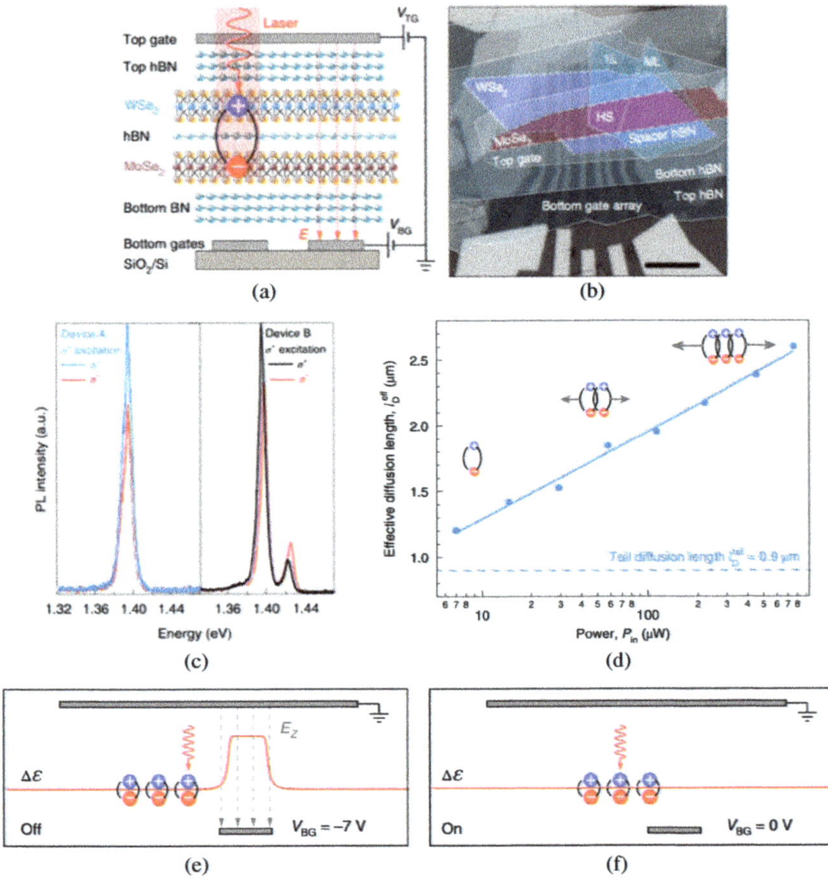

Fig. 7.34. Device for generating valley-polarised excitons. (a) Schematic of the device showing the top and bottom gates, and the interlayer excitons generated in the heterotrilayer device. (b) Optical image of fabricated device. Scale bar 10 μm. (c) Polarisation-resolved μ-PL spectrum from the WSe_2/hBN/$MoSe_2$ heterotrilayer (Device A) and the WSe_2/$MoSe_2$ heterobilayer (Device B), excited with right circularly polarised light showing that both devices have valley polarisation. (d) Extracted effective diffusion length for Device A as a function of incident power. The inset schematics demonstrate density-dependent diffusion driven by exciton-exciton repulsion. (e, f) Numerically simulated exciton energy profile (red line) in the Off state (e, $V_{TG} = 0$ V, $V_{BG} = -7$ V) and On state (f, $V_{TG} = 0$ V, $V_{BG} = 0$ V) states of the excitonic transistor where the V_{BG} is used to trap the exciton. Reproduced with permission from Ref. [82].

aligned heterobilayers. Fig. 7.33(d) shows that the intralayer excitons in the heterotrilayer device can diffuse up to 2.7 μm without recombining and the spatial exciton diffusion can be tuned on and off by electrostatic potential applied using a back gate array (Fig. 7.34(e) and (f)). Operating the gate array as electrostatic traps can also increase the exciton concentration to up to 10^{12} cm^{-2}, which paves the way for the physics demonstrating a coherent quantum state of valley-polarised excitons via Bose-Einstein condensation.

7.7.4 Magnetoelectric Device

Valleytronic devices can also enable the magnetoelectric effect based on the valley degree of freedom in 2D materials, for applications in magnetic sensing and switching. Lee *et al.* demonstrated the pure electrical generation of valley magnetisation through a charge current which generates bulk magnetisation through the valleys by breaking the three-fold rotational symmetry in monolayer MoS_2 via uniaxial stress [83]. They did this by first transferring monolayer MoS_2 onto polydimethylsiloxane (PDMS) substrates, and then stretching the PDMS along a high-symmetry axis of the MoS_2 crystal, before stamping the strained MoS_2 onto a 100 nm SiO_2 layer on a Si substrate for device fabrication. The stress-induced distortion of the lattice (see Fig. 7.35(a)) leads to ion-charge polarisation, producing a piezoelectric field E_{pz}. In the presence of E_{pz}, valley magnetisation M_v is induced by an in-plane current density J as shown in Fig. 7.35(b). The effect of straining the material can be observed via Kerr rotation imaging (recall Section 3.4).

In the unstrained material, the Kerr rotation signal was only observed at the channel edges due to the valley Hall effect, as shown in Fig. 7.35(c) and seen also in an earlier paper by Lee *et al.* [84], but it was observed

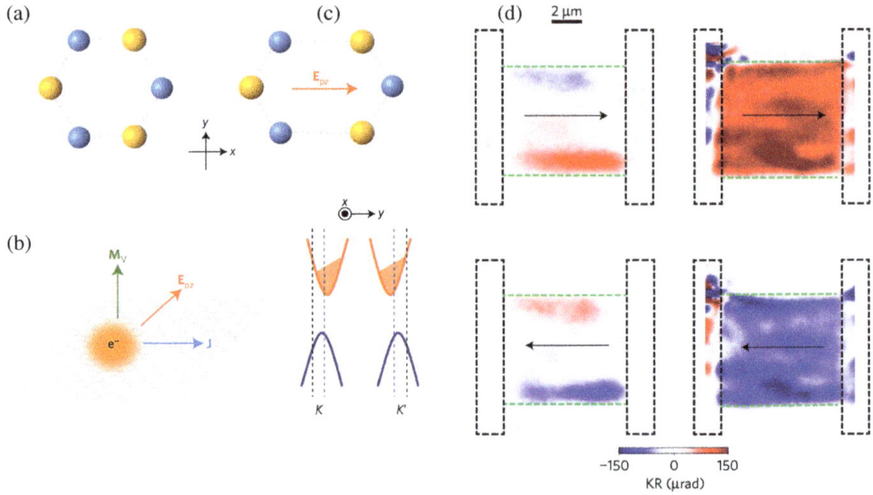

Fig. 7.35. Valley Hall effect and valley magnetoelectric effect in monolayer MoS_2. (a) Uniaxial strain producing piezoelectric field E_{pz}. (b) Out-of-plane valley magnetisation M_v induced by in-plane current density J given an orthogonal E_{pz}. Kerr rotation image of (c) unstrained and (d) strained material. Reproduced with permission from Ref. [83].

over the entire channel in the strained device (Fig. 7.35(d)) with a much stronger Kerr rotation signal per unit current, showing that the magnetoelectric effect is successfully generated in this material by applying uniaxial strain and sourcing current alone. Furthermore, the magnetoelectric effect persists up to room temperature. Overall, the observation of electric field-induced valley magnetisation at room temperature is expected to enable valley-based applications such as magnetic switching in flexible devices, and suggests a new possibility for integrating valleytronics with magnetic information storage.

7.7.5 Valley-enabled Spin Transistor

The last device to be showcased in this section is a spin FET enabled by valleytronics. Ingla-Aynés *et al.* demonstrated control of the polarisation direction of a spin current in a BLG-WSe_2 heterostructure using only the drift current (I_{DC}) and the gate voltage (V_g) in the absence of magnetic field at both 50 K and at room temperature [85]. As BLG alone has weak spin-orbit coupling, the main control comes from the presence of strong valley Zeeman spin-orbit coupling in the WSe_2 layer, which occurs due to the crystal structure of WS_2, and the presence of tungsten, a heavy transition metal. By proximitising BLG with WSe_2, a strong spin-orbit field is imprinted onto graphene as shown in Fig. 7.36(a). This effect produces an effective magnetic field in the graphene layer that is sufficient to cause coherent spin precession. Furthermore, the sign of the precessing spin polarisation detected as R_{NL} in Fig. 7.36(e) can be tuned by a back-gate voltage (V_{bg}) and by a drift current (I_{DC}) using a device architecture shown in Fig. 7.36(d).

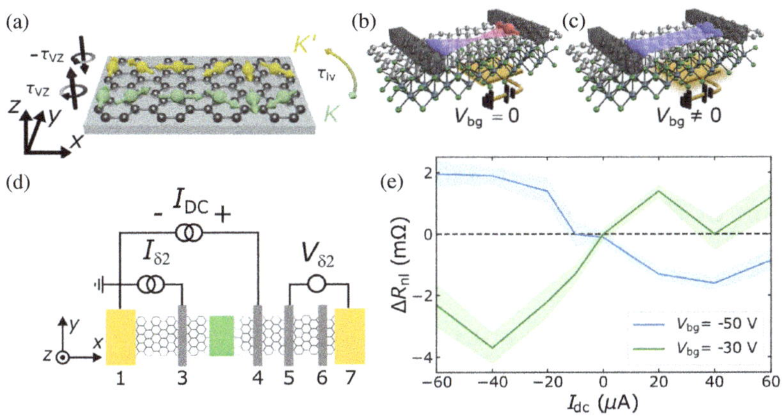

Fig. 7.36. Device working principle and BLG-WSe_2 spin transistor operation. (a) Sketch of the BLG-WSe_2 heterostructure. Out-of-plane valley Zeeman spin-orbit field (black arrows) with opposite signs at the K and K′ valleys induce in-plane spin precession with a period τ_{VZ}. Spins can scatter between the valleys via intervalley scattering (τ_{iv}). (b, c) Sketch of a spin field-effect transistor operating in the strong spin-orbit coupling regime where the valley Zeeman-induced spin precession is tuned by V_{bg} to control the sign reversal. (d) Sketch of the device configuration for spin drift measurement. (e) I_{DC} dependence of ΔR_{nl} at V_{bg} = –50 V and –30 V showing that ΔR_{nl} flips sign at different gate biases. Reproduced with permission from Ref. [85].

By tuning the carrier density using V_{bg} and spin transport time using I_{DC}, they showed the ability to control spin polarisation up to room temperature. This paves the way for energy-efficient spin-based logic applications and could complement existing spin-logic proposals which could aid in implementing energy-efficient valleytronic and spintronic devices.

7.8 Summary

In this chapter, we discussed the key engineering processes needed to realise practical valleytronics, including 2D semiconductor engineering of interfaces, heterostructures, doping, contacts, and control gates. These engineering techniques allow us to create concept valley devices including valley valves, beam splitters, and valley transistors. In addition, devices involving valleytronics also enable other novel behaviours such as valley-polarised exciton currents, valley magnetoelectricity, and valley Zeeman spin-orbit coupling-induced spin precession. Many of these devices have been demonstrated to work at room temperature, showing that valley protection of the information carriers can offer robust valleytronic devices.

Despite the advances in our understanding of valley physics and the promise that valley devices have demonstrated so far, the fabrication of high-quality scalable TMD-based valleytronic devices remains a challenge on all fronts — interface engineering, band engineering, doping consistency, clean contacts and effective gating. Since TMDs are the cornerstone of many valley devices discussed in this book, mastering the engineering of this material system remains a key priority to unlock valley-enabled benefits for devices. This will be discussed further in Chapter 8.

References

[1] A. K. Geim and I. v. Grigorieva, "Van der Waals heterostructures," *Nature*, vol. 499, no. 7459, pp. 419–425, 2013, doi: 10.1038/nature12385.

[2] X. Liu and M. C. Hersam, "Interface characterization and control of 2D materials and heterostructures," *Adv. Mater.*, vol. 30, no. 39, p. 1801586, 2018, doi: 10.1002/adma.201801586.

[3] O. V. Yazyev and Y. P. Chen, "Polycrystalline graphene and other two-dimensional materials," *Nat. Nanotechnol.*, vol. 9, no. 10, pp. 755–767, 2014, doi: 10.1038/nnano.2014.166.

[4] D. G. Purdie, N. M. Pugno, T. Taniguchi, K. Watanabe, A. C. Ferrari, and A. Lombardo, "Cleaning interfaces in layered materials heterostructures," *Nat. Commun.*, vol. 9, no. 1, p. 5387, 2018, doi: 10.1038/s41467-018-07558-3.

[5] A. v. Kretinin *et al.*, "Electronic properties of graphene encapsulated with different two-dimensional atomic crystals," *Nano Lett.*, vol. 14, no. 6, pp. 3270–3276, 2014, doi: 10.1021/nl5006542.

[6] Y. Huang *et al.*, "Universal mechanical exfoliation of large-area 2D crystals," *Nat. Commun.*, vol. 11, no. 1, p. 2453, 2020, doi: 10.1038/s41467-020-16266-w.

[7] A. Zavabeti, A. Jannat, L. Zhong, A. A. Haidry, Z. Yao, and J. Z. Ou, "Two-dimensional materials in large-areas: synthesis, properties and applications," *Nanomicro Lett.*, vol. 12, no. 1, p. 66, 2020, doi: 10.1007/s40820-020-0402-x.

[8] F. Pizzocchero *et al.*, "The hot pick-up technique for batch assembly of van der Waals heterostructures," *Nat. Commun.*, vol. 7, no. 1, p. 11894, 2016, doi: 10.1038/ncomms11894.

[9] K. Kinoshita *et al.*, "Dry release transfer of graphene and few-layer h-BN by utilizing thermoplasticity of polypropylene carbonate," *NPJ 2D Mater. Appl.*, vol. 3, no. 1, p. 22, 2019, doi: 10.1038/s41699-019-0104-8.

[10] Y. Kim, P. Herlinger, T. Taniguchi, K. Watanabe, and J. H. Smet, "Reliable postprocessing improvement of van der waals heterostructures," *ACS Nano*, vol. 13, no. 12, pp. 14182–14190, 2019, doi: 10.1021/acsnano.9b06992.

[11] S. Fan, Q. Vu, M. Tran, S. Adhikari, and Y. H. Lee, "Transfer assembly for two-dimensional van der Waals heterostructures," *2D Mater.*, vol. 7, no. 2, p. 022005, 2020, doi: 10.1088/2053-1583/ab7629.

[12] M. R. Rosenberger, H.-J. Chuang, K. M. McCreary, A. T. Hanbicki, S. v. Sivaram, and B. T. Jonker, "Nano-'squeegee' for the creation of clean 2D material interfaces," *ACS Appl. Mater. Interfaces*, vol. 10, no. 12, pp. 10379–10387, 2018, doi: 10.1021/acsami.8b01224.

[13] A. M. Goossens, V. E. Calado, A. Barreiro, K. Watanabe, T. Taniguchi, and L. M. K. Vandersypen, "Mechanical cleaning of graphene," *Appl. Phys. Lett.*, vol. 100, no. 7, p. 073110, 2012, doi: 10.1063/1.3685504.

[14] C. R. Dean *et al.*, "Boron nitride substrates for high-quality graphene electronics," *Nat. Nanotechnol.*, vol. 5, no. 10, pp. 722–726, 2010, doi: 10.1038/nnano.2010.172.

[15] X. Han, J. Lin, J. Liu, N. Wang, and D. Pan, "Effects of hexagonal boron nitride encapsulation on the electronic structure of few-layer MoS_2," *J. Phys. Chem. C*, vol. 123, no. 23, pp. 14797–14802, 2019, doi: 10.1021/acs.jpcc.9b02549.

[16] Y. Shimazaki, M. Yamamoto, I. V. Borzenets, K. Watanabe, T. Taniguchi, and S. Tarucha, "Generation and detection of pure valley current by electrically induced Berry curvature in bilayer graphene," *Nat. Phys.*, vol. 11, no. 12, pp. 1032–1036, 2015, doi: 10.1038/nphys3551.

[17] M. Sui *et al.*, "Gate-tunable topological valley transport in bilayer graphene," *Nat. Phys.*, vol. 11, no. 12, pp. 1027–1031, 2015, doi: 10.1038/nphys3485.

[18] W. Yao, B. Wu, and Y. Liu, "Growth and grain boundaries in 2D materials," *ACS Nano*, vol. 14, no. 8, pp. 9320–9346, 2020, doi: 10.1021/acsnano.0c03558.

[19] H. S. Song *et al.*, "Origin of the relatively low transport mobility of graphene grown through chemical vapor deposition," *Sci. Rep.*, vol. 2, no. 1, p. 337, 2012, doi: 10.1038/srep00337.

[20] K. W. Clark, X.-G. Zhang, I. v. Vlassiouk, G. He, R. M. Feenstra, and A.-P. Li, "Spatially resolved mapping of electrical conductivity across individual domain (grain) boundaries in graphene," *ACS Nano*, vol. 7, no. 9, pp. 7956–7966, 2013, doi: 10.1021/nn403056k.

[21] A. M. van der Zande *et al.*, "Grains and grain boundaries in highly crystalline monolayer molybdenum disulphide," *Nat. Mater.*, vol. 12, no. 6, pp. 554–561, 2013, doi: 10.1038/nmat3633.

[22] Y. L. Huang *et al.*, "Bandgap tunability at single-layer molybdenum disulphide grain boundaries," *Nat. Commun.*, vol. 6, no. 1, p. 6298, 2015, doi: 10.1038/ncomms7298.
[23] Y. Zhang *et al.*, "Invisible growth of microstructural defects in graphene chemical vapor deposition on copper foil," *Carbon*, vol. 96, pp. 237–242, 2016, doi: 10.1016/j.carbon.2015.09.041.
[24] J. Zhu *et al.*, "Boundary activated hydrogen evolution reaction on monolayer MoS_2," *Nat. Commun.*, vol. 10, no. 1, p. 1348, 2019, doi: 10.1038/s41467-019-09269-9.
[25] P. Yasaei *et al.*, "Bimodal phonon scattering in graphene grain boundaries," *Nano Lett.*, vol. 15, no. 7, pp. 4532–4540, 2015, doi: 10.1021/acs.nanolett.5b01100.
[26] G.-H. Lee *et al.*, "High-strength chemical-vapor–deposited graphene and grain boundaries," *Science*, vol. 340, no. 6136, pp. 1073–1076, 2013, doi: 10.1126/science.1235126.
[27] D. Gunlycke and C. T. White, "Graphene valley filter using a line defect," *Phys. Rev. Lett.*, vol. 106, no. 13, p. 136806, 2011, doi: 10.1103/PhysRevLett.106.136806.
[28] R. V. Gorbachev *et al.*, "Detecting topological currents in graphene superlattices," *Science*, vol. 346, no. 6208, pp. 448–451, 2014, doi: 10.1126/science.1254966.
[29] M. Houssa, R. Meng, V. Afanas'ev, and A. Stesmans, "First-principles study of the contact resistance at 2D Metal/2D semiconductor heterojunctions," *Appl. Sci.*, vol. 10, no. 8, p. 2731, 2020, doi: 10.3390/app10082731.
[30] Y. Liu, P. Stradins, and S. H. Wei, "Van der Waals metal-semiconductor junction: Weak Fermi level pinning enables effective tuning of Schottky barrier," *Sci. Adv.*, vol. 2, no. 4, p. e1600069, 2016, doi: 10.1126/sciadv.1600069.
[31] A. Chaves *et al.*, "Bandgap engineering of two-dimensional semiconductor materials," *NPJ 2D Mater. Appl.*, vol. 4, no. 1, p. 29, 2020, doi: 10.1038/s41699-020-00162-4.
[32] R. Kappera *et al.*, "Phase-engineered low-resistance contacts for ultrathin MoS_2 transistors," *Nat. Mater.*, vol. 13, no. 12, pp. 1128–1134, 2014, doi: 10.1038/nmat4080.

[33] Y. Zhang *et al.*, "Direct observation of a widely tunable bandgap in bilayer graphene," *Nature,* vol. 459, no. 7248, pp. 820–823, 2009, doi: 10.1038/nature08105.
[34] E. McCann and V. I. Fal'ko, "Landau-level degeneracy and quantum Hall effect in a graphite bilayer," *Phys. Rev. Lett.*, vol. 96, no. 8, p. 086805, 2006, doi: 10.1103/PhysRevLett.96.086805.
[35] T. Ohta, A. Bostwick, T. Seyller, K. Horn, and E. Rotenberg, "Controlling the electronic structure of bilayer graphene," *Science,* vol. 313, no. 5789, pp. 951–954, 2006, doi: 10.1126/science.1130681.
[36] J. Liu, C. Luo, H. Lu, Z. Huang, G. Long, and X. Peng, "Influence of hexagonal boron nitride on electronic structure of graphene," *Molecules*, vol. 27, no. 12, p. 3740, 2022, doi: 10.3390/molecules27123740.
[37] Y. Cao *et al.*, "Correlated insulator behaviour at half-filling in magic-angle graphene superlattices," *Nature*, vol. 556, no. 7699, pp. 80–84, 2018, doi: 10.1038/nature26154.
[38] Z.-Y. Zhao and Q.-L. Liu, "Study of the layer-dependent properties of MoS_2 nanosheets with different crystal structures by DFT calculations," *Catal. Sci. Technol.*, vol. 8, no. 7, pp. 1867–1879, 2018, doi: 10.1039/C7CY02252B.
[39] G. Zhang *et al.*, "Optical and electrical properties of two-dimensional palladium diselenide," *Appl. Phys. Lett.*, vol. 114, no. 25, p. 253102, 2019, doi: 10.1063/1.5097825.
[40] X. Yu *et al.*, "Atomically thin noble metal dichalcogenide: a broadband mid-infrared semiconductor," *Nat. Commun.*, vol. 9, no. 1, p. 1545, 2018, doi: 10.1038/s41467-018-03935-0.
[41] S. Cha *et al.*, "Generation, transport and detection of valley-locked spin photocurrent in WSe_2–graphene–Bi_2Se_3 heterostructures," *Nat. Nanotechnol.*, vol. 13, no. 10, pp. 910–914, 2018, doi: 10.1038/s41565-018-0195-y.
[42] O. L. Sanchez, D. Ovchinnikov, S. Misra, A. Allain, and A. Kis, "Valley polarization by spin injection in a light-emitting van der Waals heterojunction," *Nano Lett.*, vol. 16, no. 9, pp. 5792–5797, 2016, doi: 10.1021/acs.nanolett.6b02527.
[43] H. Yoo, K. Heo, M. H. Ansari, and S. Cho, "Recent advances in electrical doping of 2D semiconductor materials: Methods, analyses, and

applications," *Nanomaterials,* vol. 11, no. 4, p. 832, 2021, doi: 10.3390/nano11040832
[44] M. R. Laskar *et al.*, "p-type doping of MoS_2 thin films using Nb," *Appl. Phys. Lett.*, vol. 104, no. 9, p. 092104, 2014, doi: 10.1063/1.4867197.
[45] T. Hallam *et al.*, "Rhenium-doped MoS_2 films," *Appl. Phys. Lett.*, vol. 111, no. 20, p. 203101, 2017, doi: 10.1063/1.4995220.
[46] J. Suh *et al.*, "Doping against the native propensity of MoS_2: Degenerate hole doping by cation substitution," *Nano Lett.*, vol. 14, no. 12, pp. 6976–6982, 2014, doi: 10.1021/nl503251h.
[47] S. Tongay *et al.*, "Broad-range modulation of light emission in two-dimensional semiconductors by molecular physisorption gating," *Nano Lett.*, vol. 13, no. 6, pp. 2831–2836, 2013, doi: 10.1021/nl4011172.
[48] S. Mouri, Y. Miyauchi, and K. Matsuda, "Tunable photoluminescence of monolayer MoS_2 via chemical doping," *Nano Lett.*, vol. 13, no. 12, pp. 5944–5948, 2013, doi: 10.1021/nl403036h.
[49] H. Fang *et al.*, "Degenerate n-doping of few-layer transition metal dichalcogenides by potassium," *Nano Lett.*, vol. 13, no. 5, pp. 1991–1995, 2013, doi: 10.1021/nl400044m.
[50] Y. Gong *et al.*, "Spatially controlled doping of two-dimensional SnS_2 through intercalation for electronics," *Nat. Nanotechnol.*, vol. 13, no. 4, pp. 294–299, 2018, doi: 10.1038/s41565-018-0069-3.
[51] J. Yang, F. Bussolotti, H. Kawai, and K. E. J. Goh, "Tuning the conductivity type in monolayer WS_2 and MoS_2 by sulfur vacancies," *Phys. Status Solidi RRL*, vol. 14, no. 9, p. 2000248, 2020, doi: 10.1002/pssr.202000248.
[52] C. Lei, Y. Ma, T. Zhang, X. Xu, B. Huang, and Y. Dai, "Valley polarization in monolayer CrX_2 (X = S, Se) with magnetically doping and proximity coupling," *New J. Phys.*, vol. 22, no. 3, p. 033002, 2020, doi: 10.1088/1367-2630/ab7256.
[53] K. R. Sahoo *et al.*, "Enhanced room-temperature spin-valley coupling in V-doped MoS_2," *Phys. Rev. Mater.*, vol. 6, no. 8, p. 085202, 2022, doi: 10.1103/PhysRevMaterials.6.085202.
[54] A. Allain, J. Kang, K. Banerjee, and A. Kis, "Electrical contacts to two-dimensional semiconductors," *Nat. Mater.*, vol. 14, no. 12, pp. 1195–1205, 2015, doi: 10.1038/nmat4452.

[55] T. Roy *et al.*, "Field-effect transistors built from all two-dimensional material components," *ACS Nano*, vol. 8, no. 6, pp. 6259–6264, 2014, doi: 10.1021/nn501723y.
[56] D. Voiry, A. Mohite, and M. Chhowalla, "Phase engineering of transition metal dichalcogenides," *Chem. Soc. Rev.*, vol. 44, no. 9, pp. 2702–2712, 2015, doi: 10.1039/C5CS00151J.
[57] Y. Liu *et al.*, "Approaching the Schottky–Mott limit in van der Waals metal–semiconductor junctions," *Nature*, vol. 557, no. 7707, pp. 696–700, 2018, doi: 10.1038/s41586-018-0129-8.
[58] Y. Wang *et al.*, "Van der Waals contacts between three-dimensional metals and two-dimensional semiconductors," *Nature*, vol. 568, no. 7750, pp. 70–74, 2019, doi: 10.1038/s41586-019-1052-3.
[59] C. S. Lau *et al.*, "Quantum transport in two-dimensional WS_2 with high-efficiency carrier injection through indium alloy contacts," *ACS Nano*, vol. 14, no. 10, pp. 13700–13708, 2020, doi: 10.1021/acsnano.0c05915.
[60] J. Kang, W. Liu, D. Sarkar, D. Jena, and K. Banerjee, "Computational study of metal contacts to monolayer transition-metal dichalcogenide semiconductors," *Phys. Rev. X*, vol. 4, no. 3, p. 031005, 2014, doi: 10.1103/PhysRevX.4.031005.
[61] C. S. Lau *et al.*, "Gate-defined quantum confinement in CVD 2D WS_2," *Adv. Mater.*, vol. 34, no. 25, p. 2103907, 2022, doi: 10.1002/adma.202103907.
[62] X. Cai *et al.*, "Bridging the gap between atomically thin semiconductors and metal leads," *Nat. Commun.*, vol. 13, no. 1, p. 1777, 2022, doi: 10.1038/s41467-022-29449-4.
[63] Y. Wang *et al.*, "P-type electrical contacts for 2D transition-metal dichalcogenides," *Nature,* vol. 610, no. 7930, pp. 61–66, 2022, doi: 10.1038/s41586-022-05134-w.
[64] L. Wang *et al.*, "One-dimensional electrical contact to a two-dimensional material," *Science*, vol. 342, no. 6158, pp. 614–617, 2013, doi: 10.1126/science.1244358.
[65] L. Cheng and Y. Liu, "What limits the intrinsic mobility of electrons and holes in two dimensional metal dichalcogenides?," *J. Am. Chem. Soc.*, vol. 140, no. 51, pp. 17895–17900, 2018, doi: 10.1021/jacs.8b07871.

[66] X. Cui *et al.*, "Multi-terminal transport measurements of MoS_2 using a van der Waals heterostructure device platform," *Nat. Nanotechnol.*, vol. 10, no. 6, pp. 534–540, 2015, doi: 10.1038/nnano.2015.70.

[67] P. Zhang, Y. Zhang, Y. Wei, H. Jiang, X. Wang, and Y. Gong, "Contact engineering for two-dimensional semiconductors," *J. Semicond.*, vol. 41, no. 7, p. 071901, 2020, doi: 10.1088/1674-4926/41/7/071901.

[68] C.-S. Lee *et al.*, "Epitaxial van der Waals contacts between transition-metal dichalcogenide monolayer polymorphs," *Nano Lett.*, vol. 19, no. 3, pp. 1814–1820, 2019, doi: 10.1021/acs.nanolett.8b04869.

[69] M. Zhao *et al.*, "Large-scale chemical assembly of atomically thin transistors and circuits," *Nat. Nanotechnol.*, vol. 11, no. 11, pp. 954–959, 2016, doi: 10.1038/nnano.2016.115.

[70] Y. Liu *et al.*, "Toward barrier free contact to molybdenum disulfide using graphene electrodes," *Nano Lett.*, vol. 15, no. 5, pp. 3030–3034, 2015, doi: 10.1021/nl504957p.

[71] S. B. Mitta *et al.*, "Electrical characterization of 2D materials-based field-effect transistors," *2D Mater.*, vol. 8, no. 1, p. 012002, 2021, doi: 10.1088/2053-1583/abc187.

[72] D. Vaquero, V. Clericò, J. Salvador-Sánchez, J. Quereda, E. Diez, and A. M. Pérez-Muñoz, "Ionic-liquid gating in two-dimensional TMDs: The operation principles and spectroscopic capabilities," *Micromachines (Basel)*, vol. 12, no. 12, p. 1576, 2021, doi: 10.3390/mi12121576.

[73] D. Domaretskiy, M. Philippi, M. Gibertini, N. Ubrig, I. Gutiérrez-Lezama, and A. F. Morpurgo, "Quenching the bandgap of two-dimensional semiconductors with a perpendicular electric field," *Nat. Nanotechnol.*, vol. 17, no. 10, pp. 1078–1083, 2022, doi: 10.1038/s41565-022-01183-4.

[74] K. S. Novoselov *et al.*, "Electric field effect in atomically thin carbon films," *Science*, vol. 306, no. 5696, pp. 666–669, 2004, doi: 10.1126/science.1102896.

[75] Z.-E. Ooi, J. Y. Chee, S. L. Wong, C. S. Lau, and K. E. J. Goh, "Isoemissive photoluminescence from a quaternary system of valley-polarized, defect-bound excitons and trions in two-dimensional transition metal dichalcogenides," *J. Phys. Chem. C*, vol. 125, no. 23, pp. 12721–12729, 2021, doi: 10.1021/acs.jpcc.1c00535.

[76] X. Liu *et al.*, "High performance field-effect transistor based on multilayer tungsten disulfide," *ACS Nano*, vol. 8, no. 10, pp. 10396–10402, 2014, doi: 10.1021/nn505253p.

[77] A. Sebastian, R. Pendurthi, T. H. Choudhury, J. M. Redwing, and S. Das, "Benchmarking monolayer MoS_2 and WS_2 field-effect transistors," *Nat. Commun.*, vol. 12, no. 1, p. 693, 2021, doi: 10.1038/s41467-020-20732-w.

[78] M. Eich *et al.*, "Spin and valley states in gate-defined bilayer graphene quantum dots," *Phys. Rev. X*, vol. 8, no. 3, p. 031023, 2018, doi: 10.1103/PhysRevX.8.031023.

[79] M. Hamer *et al.*, "Gate-defined quantum confinement in InSe-based van der Waals heterostructures," *Nano Lett.*, vol. 18, no. 6, pp. 3950–3955, 2018, doi: 10.1021/acs.nanolett.8b01376.

[80] J. Li *et al.*, "A valley valve and electron beam splitter," *Science*, vol. 362, no. 6419, pp. 1149–1152, 2018, doi: 10.1126/science.aao5989.

[81] L. Li *et al.*, "Room-temperature valleytronic transistor," *Nat. Nanotechnol.*, vol. 15, no. 9, pp. 743–749, 2020, doi: 10.1038/s41565-020-0727-0.

[82] D. Unuchek *et al.*, "Valley-polarized exciton currents in a van der Waals heterostructure," *Nat. Nanotechnol.*, vol. 14, no. 12, pp. 1104–1109, 2019, doi: 10.1038/s41565-019-0559-y.

[83] J. Lee, Z. Wang, H. Xie, K. F. Mak, and J. Shan, "Valley magnetoelectricity in single-layer MoS_2," *Nat. Mater.*, vol. 16, no. 9, pp. 887–891, 2017, doi: 10.1038/nmat4931.

[84] J. Lee, K. F. Mak, and J. Shan, "Electrical control of the valley Hall effect in bilayer MoS_2 transistors," *Nat. Nanotechnol.*, vol. 11, no. 5, pp. 421–425, 2016, doi: 10.1038/nnano.2015.337.

[85] J. Ingla-Aynés, F. Herling, J. Fabian, L. E. Hueso, and F. Casanova, "Electrical control of valley-Zeeman spin-orbit-coupling–induced spin precession at room temperature," *Phys. Rev. Lett.*, vol. 127, no. 4, p. 047202, 2021, doi: 10.1103/PhysRevLett.127.047202.

https://doi.org/10.1142/9789811229107_0008

Chapter 8

Opportunities for Applications

Kuan Eng Johnson Goh

8.1 Back to the Future

We have taken a whirlwind tour of the development of valleytronics with the intent of providing the reader with a fast entry into this exciting subject with potentially disruptive device concepts ahead and perhaps some novel technologies for the near future. Our purpose was to excite, enable and bring a generation of physicists, material scientists, chemists, engineers and more to partner in the research and development of valleytronics. We acknowledge that this attempt at brevity will not do justice to the richness of the subject, and we encourage readers to dive deeper beyond the chapters herein to the references provided. Certainly, more books will be written on this subject in the coming decades, and we hope that this introductory volume could serve a foundational role.

We conclude this volume by providing a perspective summary of the field to date, discussing challenges yet to be overcome, and opportunities for applications in the near term as well as the future.

Our brief historical review in Chapter 1 revealed that the quest for valley control is not a recent idea but a longing from decades before, almost soon after the advent of modern semiconductor technology. The renewed interest mainly arose from the discovery that in isolated 2D semiconductors with broken inversion symmetry like TMDs (monolayer transition dichalcogenides, e.g., MoS_2, WS_2, $MoSe_2$, WSe_2) or bilayer graphene, the valley degree of freedom is distinguishably coupled to the opposite spins of charge carriers, and this opens the door for extracting valley information from these materials, and importantly control valley electronics directly via external fields (electrostatic or magnetic, or light).

We attempted to cover the key physical concepts necessary to have a fundamental grasp (perhaps intuitively) of the required valley physics in Chapter 2. In the main, we explained that the key concept of the geometric phase or Berry phase leads to the anomalous Hall effect in inversion-asymmetric 2D materials. The concept of the Berry phase is crucial because, in a system with broken inversion symmetry, the related observable quantity known as Berry curvature is non-zero and acquires opposite signs in the K and opposite K′ valleys. This leads to valley-distinguishable spin signatures which may be detected via selective excitation by circularly polarised light. It also manifests as the anomalous Hall effect, sometimes called the Berry curvature effect, wherein an in-plane electric field drives opposite spins to the opposite edges of the inversion-asymmetric 2D material in the absence of an external magnetic field. The ability to further manipulate/control the valley degree of freedom is also shown in the cases of an external electric field (AC Stark effect) and external magnetic field (valley Zeeman effect). This is a subject that deserves a whole book in itself, hence the initiated reader is referred to the key references in Chapter 2 if further depth and details are desired.

In Chapter 3, we discussed the two requisite engineering concepts of valley accessibility (or addressability) and control that are crucial to realise practical valleytronics and provided a survey of the key literature that supports the feasibility of these two capabilities in 2D transition metal dichalcogenides (TMDs). The concept of valley selectivity arising from opposite Berry curvatures in the K and K′ valleys is now well established by this body of works, thanks to the variety of techniques employed ranging from circular dichroic photoluminescence, spin-resolved photoemission spectroscopy, Kerr-rotation measurements, etc. Many of these techniques, however, remain in the experimental research domain and there is no ready-for-use commercial equipment to screen the quality of 2D materials *en masse* at different stages of production. Whilst this represents an urgent gap that impedes development, it opens opportunities for the development of a new class of material screening tools to support the production of 2D materials for valleytronics.

Towards realising a true valley device, direct valley controllability or manipulation remains a rather elusive endeavour. Concrete demonstrations of the ability to manipulate valley population and dynamics are few, but some key experiments using electrically gated devices show such potential in Chapter 3. Indeed, the ability to exploit valley physics to switch the spins in a layer of MoS_2 [1] reproducibly by an electric field is a promising demonstration that could spark ideas, for example, for memory, tunable magnetic or spintronic devices. There remains a lot of room for exploration and innovation in valley control. The realisation of a valley qubit in 2D materials remains a holy grail despite several theoretical proposals [2]–[5]. However, steady progress in this direction has been reported [6]–[13]. This is fertile ground for further research and development.

Importantly, to actualise any valleytronic device, we would need to engineer reproducible materials and develop processes such as encapsulation to keep the active device layers from degradation and provide consistent performance over time. This was covered in Chapter 4 whereby the traditional methods of exfoliation and stamp transfer are described and their inadequacies for scale-up have been discussed. We also covered the recent attempts to use more scalable production methods such as chemical vapour deposition and molecular beam epitaxy. Despite these, the dearth of consistent quality materials for valleytronic devices remains a key challenge, and more materials development work remains before device engineering can be meaningful. Thus high-quality production of scalable 2D materials, and related dielectrics development both for substrate and encapsulation of the active device materials are all fertile grounds for future research.

Naively, one might assume that like the transistor, full electrical operation of a valley device would be the most practical pathway for development, and reported works have alluded to this direction [14]–[17]. Chapter 5 covered the attempts at valley current generation and detection towards achieving controllable and readable valleytronics devices. However, more frequently, we find that it is indeed still difficult to do everything electrically in most reported studies. Hence, some groups attempted to generate valley currents by optical excitation and detect the current electrically [18], while others drive the currents electrically and detect the flow optically [19]. More recently, Li *et al.* reported a device with fully electrically driven and detected valley currents [17]. This is indeed encouraging, but many more such device demonstrations will be required before performance metrics can be established. The quest for full electrical operability should not preclude further experimentation and development of devices involving hybrid operations (electrical, optical, magnetic, strain, micromechanical resonators, etc.) that are valley relevant. Further work in such areas might find productive efforts for sensing devices, spin-to-photon or photon-to-spin converters, valley-phonon coupled devices and possibly more.

Chapter 6 covered an exciting possibility to exploit valley physics for building single-photon emitters highly sought for quantum information and computing technologies. In 2D materials, exciton transitions from defects have not only manifested such non-classical photon antibunching behaviour but do so with optical linear polarisation, thanks to the superposition of valley pseudospin states. We covered the photophysics of single-photon generation, valley-based origin of optical linear polarisation, how these properties can be useful in quantum information applications such as the BB84 quantum key distribution protocol, and current experimental efforts toward these goals. We note that many current endeavours exploit the intrinsic material defects in 2D TMD materials that localise the single-photon emissions. Further developments for feasible production and device manufacturing would need to be able to deterministically place the defect-localised emission sites according to the device design and ensure the surrounding material quality (substrate, cap layer, 2D materials, gates) is sufficiently well-controlled to ensure spectral stability. The achievements of the research community in this area over the last several years are noteworthy: the realisation of high single-photon purity with $g^{(2)}(0) < 0.1$, electrically driven single-photon emission, and scalable site-controlled nanopillar fabrication for inducing single-photon emitters. These are promising developments that could see 2D TMD materials play an important future role in quantum information and communication.

Finally, Chapter 7 is an in-your-face reality check that brings out perhaps the less glamorous and often understated need for practical engineering of valleytronic devices. We cannot escape the need for many systematic studies and development that deal with interfacing materials for valleytronics, fabrication of heterostructures, doping, contacts and electrostatic gating strategies. Several groups [20]–[27] are actively working in these areas, and more such developments will be needed. There will be opportunities within such endeavours to innovate in both materials and instrumentation. Realising the unique properties of the novel engineered material stack [28], [29] may also provide the impetus for new device concepts for valleytronics, electronics, optoelectronics, photonics, phononics, plasmonics, thermoelectrics, qubits, or a combination thereof.

8.2 Light in the Valley

The opportunity to exploit the valley degree of freedom for real applications is greater now than ever. We are no longer stuck in the valleys to which practical access is constrained by the difficult manipulation of strain in materials [30]. The discovery of inversion-asymmetric 2D materials brought chiral light into the valley so that we may selectively excite or tune carrier populations in specific valleys. Control can be by electric or magnetic fields, electromagnetic waves, or perhaps some yet-to-be-demonstrated methods, e.g., phononic coupling. We conclude the volume with our perspective of what are the key challenges remaining in the categories of materials, device engineering, instrumentation, and practical applications of valleytronics.

8.2.1 Materials

At the fundamental science level, we have made significant advances in terms of understanding van der Waals materials and their potential for device applications [31]. We have attempted to condense the key learnings relevant to valleytronics in the previous chapters. The change in electrical properties as one goes from bulk to few or monolayer TMDs presents an alluring thesis for novel applications which are yet to be seen in the form of products on the shelf. Nevertheless, there remain several unanswered questions, and perhaps more to come in time. Indeed, TMD heterolayers [20], [28], [31]–[33], and twisted layers [33]–[40] are just recent examples of the permutations and combinations that such van der Waals systems afford the research community. We believe there is ample ground for many more scientific discoveries in TMDs and valleytronics.

For practical device realisation, the availability of consistent high-quality material substrates that host the active device materials/interfaces will be an important prerequisite. Without such quality control over the starting substrates, it would be impossible to begin the work of optimising fabrication processes, electrical contacts, gating strategies, and comparing one device architecture against another. Achieving reproducible benchmarks for device performance would also be impractical. Whilst significant research and development are ongoing in the field of 2D van der Waals materials, including TMDs, the issue of what is a good starting substrate for hosting the active device materials for valleytronics remains in debate. Common substrates for growing TMDs include oxidised silicon wafers and sapphire. However, these substrates are often not suitable as the host substrates upon which to fabricate high-quality devices as the substrates could suffer degradation during the growth process, or may be unfeasible for further processing, or perhaps unsuitable for the device architecture. For example, the silicon oxide layer on commercial silicon wafers typically becomes more electrically leaky after being exposed to growth precursors at high temperatures in a chemical vapour deposition process [41]. These constraints force researchers to transfer the TMDs from the substrates upon which they were grown onto other more appropriate substrates for the device fabrication using various novel lift-off and stamping methods [42]–[45]. Inevitably, these transfer processes likely contaminate the TMD and make it even more challenging to maintain consistently high-quality TMD production. *A scalable method for achieving high-quality, large-scale TMDs on a technologically relevant substrate remains unfulfilled.* This is a fundamental technology area that requires significant further investment before useful optimisation of valleytronic devices, and any practical TMD-based devices, can be achieved. The closest analogy would be the invention of the Czochralski crystal growth process that produces the monocrystalline silicon ingots for dicing into the wafer substrates that form the bedrock for the semiconductor industry. Achieving equivalent wafer technologies for TMDs will be a critical key to unlocking valleytronics and other TMD-based benefits for devices. There remains therefore enormous opportunity for material growers in TMDs.

Beyond needing consistent quality in wafer-scale TMD, the processes required to fabricate devices on them in a reproducible manner remain outstanding. Fabrication ideas abound in literature, but development lags behind significantly because researchers have to overcome the limitations of inconsistent starting qualities of TMDs, challenges of transfer contaminations, and poor control over material interfaces, to just name a few bugbears.

A number of groups have been actively focused on solving the issues of making good ohmic contacts to TMDs from room temperature [21], [27] to cryogenic temperatures [22] for various device applications. Some of these works exploit the novel phase-change control of materials [46] or "gentle" stacking of metal contacts on TMDs [27] in order to achieve remarkable materials interface engineering that attains good control over electrical contact characteristics. Yet, demonstrations of scalable processing with good electrical contacts [21], [27], [47], [48] remain few and far between. To make further progress in TMD device fabrication, and in particular for valleytronics, this foundational technology of good electrical contacts is urgently required. Without which, the inefficient charge injection into TMD devices will detrimentally dominate the device behaviour and likely negate the benefits we hope to elicit from valleytronics.

Clean interfaces between the substrate dielectric, the sandwiched TMD (active device material), and the encapsulating dielectric layer is another critical process technology that requires urgent development. Intentional doping with relevant dopant atoms could enhance device performance or produce novel effects. Unintended contaminants, intrinsic or induced crystal defects (vacancies, interstitials, grain boundaries, etc.) can cause the devices to behave in unexpected/undesirable ways. For example, unintentional doping can produce unwanted defect states in the TMD bandgap, resulting in Fermi level pinning which hinders contact engineering [23], [27], [49]. For the realisation of spin-based devices (e.g., spin-valley qubits), the presence of nearby non-zero spin nuclei can result in spin decoherence. More recently, it has also been shown that interfacial roughness caused by standard ALD growth of the dielectric cap layer can adversely impact the carrier mobility in TMD devices [12]. This multitude of materials-related issues remain unresolved and present a fertile ground for many new materials engineering endeavours and perhaps materials science discoveries.

8.2.2 Device Engineering

Device engineering is very much intertwined with the aforesaid materials engineering to a large extent — primarily because the materials need to be interfaced correctly to achieve the desired device behaviour (see Fig. 8.1 and the detailed discussions on device fabrication in Chapter 7). Decisions range from what type of substrate to use,

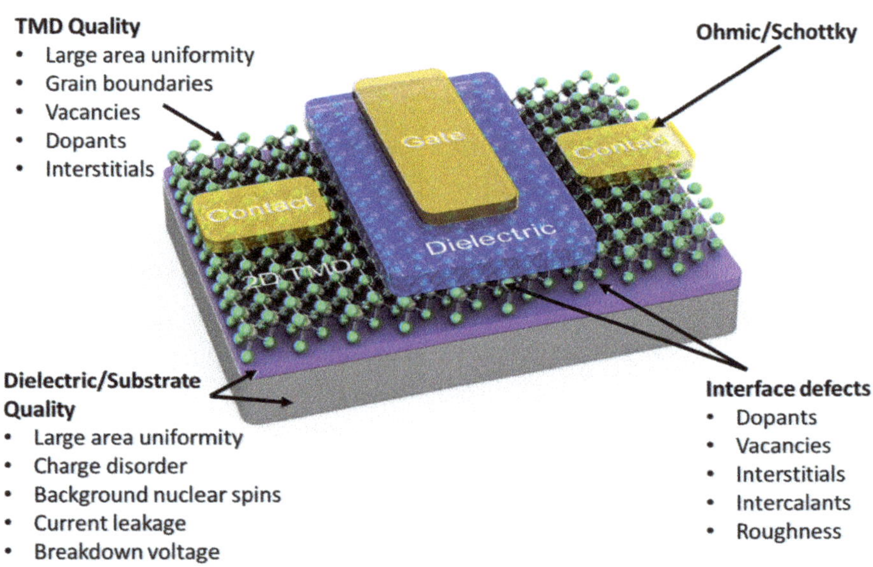

Fig. 8.1. Anatomy of a TMD device material stack, showing the materials challenges that must be controlled or overcome for device engineering to be feasible.

what bottom and top dielectrics are compatible for fabrication and operation, what 2D TMD stack (homo- or heterostructure) to use, what contact strategy to employ, and what gating strategy (top and/or bottom) for operation. These choices are incumbent upon the material quality (TMDs, dielectrics, metal contact, and their concomitant interfaces) and processability (doping, controlled deposition, selective etching, compatibility with lithographic patterning, scalability, etc.) in order to achieve the desired device function. Fig. 8.1 highlights some of these pertinent concerns that must be addressed.

The key in device engineering is always to minimise the parasitic effects and maximise the sought-after device action. An analogy would be to minimise fumbling with the door lock when the main action is to open a door. In the case of valleytronics, it is key to ensure at each step of the fabrication process, the advantage of the spin-valley coupling is preserved. As we saw in Chapter 2, this coupling is in essence the key by which we can access the valley information and also control the valley device. Without which, we lose control of the valley degree of freedom. Once we master the materials-related processes above and retain the spin-valley coupling, the exploration of different device architectures becomes meaningful.

Several device architectures that exploit the valley degree of freedom have been described in Chapters 6 and 7. These range from switchable 2D magnets, electron beam splitters, valley filters/valves, to the full demonstration of a room-temperature valley-based field-effect transistor. Nevertheless, these device demonstrations remain only a handful largely because of the low device yields arising from difficulties in obtaining good-quality materials and limited control in processing — the best-quality devices are still mainly obtained via manual exfoliation of odd-sized 2D materials from bulk crystals. Although these are promising first proof-of-concept valleytronic devices, increasing device yields and reproducibility would be crucial to optimise device architecture and allow augmentation of device capabilities.

8.2.3 Instrumentation

Important characterisation techniques tailored to studying valley selectivity and controllability have been well summarised in Chapter 3. These include techniques such as circular dichroic photoluminescence, spin- and angle-resolved photoemission spectroscopy, Kerr-rotation measurements, and time-resolved techniques for the determination of valley lifetimes. No doubt these experimental demonstrations are important first observations to validate our understanding of TMDs. However, such experimental setups are highly specialised and often time consuming and costly to establish and would typically be dismantled and reconfigured for other experiments in a research laboratory. Commercial setups for screening valleytronic materials remain unavailable. In part, this impedes the materials engineering required to achieve the consistent high-quality TMDs required for the development of valleytronics and other 2D TMD devices. A feasible engineering development for valleytronic materials (e.g., TMDs) would require timely and ideally in-line screening and feedback. Hence the development of key instrumentation for providing the necessary material screening with timely feedback will be critical. Synchrotron-based characterisation tools and typical experimental setups for circular dichroic photoluminescence or Kerr-rotation measurements may not easily lend themselves to being incorporated into a material process line. In addition, some measurements could be time consuming, and a scientific rationalisation will be needed to determine if proxy methods could be developed for faster screening. In anticipation of valleytronic devices being manufactured, new tools for testing, stressing, and benchmarking will be required. There is therefore much room for instrument builders to participate in engineering the processes for valleytronic materials development and device testing. The new instruments developed are likely to benefit the wider community of 2D materials research and development beyond valleytronics.

8.2.4 Practical Applications

Valleytronics represent an exciting paradigm for devices that we have recently learnt how to control in a practical sense. If production is realised, we expect novel implementations that bring spintronics and qubit devices to new heights [11], [50], [51], and perhaps a disruption in the information technology sector. There could also be new applications in optoelectronics that we have yet to envision since the TMD materials we are using provide the single platform to combine both optical and electronic effects for device realisation. Once the materials and device-related processes are honed, we may indeed also realise the touted advantages of ultrascaled and ultralow power 2D TMD devices without the common device parasitics and fundamental constraints that limit the conventional semiconductor devices. This, in combination with full control over the valley degree of freedom, would open new vistas for device aficionados and possibly unexpected new applications. The field is young and the grass is green in the valley.

Welcome to Valleytronics.

References

[1] D. Xiao, G.-B. Liu, W. Feng, X. Xu, and W. Yao, "Coupled spin and valley physics in monolayers of MoS_2 and other group-VI dichalcogenides," *Phys. Rev. Lett.*, vol. 108, no. 19, p. 196802, 2012, doi: 10.1103/PhysRevLett.108.196802.

[2] Z. Gong *et al.*, "Magnetoelectric effects and valley-controlled spin quantum gates in transition metal dichalcogenide bilayers," *Nat. Commun.*, vol. 4, no. May, pp. 1–6, 2013, doi: 10.1038/ncomms3053.

[3] A. Kormányos, V. Zólyomi, N. D. Drummond, and G. Burkard, "Spin-orbit coupling, quantum dots, and qubits in monolayer transition metal dichalcogenides," *Phys. Rev. X*, vol. 4, no. 1, p. 011034, 2014, doi: 10.1103/PhysRevX.4.011034.

[4] J. Pawłowski, D. Zebrowski, and S. Bednarek, "Valley qubit in a gated MoS_2 monolayer quantum dot," *Phys. Rev. B*, vol. 97, no. 15, p. 155412, 2018, doi: 10.1103/PhysRevB.97.155412.

[5] G. Széchenyi, L. Chirolli, and A. Pályi, "Impurity-assisted electric control of spin-valley qubits in monolayer $MoS_{2,}$" *2D Mater.*, vol. 5, no. 3, p. 035004, 2018, doi: 10.1088/2053-1583/aab80e.

[6] Z. Zhang *et al.*, "Electrotunable artificial molecules based on van der Waals heterostructures," *Sci. Adv.*, vol. 3, no. 10, pp. 1–7, 2017, doi: 10.1126/sciadv.1701699.

[7] R. Pisoni *et al.*, "Gate-defined one-dimensional channel and broken symmetry states in MoS_2 van der waals heterostructures," *Nano Lett.*, vol. 17, no. 8, pp. 5008–5011, 2017, doi: 10.1021/acs.nanolett.7b02186.

[8] R. Pisoni *et al.*, "Gate-tunable quantum dot in a high quality single layer MoS_2 van der Waals heterostructure," *Appl. Phys. Lett.*, vol. 112, no. 12, p. 123101, 2018, doi: 10.1063/1.5021113.

[9] K. Wang *et al.*, "Electrical control of charged carriers and excitons in atomically thin materials," *Nat. Nanotechnol.*, vol. 13, no. 2, pp. 128–132, 2018, doi: 10.1038/s41565-017-0030-x.

[10] D. Kotekar-Patil, J. Deng, S. L. Wong, and K. E. J. Goh, "Coulomb blockade in etched single and few layer MoS_2 nanoribbons," *ACS*

Appl. Electron. Mater., vol. 1, no. 11, pp. 2202–2207, 2019, doi: 10.1021/acsaelm.9b00390.

[11] K. E. J. Goh, F. Bussolotti, C. S. Lau, D. Kotekar-Patil, Z. E. Ooi, and J. Y. Chee, "Toward valley-coupled spin qubits," *Adv. Quantum Technol.*, vol. 3, no. 6, p. 1900123, 2020, doi: 10.1002/qute.201900123.

[12] C. S. Lau *et al.*, "Gate-defined quantum confinement in CVD 2D WS_2," *Adv. Mater.*, vol. 34, no. 25, p. 2103907, 2021, doi: 10.1002/adma.202103907.

[13] J. Boddison-Chouinard *et al.*, "Gate-controlled quantum dots in monolayer WSe_2," *Appl. Phys. Lett.*, vol. 119, no. 13, p. 133104, 2021, doi: 10.1063/5.0062838.

[14] Y. Shimazaki, M. Yamamoto, I. V. Borzenets, K. Watanabe, T. Taniguchi, and S. Tarucha, "Generation and detection of pure valley current by electrically induced Berry curvature in bilayer graphene," *Nat. Phys.*, vol. 11, no. 12, pp. 1032–1036, 2015, doi: 10.1038/nphys3551.

[15] F. Bussolotti *et al.*, "Roadmap on finding chiral valleys: screening 2D materials for valleytronics," *Nano Futur.*, vol. 2, no. 3, p. 032001, 2018, doi: 10.1088/2399-1984/aac9d7.

[16] T. Y. T. Hung, K. Y. Camsari, S. Zhang, P. Upadhyaya, and Z. Chen, "Direct observation of valley-coupled topological current in MoS_2," *Sci. Adv.*, vol. 5, no. 4, p. eaau6478, 2019, doi: 10.1126/sciadv.aau6478.

[17] L. Li *et al.*, "Room-temperature valleytronic transistor," *Nat. Nanotechnol.*, vol. 15, pp. 743–749, 2020, doi: 10.1038/s41565-020-0727-0.

[18] L. Xie and X. Cui, "Manipulating spin-polarized photocurrents in 2D transition metal dichalcogenides," *Proc. Natl. Acad. Sci.*, vol. 113, no. 14, pp. 3746–3750, 2016, doi: 10.1073/pnas.1523012113.

[19] J. Lee, K. F. Mak, and J. Shan, "Electrical control of the valley Hall effect in bilayer MoS_2 transistors," *Nat. Nanotechnol.*, vol. 11, no. 5, pp. 421–425, 2016, doi: 10.1038/nnano.2015.337.

[20] Z. Zhang, P. Chen, X. Duan, K. Zang, J. Luo, and X. Duan, "Robust epitaxial growth of two-dimensional heterostructures, multiheterostructures, and superlattices," *Science*, vol. 357, no. 6353, pp. 788–792, 2017, doi: 10.1126/science.aan6814.

[21] Y. Wang *et al.*, "Van der Waals contacts between three-dimensional metals and two-dimensional semiconductors," *Nature*, vol. 568, no. 7750, pp. 70–74, 2019, doi: 10.1038/s41586-019-1052-3.

[22] C. S. Lau *et al.*, "Quantum transport in two-dimensional WS_2 with high-efficiency carrier injection through indium alloy contacts," *ACS Nano*, vol. 14, no. 10, pp. 13700–13708, 2020, doi: 10.1021/acsnano.0c05915.

[23] F. Bussolotti, J. Yang, H. Kawai, C. P. Y. Wong, and K. E. J. Goh, "Impact of S-vacancies on the charge injection barrier at the electrical contact with the MoS_2 monolayer," *ACS Nano*, vol. 15, no. 2, pp. 2686–2697, 2021, doi: 10.1021/acsnano.0c07982.

[24] F. Bussolotti *et al.*, "Electronic properties of atomically thin MoS_2 layers grown by physical vapour deposition: band structure and energy level alignment at layer/substrate interfaces," *RSC Adv.*, vol. 8, no. 14, pp. 7744–7752, 2018, doi: 10.1039/C8RA00635K.

[25] J. Yang, H. Kawai, C. P. Y. Wong, and K. E. J. Goh, "Electrical doping effect of vacancies on monolayer MoS_2," *J. Phys. Chem. C*, vol. 123, no. 5, pp. 2933–2939, 2019, doi: 10.1021/acs.jpcc.8b10496.

[26] J. Yang, F. Bussolotti, H. Kawai, and K. E. J. Goh, "Tuning the conductivity type in monolayer WS_2 and MoS_2 by sulfur vacancies," *Phys. Status Solidi RRL*, vol. 14, no. 9, p. 2000248, 2020, doi: 10.1002/pssr.202000248.

[27] Y. Liu *et al.*, "Approaching the Schottky-Mott limit in van der Waals metal-semiconductor junctions," *Nature*, vol. 557, no. 7707, pp. 696–700, 2018, doi: 10.1038/s41586-018-0129-8.

[28] K. S. Novoselov, A. Mishchenko, A. Carvalho, and A. H. Castro Neto, "2D materials and van der Waals heterostructures," *Science*, vol. 353, no. 6298, 2016, doi: 10.1126/science.aac9439.

[29] K. F. Mak and J. Shan, "Semiconductor Moiré materials," *Nat. Nanotechnol.*, vol. 17, no. 7, pp. 686–695, 2022, doi: 10.1038/s41565-022-01165-6.

[30] O. Gunawan, Y. P. Shkolnikov, K. Vakili, T. Gokmen, E. P. De Poortere, and M. Shayegan, "Valley susceptibility of an interacting two-dimensional electron system," *Phys. Rev. Lett.*, vol. 97, no. 18, pp. 1–4, 2006, doi: 10.1103/PhysRevLett.97.186404.

[31] Y. Liu, N. O. Weiss, X. Duan, H.-C. Cheng, Y. Huang, and X. Duan, "Van der Waals heterostructures and devices," *Nat. Rev. Mater.*, vol. 1, no. 9, p. 16042, 2016, doi: 10.1038/natrevmats.2016.42.

[32] Z. Huang *et al.*, "Robust room temperature valley Hall effect of interlayer excitons," *Nano Lett.*, vol. 20, no. 2, pp. 1345–1351, 2020, doi: 10.1021/acs.nanolett.9b04836.

[33] W. Yang *et al.*, "Interlayer interactions in 2D WS_2/MoS_2 heterostructures monolithically grown by: In situ physical vapor deposition," *Nanoscale*, vol. 10, no. 48, pp. 22927–22936, 2018, doi: 10.1039/c8nr07498d.

[34] S. Carr, D. Massatt, S. Fang, P. Cazeaux, M. Luskin, and E. Kaxiras, "Twistronics: Manipulating the electronic properties of two-dimensional layered structures through their twist angle," *Phys. Rev. B*, vol. 95, no. 7, p. 075420, 2017, doi: 10.1103/PhysRevB.95.075420.

[35] S. Zheng *et al.*, "Coupling and interlayer exciton in twist-stacked WS_2 bilayers," *Adv. Opt. Mater.*, vol. 3, no. 11, pp. 1600–1605, Nov. 2015, doi: 10.1002/adom.201500301.

[36] P. C. Yeh *et al.*, "Direct measurement of the tunable electronic structure of bilayer MoS_2 by interlayer twist," *Nano Lett.*, vol. 16, no. 2, pp. 953–959, 2016, doi: 10.1021/acs.nanolett.5b03883.

[37] C. G. Péterfalvi, A. David, P. Rakyta, G. Burkard, and A. Kormányos, "Quantum interference tuning of spin-orbit coupling in twisted van der Waals trilayers," *Phys. Rev. Res.*, vol. 4, no. 2, 2022, doi: 10.1103/PhysRevResearch.4.L022049.

[38] K. L. Seyler *et al.*, "Signatures of Moiré-trapped valley excitons in $MoSe_2/WSe_2$ heterobilayers," *Nature*, vol. 567, no. 7746, pp. 66–70, 2019, doi: 10.1038/s41586-019-0957-1.

[39] Z. Hennighausen and S. Kar, "Twistronics: a turning point in 2D quantum materials," *Electron. Struct.*, vol. 3, no. 1, p. 014004, 2021, doi: 10.1088/2516-1075/abd957.

[40] N. P. Wilson, W. Yao, J. Shan, and X. Xu, "Excitons and emergent quantum phenomena in stacked 2D semiconductors," *Nature*, vol. 599, no. 7885, pp. 383–392, 2021, doi: 10.1038/s41586-021-03979-1.

[41] U. Cho, S. Kim, C. Y. Shin, and I. Song, "Tabletop fabrication of high-performance MoS_2 field-effect transistors," *ACS Omega*, vol. 7, no. 24, pp. 21220–21224, 2022, doi: 10.1021/acsomega.2c02188.

[42] M. C. F. Da Costa, H. B. Ribeiro, F. Kessler, E. A. T. De Souza, and G. J. M. Fechine, "Micromechanical exfoliation of two-dimensional materials by a polymeric stamp," *Mater. Res. Express*, vol. 3, no. 2, p. 025303, 2016, doi: 10.1088/2053-1591/3/2/025303.

[43] F. Liu, "Mechanical exfoliation of large area 2D materials from vdW crystals," *Prog. Surf. Sci.*, vol. 96, no. 2, p. 100626, 2021, doi: 10.1016/j.progsurf.2021.100626.

[44] A. Castellanos-Gomez *et al.*, "Deterministic transfer of two-dimensional materials by all-dry viscoelastic stamping," *2D Mater.*, vol. 1, no. 1, p. 011002, 2014, doi: 10.1088/2053-1583/1/1/011002.

[45] L. Wang *et al.*, "One-dimensional electrical contact to a two-dimensional material," *Science*, vol. 342, no. 6158, pp. 614–617, 2013, doi: 10.1126/science.1244358.

[46] D. Voiry, A. Mohite, and M. Chhowalla, "Phase engineering of transition metal dichalcogenides," *Chem. Soc. Rev.*, vol. 44, pp. 2702–2712, 2015, doi: 10.1039/C5CS00151J.

[47] R. Kappera *et al.*, "Phase-engineered low-resistance contacts for ultrathin MoS_2 transistors," *Nat. Mater.*, vol. 13, no. 12, p. 1128, 2014, doi: 10.1038/nmat4080.

[48] C. S. Lau *et al.*, "Quantum transport in two-dimensional WS_2 with high-efficiency carrier injection through indium alloy contacts," *ACS Nano*, vol. 14, no. 10, pp. 13700–13708, 2020, doi: 10.1021/acsnano.0c05915.

[49] M. Chhowalla, D. Jena, and H. Zhang, "Two-dimensional semiconductors for transistors," *Nat. Rev. Mater.*, vol. 1, no. 11., pp. 1–15, 2016. doi: 10.1038/natrevmats.2016.52.

[50] X. Xu, W. Yao, D. Xiao, and T. F. Heinz, "Spin and pseudospins in layered transition metal dichalcogenides," *Nat. Phys.*, vol. 10, no. 5, pp. 343–350, 2014, doi: 10.1038/nphys2942.

[51] K. E. J. Goh, L. A. Krivitsky, and D. L. Polla, "Quantum technologies for engineering: the materials challenge," *Mater. Quantum Technol.*, vol. 2, no. 1, p. 013002, 2022, doi: 10.1088/2633-4356/ac55fb.

https://doi.org/10.1142/9789811229107_bmatter

Index

www.ingramcontent.com/pod-product-compliance
Ingram Content Group UK Ltd.
Pitfield, Milton Keynes, MK11 3LW, UK
UKHW022105170926
8624IPUK00024B/88